METHODS OF SOLVING PROBLEMS

PROBLEMS

in

Elementary,
Middle, and High School
MATHEMATICS

Mihai Rosu

To order additional copies of this book, contact:
Xlibris Corporation
1-888-795-4274
www.Xlibris.com
Orders@Xlibris.com

88487

Contents

Chapter I... **11**

NATURAL AND INTEGER NUMBERS

Integer numbers (**Z**). The order of the operations and the use of brackets, Positive powers, Operations with powers, Integer exponents, Properties of the powers, Problems of divisibility, Linear equations with integer solutions, Inequations and intervals, Recapitulative tests.

Chapter II **38**

RATIONAL NUMBERS AND FRACTIONS

The set of rational numbers(**Q**); Decimal numbers and fractions; Comparition of rational numbers; Operations with fractions; Order of the operations; Negative and positive fractions; Linear equations; Proportions; Procents, Recapitulative tests.

Chapter III.. **77**

ALGEBRAIC EXPRESSIONS AND LINEAR FUNCTIONS

Algebraic expressions, Operations, Algebraic rational expressions, Complex fractions, Rational equations, Linear equations containing absolute value, Inequations, Systems of linear equations, Linear functions: slops, x-axis and y-axis intersection, graphs.

Chapter IV

POWERS AND RADICALS.. **102**

Real numbers (**R**), Positive and negative exponents negative exponents, Radicals, Properties of the radicals, Graphs of the function with integer exponent, Square root function, Algebraic irrational expressions, Irrational equations, Irrational inequations, Systems of irrational equations.

Chapter V **124**

QUADRATIC FUNCTION AND APPLICATIONS

Algebraic form of the quadratic function, Factored form, Vertex form (standard form), The roots of the quadratic equation, Axe of symmetry, Maximum or minimum of the quadratic function, Graph of the quadratic function, Relations between roots and coefficients, Solving algebraic rational inequations, Domain and range, Homogeneous, symmetric, and nonlinear systems of equations, Word problems using quadratic function.

Chapter VI **139**

COMPLEX NUMBERS

Algebraic form of a complex number, Equality in **C**, The complex plane, Operation with complex numbers, Conjugate and modulus (absolute value) of a complex number, Quadratic equations, Complex functions, Equations and systems in **C**.

Chapter VII **149**

EXPONENTIAL AND LOGARITHMIC FUNCTIONS

Definition of exponential and logarithmic functions, Domain and range, Graphs and transformations, Properties of exponents, Properties of logarithms, Exponential and logarithmic equations and inequations, Sistems of exponential and logarithmic equations, Inequations containing exponential and logarithmic expressions.

Chapter VIII **165**

ARITHMETIC AND GEOMETRIC PROGRESSIONS

Sequences and summation notation, Definition of an arithmetic and a geometric progerssion, Formula for general term, Common difference and ratio, Partial sums, Formula for S_n for an arithmetic and a geometric progerssion, Infinite series, Some considerations of *Fibonacci numbers*.

Chapter IX... **185**

POLYNOMIALS AND ALGEBRAIC EQUATIONS

Algebraic form of a polynomial, Synthetic division (*Horner's Algorithm*), Long division, Reminder theorem (*Bézout's theorem*), Fundamental theorem of algebra (*d'Alambert-Gauss*), Relation between and coefficients (*Vieta's formulas*), Factor theorem, polynomial with rational, real, or complex roots and coefficients, Reciprocal quadratic equations.

Chapter X **201**

PERMUTATIONS AND COMBINATIONS. NEWTON'S BINOMIAL.

The factorial notation, Arrangements, Combinations, The number of k-subsets of a set of n elements, *Newton*'s binomial theorem, The general term of the *Newton*'s binomial, Sums of combinations, Calculation of combination sums using complex numbers.

ANSWERS and SOLUTIONS

Chapter I.. **217**

Chapter II **227**

Chapter III.. **240**

Chapter IV **255**

Chapter V **265**

Chapter VI **278**

Chapter VII **289**

Chapter VIII... **307**

Chapter IX.. **315**

Chapter X **329**

Bibliography **337**

I dedicate this book to my truly inspiring sons,
Alexandru and **Vlad**,
whose help has been essential
in finalizing chapters I-VI
and to whom I owe a debt of gratitude
I shall never be able to repay.

Preface to the First Edition

I have developed this book to help elementary, middle and high school students acquire a deeper understanding of mathematics and prepare themselves for Olympiads and contests in this field. My wish is to instill in my students the art of reasoning and the power of evaluating and solving mathematics issues at a higher level.

The first section of this book covers almost the entire North American curriculum with over 3,200 problems and exercises organized into ten chapters. Each chapter contains excercises that increase in difficulty from the simplest ones to problems used in major mathematics competitions. At the end of each chapter, I have included problems I have published in the "Mathematical Olympiads Correspondence Program" (*http://www.math.ca/Competitions/MOCP/*) in Ontario. Some problems are accompanied by at least one full solution; others come only with the answer, leaving the students the joy of solving them on their own.

Occasionally, some problems at the end of the chapter appear impossible to solve. However, my thirty-year experience in teaching mathematics and my interactions with students enable me to demonstrate the solutions in a clear, easy to understand manner.

The second part of the book is a presentation of several problems solving, alongside a series of guiding ways and answers to other categories of problems.

I am confident that any students who pick up this book and solve one set of excercises can expand their knowledge, fuel their interest, and excel in their mathematics courses. I also hope this paper will become instrumental to mathematics teachers working both in the private and public education systems.

Mihai Rosu
rosumihai@yahoo.ca

Chapter I

NATURAL AND INTEGER NUMBERS

$\mathbf{N} = \{0,1,2,...\}$ Natural number, $\mathbf{N}^* = \{1,2,3,...\}$

$\mathbf{Z} = \{-2,-1,0,1,2,...\}$ Integer numbers,

Properties of the powers $\quad a^n = \underbrace{a \cdot a \cdot ... \cdot a}_{n \ times}, \quad n \in \mathbf{N}^*.$

a) $a^0 = 1,$

b) $(ab)^n = a^n b^n,$

c) $\left(\dfrac{a}{b}\right)^n = \dfrac{a^n}{b^n},$

d) $a^n a^m = a^{n+m},$

e) $\dfrac{a^n}{a^m} = a^{n-m},$

f) $\left(a^m\right)^n = a^{mn}, \quad a,b \in \mathbf{Z}, \ b \neq 0, \ m,n \in \mathbf{N}.$

Base ten numeration system $\quad \overline{abcd} = a \cdot 10^3 + b \cdot 10^2 + c \cdot 10 + d.$

Integer numbers. Order of the operations

I. 1. Replace the □ with "=", "<", ">" such that the following relations are true:

a) $4 + 3 \ \square \ 2 + 3$;

b) $7 - 3 \ \square \ 2 + 3$;

c) $9 - 3 \ \square \ 3 + 3$;

d) $9 - 4 \ \square \ 5 + 2$;

e) $8 - 1 \ \square \ 5 + 2$;

f) $10 - 3 \ \square \ 6 + 3$;

g) $8 + 1 \ \square \ 8 + 1$;

h) $7 + 2 \ \square \ 6 + 3$;

i) $9 - 1 \ \square \ 5 + 3$;

j) $7 + 3 \ \square \ 6 + 4$.

I. 2. Calculate:

a) $1 + 3$;

b) $22 + 30$;

c) $17 - 2$;

d) $50 - 22$;

e) $65 + 47$;

f) $57 - 15$;

g) $55 + 34$;

h) $88 - 55$.

i) From the product of the numbers 27 and 22 subtract their difference;

j) I think to one number. I add 20 to this number. From the result I subtract 43 and I get 10. What is number?

I. 3. Complete the following tables:

x	18	59	76
y	22	12	15
$x+y$			

a	b	c	$a+b$	$a+c$	$a+b+c$
32	14	19			
14	15	13			

I. 4. Fill out a correct number in each square:

a) $\square +3 < 5$; b) $4+\square = 9$; c) $7-\square < 4$;

d) $3 + 5 = \square + 3$; e) $5+ \square > 8 - 2$.

Find the unknown terms in the operations:

f) $45 - 12 + b = 56+17$; g) $50 + 0 = x + 49$;

h) $65 - 22+b = 58 + 37$; i) $45 + 32 - a = 6 + 31$;

j) $a + 85-12 = 86 + 27$.

I. 5. a) Calculate:

$43 + 18 =$	$55 - 25 - 8 =$
$35 + 29 =$	$84 - 56 - 8 =$
$83 - 27 =$	$38 + 29 +17 =$
$84 - 36 =$	$46 + 9 - 16 =$

b) Calculate grouping the factors:

$2 + 13 + 7 + 18 =$

$3 + 24 + 6 + 17 =.$

c) Find the unknown numbers:

$n - 26 = 25$	$35 + a = 71$
$75 + d = 79$	$35 - a = 11$
$12 + 13 - x = 14$	$78 - (x + 21) = 46$

d) From the sum of the numbers 9, 18, 27 and 36 subtract the difference between 36 and 17.

e) In a garden there are 48 roses, in another garden there are 24 more roses than the other. How many roses are in both gardens?

f) Vlad has 48 stamps and bought another 56 stamps. He gave to Alexandru 29 stamps. How many stamps has Vlad left?

g) Fill in:

0, 5, 10, __, __, __, __, __, __, __, 50.
0, 3, 6, __, __, __, __, __, __, __,30.
0, 10, 20, __, __, __, __, __, __, __, 100.

h)   i) Multiply 767 by 18.
ii) Find the number 6 times larger than 97.
iii) Add 25 to the triple of the number 182.
iv) Substract 100 from the triple of the number 99.

i) Vlad is 7 years old, his mothers age is 6 times more and father's age is 5 years older than his mother. How many times is Vlad's age less than his father's age?

j) Find "a" from the equalities:

$a \div 6 = 7$; $9 \cdot a = 54$; $27 \div a = 3$.

I. 6. Fill out the table

a	b	c	d	c-b	$3a$-b	$3c$-d+1	a+$3b$-c	$3b$+$2c$-d+2
7		63	189	42				
10		16	48	0				
7		15	30	0				
5	15		135	30				
4		36	144	17				

I. 7. Calculate the expressions when
$a = 3, b = 9, c = 27, d = 81$:

a) $d - 3c + b - 3a$;

b) $ad - 3ac + ab - 3a^2$;

c) $ad - c + 1$;

d) $a^2 b - ac + a$

e) $bc - 3d + 1$;

f) $a^3 b - 3ac + 27 - a^3$;

g) $b^2 - d$;

h) $b^3 - bd$;

i) $a^2 b - d$;

j) $a^3 b - ad$.

I. 8. Calculate:

a) $-8 + 5 - 2$; b) $13 + (-8) + (-7) - (-3)$; c) $(-8) + (+6) - (-3)$;

d) $-5 - (-4) + 3 - 10$; e) $8 - (7-9) - 2$; f) $2 - 3 - 5 + 6 - 7 + 8 - 9$;

g) $(-10+8) + (4+6) - (-5-3)$.

I. 9. Calculate:

a) $(-4) + (-25) + 10 + (-25) + (+7) + 10 + 8 + (-30)$;

b) $(-27) + 32 + (-70) + (+40) + (-40)$;

c) $(-35) + 86 + 104 + 18 + (-20) + (+51) + (-15)$;

d) $45 + 31 + (-16) + (-31) + 22$;

e) $11 + 2 + 3 + (-4) + (-5) + (-6)$;

f) $1 + (-2) + 3 + (-4) + 5 + (-6)$;

g) $-8 - 7 - 6 - 5 - 4 - 3 - 2 - 1 + 0 - 1 + 2 - 3 + 4 - 5 + 6 - 7 + 8$.

I. 10. Calculate:

a) $(2 + 3 - 5 + 7 - 5) + (3 - 5 + 8 - 12) - (-3 + 5 - 7)$;

b) $(4 + 5 - 7) + (-4 - 9 + 13) - (-7 + 12 - 75)$;

c) $(42 + 23) - (23 + 42) + (-41 + 40 - 0)$;

d) $2 - [3 - (5 + 6 + 9 - 20 + 1)]$;

e) $(3 - 4 + 7) - [(5 - 4 + 9) - (-3 + 9) - 2]$;

f) $2 - \{3 - [+9 - (4 - 5)] + 1\}$;

g) $(-5) - [(-34) + (+16) - (-19) + (-1)] - (+11)$;

h) $(-11 + 13) - \{(-15) + (+7) - [-(-23) - (+29) -14 - (+13)]-$
$-(-32 + 45)\}$.

I. 11. Calculate:

a) $-39 - [-17 - (-15 + 39 - 26) - 20] - (-14 + 22)$;

b) $-(-52) - \{-24 + 31 - [(-37) + (+26) - (+15) + (-51)] - (-31)\}$;

c) $-32 + \{-33 - [-(-41) + (-19) + (-6)] - (-15 - 8)\}$;

d) $6 - \{-27 - [-4 - (-18 + 37) - (43 - 50)]\}$.

I. 12. Fill out the table

a	b	c	a-b	b-c	$(a$-$b)+2c$	$(2a$-$2b)+(b$-$c)$
26			11	1		
	25		1	1		
125			25	0		
45			5	10		
60		6	30	10		
	7		1	6		

I. 13. Calculate:

a) $13 - \{(-32) + [(-45) + (+43) + (-24)] - (+71)\} - 141$;

b) $44 - \{-43 - [-17 + (-59) + (+38)] + (+77)\} - 18$;

c) $-26 - \{-49 + [-37 + (+55) + (-32)] + (-33)\} - 30$;

d) $-35 + \{-22 - [-39 + (-52) + (+65)] + (-69)\} + 61$;

e) $-21 - \{-15 + [-30 + 3 + (-50)] + 16\} + 10$.

I. 14. Calculate:

a) –99 – {–93 – [–21 + 65 + (–50)]} + 18;

b) 24 – {(–58) + 67 – [(–19) + 43] + 54} + 3;

c) 35 + {–44 – [–24 + (–35 + 50) + 16] + (–40)} + 62;

d) –32 + {–10 – [71 – (–71 + 80) + (–40)] + (–50)} + 83;

e) 69 – {–43 – [–35 + (–39) + 52] + 88} + 33;

f) –5 – (–2) – {(–5) – [(–8) + (–3) – (–7)]+5}.

I. 15. Calculate:

a) (–9) · (+2) · (–2) · (–4); **b)** (+4) · (+1) · (+3) · (–2) · (–2);

c) –(–2)·(+3)· (–1) · (–8) · (–4); **d)** (108) ÷ (–3); **e)** (–35) ÷ (–5);

f) 45 ÷ (–9); **g)** (–75) ÷ (+25); **h)** (–92) ÷ (–23).

I. 16. Calculate:

a) (–2) · [27 + 108 ÷ (–9)];

b) 24 + (–81) ÷ (+3) – (–2) · (+6); **c)** 21 + (–8) · (+4) ÷ (–16);

d) –55 ÷ (–11) · (–7) ÷ 35; **e)** –4 · {(18 – 10) ÷ [6 – (–2)]};

f) 34 ÷ (–17) + (–24) ÷ (–8); **g)** 64 ÷ 32 ÷ (–2); **h)** 54 ÷ [27 ÷ (–3)];

i) (74 – 102 ÷ 3) · 2 – 42÷6; **j)** (945 ÷ 9) ÷ (–5) – (–32) ÷ (+8).

I. 17. Calculate:

a) –2 · [(–3) · (+8) + (–3) · (+2)]; **b)** –2 · [(–6 · 8) + (–6 · 2)];

c) (–13) + (–6) ÷ 2 – {(–5) · (–4) – [2 – (3 – 5)]};

d) (–3) · [(–15) + 4 ÷ (5 – 3)]; **e)** (–5) · [17 + (–3) · (–5) – 6];

f) (–2) · [(–40) · (–1) + (–28) ÷ (+7);

g) –[(–5) – (–48) ÷(–16) + (–2) · (+3)].

I. 18. Calculate:

a) {80 + [10 – 5 + (7 – 3)] – 9} ÷ (–40);

b) 50 · {59 – [7 · 8 + (13 – 2 · 5)]};

c) 22 ÷ {17 – [15 – 3(7 – 4)]};

d) 3 · {18 + [3 + 4 · (4 – 2)]};

e) (15 – 8) · {8 + [3 + (7 – 1)]};

f) 10 + [1 + (15 – 2) · 4];

g) 30 – {121 ÷ (–11) · (–11) – 4 · (–4) – 3 · [3 – (9 – 2)]};

h) (–1) · [(–7) + (–23)] · (–7) – (8 – 10) · (–4);

i) $5 - (-3)[4 - (7 - 3)] \div 3 \cdot [5 + (-3) \cdot (-6)]$;

j) $(-3)(-4) + [(-5) - (-6)] \cdot (-7)$;

k) $(-3) + (-4)[(-5) - (-6)] - (-7)$;

l) $[(-1) + (-6) + (-5)] \div (-6) - (+7)$.

I. 19. Calculate:

a) $(-5) \cdot \{(-14) \cdot (-3) - (-25) \cdot (-2) - [(-4) \cdot (+3) - (-5) \cdot (+1)]\}$;

b) $(-3) \cdot \{(-24) \cdot (-3) - (-5) \cdot (-13) - [-(-34) - 3 \cdot (-21 + 36)]\}$;

c) $(-4) \cdot [(-32) \div (-8) - (-48) \div (-4) + (-25 + 10)]$;

d) $(-9) \cdot \{-5 - [(-24) \div (-2) - (-54) \div (-6)] \cdot (-4)$.

I. 20. Calculate:

a) $[(-2) \cdot (+27) - (-9) \cdot (-4) - (-7) \cdot (+16)] \div (-10 + 8)$;

b) $[(-2) \cdot (-32) + (+3) \cdot (-24) - (-5)(-9)] \div (-65 + 12)$;

c) $\{-36 - [(-8) \cdot (+7) - (-23) \div (-1) + (-36) \div (-3) -$
   $-(-38 + 19)]\} \div (-4)$;

d) $[(-6) \cdot (-5) - (-24) \div (-3) - (+27) \div (-3)] \cdot [(-2) \cdot (-7) - (-2) \cdot (-9)]$;

e) $[(+18) \div (-3) - (-45) \div (-3) - (-32) \div (+2)] \cdot [-24 + (-3) \cdot (-6)]$;

f) $[(-36) \div (-12) \cdot (+3) - (-18) \div (+2) \cdot (-2)] \cdot [-52 + (-25) \cdot (-2)]$.

I. 21. Fill out the table

a	b	c	d	a²c-3a²b+a+b	a²+3ab²-abc
6	18	54	162		
3	9	27	81		
11	33	99	297		
9	27	81	243		
17	51	153	459		

I. 22. Calculate:

a) $(-2)^3$; b) $(-5)^1$; c) $(-15)^2$; d) $(-1)^{203}$; e) $(-1)^{600}$; f) 17^1;

g) 1^{500}; **h)** 451^0; **i)** 0^{801}; **j)** $(-2)^3 \cdot (-3)^3 \cdot (-1) + 2^3 \cdot 3^2$;

k) $(-1)^3 \cdot (-5)^2 + 2^3 \cdot (-3)^2 \cdot 3 - 5^0$; **l)** $2^{23} \cdot (-5)^{25} + 2^{25} \cdot (-5)^{23}$;

m) 2^{3^4}; **n)** $(-3)^{2^3}$; **o)** $\left(2^1 \cdot 2^3 \cdot 2^5\right)^9 + [(-2)^1 \cdot (-2)^3 \cdot (-2)^5]^9$;

p) $(-2) \cdot (2^2)^3 \cdot (-2^4)^4 \div [2 \cdot (-2)^6 \cdot 2^{16}]$.

I. 23. Calculate:

a) $\left[(-2)^3 \cdot 2^3 \cdot 2 \cdot (-2)^2\right]^3 \div 8^5$; **b)** $\left[(-3)^4 \cdot 3^2 \cdot 3^3 \cdot (-3)^5\right]^2 \div 9^8$;

c) $2^3 \cdot \left[(-8)^1 \cdot (-4)^2 \cdot (-2)^3\right]^2 \div \left[(-2)^2\right]^9$;

d) $\left[(-9)^2 \cdot 27^4 \cdot (-81)^2\right] \div (-81^2)^3$.

I. 24. Calculate:

a) $\left\{\left[(-2)^4 \cdot 2^3 \cdot (-2)^3\right] \div \left[2 \cdot (-2)^2 \cdot (-2)^3\right]\right\} \div 4^2$;

b) $\left\{\left[3^3 \cdot (-3)^4 \cdot (-3)^5\right] \div \left[(-3)^4 \cdot (-3)^2 \cdot (-3)^3\right]\right\} \div (-3)^2$

c) $\left\{\left[(-2)^6 \cdot 4^2 \cdot (-8)^5\right] \div \left[(-4)^3 \cdot (-64)^2\right]\right\} \div (-8)$;

d) $\left\{\left[(-3)^4 \cdot 3^2 \cdot (-27)^2\right] \div \left[(-9)^3 \cdot (+81)\right]\right\} \div (-3)$.

I. 25. Calculate:

a) $-3^9 \div (-3)^9 + (-3)^5 \div (-3^5) + (-5 \cdot 4)^7 \div \left(2^{11} \cdot 5^3\right)$;

b) $(-2)^{45} \div (2^8)^5 \cdot \left[(-3)^{21} + 5^{21} \div (-5)^{10} - (-8)^2 \cdot (-2)^9\right] \div$
$\div \left(-3^5 + 5^{11} + 2^{15}\right)$;

c) $\left[3 + (-3)^2 \cdot (-3^4)^6 + 3^3 \cdot 81^3 + 3 \cdot (2^4)^{11}\right] \div (1 + 3^{25} + 3^{14} + 2^{44}) - 2$.

I. 26. Calculate:

a) $(-3^3)^3 \cdot (-3^3)^2 \div (-3^{3^2}) + 2^{3^4} \div (-2)^{9^2} + (-2^4)^{2^2} \div 2^{2^{2^2}}$;

b) $\left[(-8)^4 \cdot 32^2 \cdot 2^2 \div (4^7)\right]^3 \div 8^{10} - \left[8^5 \cdot (-4)^2 \div (-8)^5\right]^3 \div (-4)^5$;

c) $\left[(-8^2)^2 \div (4^{10} \div 32^2)\right]^5 \div (2^3)^3 - \left[(-81^2)^2 \div (9^3 \cdot 3^4)\right]^3 \div$
$\div \left[81^2 \div (-3^3)^2\right]^9$;

d) $\left\{\left[(-25)^3 \cdot 5^7 \div (5^3)^4\right]^5 \div (-25)^2\right\}^4 \div \left[-625^2 \div 25^3\right]^2$;

e) $(-27)^6 \div \left[(-3)^5 \cdot (-3)^3\right]^2 + (-3)^2 \cdot (-3)^8 \div (-3^3)^3 + 3^3 \cdot 3 \div 3^4$;

f) $\left[(-2)\cdot(-2)^2\cdot(-2)^3\cdot...\cdot(-2)^{111}\right]\div\left[(-2)\cdot(-2)^2\cdot(-2)^3\cdot...\cdot(-2)^{110}\right].$

I. 27. Calculate:

a) $2^3\cdot\left\{10+1^{20}\cdot\left[\left(2^2\right)^2\cdot2^{2^2}\div2^4-5\cdot\left(3^3-5^2\right)\right]\div3+2\right\};$

b) $\left[5^{62}\div\left(5^2\right)^{30}+\left(3^2\right)^3\div3^{2^2}\right]\div2\,;$

c) $\left(2^{n+1}\cdot3^{n+2}\right)\div\left(2^n\cdot3^n\right)+1^{n+1},n\in N\,;$

d) $\left(3^2+2\cdot2^3+5^9\right)\div\left[9+2^{12}\div2^8+\left(5^3\right)^3\right].$

I. 28. Calculate:

a) $(-1)^7+(-1)^{86}-(-1)^{101}+(-1)^{37};$ **b)** $(-1)^n+2(-1)^{n+1}-3(-1)^{n+2};$

c) $(-1)^n+3\cdot(-1)^{2n+1}-2(-1)^{2n};$

d) $1\cdot(-1)^{2n}+2(-1)^{2n+1}+3(-1)^{2n+2}+4(-1)^{2n+3}+...+100\cdot(-1)^{2n+99}.$

I. 29. Calculate:

a) i) $1-2^3+3^2$, **ii)** $\left(5^2+5+1\right)\div\left(3^3+2^2\right);$

b) $\left[10^3\div5^3+\left(2^2\right)^3\div2^{2^2}-3^3\right]\cdot2^{1^2}+1^{2010}+2010^0;$

c) $968-\left\{14\cdot13-5\cdot\left[2^3\cdot5\div\left(3^2\cdot5\div15+5\right)+2\cdot\left(13^4\div13^3-12\right)\right]\right\}\cdot5;$

d) i) $4^2\cdot15-4^2\cdot14$, **ii)** $11^2\cdot20-11^2\cdot18-11^2;$

iii) $30\cdot11^2-11^2\cdot28;$ **iv)** $\left(2^{2007}+2^{2007}\right)\div2^{2006};$

v) $\left(3^{2007}+3^{2005}\right)\div3^{2005};$ **vi)** $2^7-2^6-2^5-2^4+2^3;$

e) $\left\{\left[\left(7^2-14\right)+2^5\div4\right]\div\left(3^3+2^4\right)+3^2\right\}\cdot11-\left[\left(4+3^2\right)\cdot5-55\right];$

f) $\left\{\left[\left(5^4-5^2\right)\div5^2+\left(2^5+4^3\right)\cdot2^2\right]-3^4\right\}-2010^0.$

I. 30. Calculate:

a) $3x^3\cdot(-3x^2);$ **b)** $-4m^4\cdot(-3m^2);$ **c)** $2x^4\cdot(-4x^3);$

d) $5a^2b\cdot(8a^2b^3);$ **e)** $2x^3y\cdot(7x^3y^2);$ **f)** $3x\cdot4xy\cdot(2\cdot3x^2);$

g) $2ab^2c\cdot(-6a^2b^2);$ **h)** $5m^2n\cdot(2m^2n^5);$

i) $5a^2b\cdot(-a^2b^3)\cdot(4a^4b^3).$

I. 31. The decreasing order of the numbers
aaa, bbb, aba, and *bab* when $a < b$ is:
a) *bbb, aba, bab, aaa*
b) *aba, aaa, bbb, bab*
c) *bbb, bab, aba, aaa*
d) *bab, aaa, bbb, aba*
e) *aaa, bbb, aba, bab.*

I. 32. Collect the terms:
a) $5a + 6a - 3a$; **b)** $4x^2y - 3x^2y + 2x^2y$; **c)** $4mn - 3mn - 6mn$;
d) $m^2 - 4 + 3m^2 + 1$; **e)** $2x^3 + 6x^3y - 4x^3 - 2x^3y$;
f) $2rs^2 + rs^2 + rs + 5rs - 4rs^2$; **g)** $-4a^2b + 2a^2b + 2ab - 5a^2b$;
h) $6x^2yz - 3x^2yz - 2x^2yz$; **i)** $2x^2y - 4x^2y + 5xy^2$.

I. 33. Fill out the table

a	b	c	d	6a³c÷a−18a³b÷a+b	3a³c²÷b−a³cd÷b+b÷a
5	15	45	135		
3	9	27	81		
1	3	9	27		
9	27	81	243		
17	51	153	459		

Equations and inequtions in the set **Z** (integer numbers)

I. 34. Prove that for any $k \in$ **Z**,
a) $3 \mid 3k$; **b)** $7 \mid 14k$; **b)** $5 \mid 5x + 5y$ for any $x, y \in$ **Z**;
c) $3 \mid a$ and $5 \mid a$, then $15 \mid a$, $a \in$ **Z**;

d) $3 \mid (12a + 21b)$, for any $a, b \in \mathbf{Z}$;

e) $31 \mid 3^{n+3} + 4 \cdot 3^n$, for any $n \in \mathbf{N}$;

f) $17 \mid 2^{2n+3} \cdot 3^{n+1} + 3^{n+2} \cdot 4^{n+1} + 2^{n+3} \cdot 6^n$, for any $n \in \mathbf{N}$.

I. 35. Prove that if $7 \mid (5x + y)$, then $7 \mid (2x - y)$, where $x, y \in \mathbf{Z}$.

I. 36. Solve for a the equations:

a) $5 + a = 8$; **b)** $55 + a = 59$; **c)** $5 - a = 2$; **d)** $78 - a = 71$;

e) $a - 3 = 6$; **f)** $38 - a = 24$; **g)** $a - 54 = 23$; **h)** $31 + a = 69$;

i) $120 - 2a = 40$; **j)** $279 - 3x = 222$.

I. 37. Solve in $\mathbf{Z}$ the equations:

a) $x + 4 = 11$; **b)** $y - 8 = 1$; **c)** $z - 2 = 3$; **d)** $x + 6 = 2$;

e) $b + 4 = 1$; **f)** $-1 = x + 9$; **g)** $-20 + x = 19$;

h) $(x + 6) - 3 = -2$; **i)** $(y - 5) + 19 = 14$;

j) $(z - 4) + 6 = 21$; **k)** $(b + 1) + 9 = 10$; **l)** $x - 8 = |2 - 9|$;

m) $x - 7 = |4 - 11|$; **n)** $x + |5 - 12| = -4$.

I. 38. Solve in $\mathbf{Z}$ the equations:

a) $12x = 144$; **b)** $8y = -32$; **c)** $7a = -28$; **d)** $-4x = 64$;

e) $-7y = -35$; **f)** $-35z = -105$; **g)** $-5x = 40$;

h) $-2t = 22$; **i)** $3x = -3$.

I. 39. Solve in $\mathbf{Z}$ the equations:

a) $2x + 5 = 11$; **b)** $11y - 5 = 17$; **c)** $-13 + 4z = -37$;

d) $12 + 5t = -78$; **e)** $19 = 5 - 2x$; **f)** $17 = -3 - 5y$;

g) $-17 = -5 + 3x$; **h)** $2x + 82 = 0$;

i) $4y + 180 = 0$.

I. 40. Solve in $\mathbf{Z}$ the equations:

a) $5x - x + 2 = -30$; **b)** $5x - 4x + 2 = 5$; **c)** $y - 7 - 4y = -4$;

d) $7x - 5x + 10 = 0$; **e)** $4t - 3t - 2 = 0$; **f)** $0 = x - 14 - 3x$;

g) $0 = a + 25 + 4a$; **h)** $5 \cdot (x - 2) = -25$; **i)** $6 \cdot (y + 9) = -36$;

j) $-3 \cdot (x - 3) = 12$; **k)** $-8 \cdot (a + 5) = 56$; **l)** $3 \cdot (t + 1) + 2 = -7$;

m) $2 \cdot (y - 3) - 7 = 5$; **n)** $4 \cdot (x + 8) - 13 = -1$;
o) $5 \cdot (z - 2) + 30 = 20$; **p)** $8 + 3 \cdot (x + 3) = -4$;
q) $13 + 4 \cdot (x - 4) = -3$; **r)** $-4 - 3 \cdot (2x + 1) = -13$;
s) $-5 + 4 \cdot (3x - 6) = 7$; **t)** $(2x - 10) + (x - 2) + x = 0$;
u) $(3 - x) + (4 - x) + (5 - x) = 0$; **v)** $3(x + 4) - 2x = -5$;
w) $-(1 - 3y) + 2(y - 3) - y = -4 + 3y + 1$;
x) $-x - 2[6 - (1 - 2x)] = 0$; **y)** $-2 \cdot [z + 3 \cdot (5 - z)] - 5 \cdot (z - 7) = 0$;
z) $5 \cdot a + 2 \cdot [3 \cdot (1 - a) - 2 \cdot (1 + a)] = 22$.

I. 41. Solve in **Z** the equations:
a) $4x = x + 21$; **b)** $8x = 6x + 24$; **c)** $12y = 80 + 8y$;
d) $a = 51 - 2a$; **)** $x = 84 - 6x$; **f)** $11x - 91 = 4x$; **g)** $99x + 54 = 93x$;
h) $41y - 28 = 34y$; **i)** $49 - 5x = 2x$; **j)** $2x + 6 = 1 + x$;
k) $3y = 9 + 2y$; **l)** $13a = 55 + 2a$; **m)** $7x + 5 = 5x - 7$;
n) $4y - 3 = y + 6$; **o)** $6t - 2 = t + 13$; **p)** $3(x + 4) = 2x$;
q) $2(x - 5) = x - 1$; **r)** $5 \cdot (1 + x) = 2 \cdot (7 + x)$; **s)** $5x + 4 \cdot (1 - x) = -x$;
t) $2 \cdot (2 + x) - 3x = x + 10$; **u)** $7x - 2 = 3x + 2$; **v)** $4x - 2 \cdot (1 - x) = 4$;
w) $2 \cdot [3 \cdot (2 + x) - (-x)] = 5 \cdot (x + 3) - 2(1 - x)$;
x) $x - 1 - [2 \cdot (3 - 2x) - 3 \cdot (3 - x)] = x + 1$.

I. 42. Solve in **Z** the inequations:
a) $x + 3 > 6$; **b)** $x - 3 \leq 4$; **c)** $-2 > x - 6$; **d)** $x + 8 \leq 1$;
e) $3x < 12$; **f)** $2x \leq -16$; **g)** $-4x < 20$; **h)** $-2x \geq -10$;
i) $-x + 4 \geq 1$; **j)** $x - 5 < 13$; **k)** $x + 3 \leq 12$.

I. 43. Solve in **Z** the inequations:
a) $5x > 4x + 5$; **b)** $7x - 3 > 6x + 6$; **c)** $2x - 1 < 4x + 3$;
d) $3 + 3x > 7 - x$; **e)** $9 - 2x \geq 6 - x$; **f)** $3(x - 4) \leq 6$;
g) $8 < 4 \cdot (3 - x)$; **h)** $2 \cdot (x - 6) < 2 \cdot (x - 2)$;
i) $6 \cdot (1 - y) \geq 3 \cdot (3 - y)$.

I. 44. Solve in **Z** the inequations:
a) $3 \cdot (3 - x) - 5 \cdot (5 + x) > 0$; **b)** $2 \cdot (x - 3) > 3 \cdot (x + 2) - 3$;
c) $2 \cdot (x - 2) - 4 \leq 3(x + 1) + 3$; **d)** $2 \cdot (y - 1) - 3(y + 2) \leq 3 - 2y$;
e) $3 - a + 2 \cdot (a - 3) \geq 3 \cdot (a + 1)$;
f) $3 \cdot (2y - 1) - 2 \cdot (y + 1) > 3y + 8$.

I. 45. Solve in **Z** the inequations:

a) $-3 < x \le 4$; b) $4 \le 2x \le 8$; c) $-2 \le x + 2 < 3$; d) $-3 < x - 4 \le 2$;
e) $-3 < -2 + x \le 2$; f) $-4 < -x \le 2$; g) $-1 \le -x + 1 < 3$;
h) $-9 \le -3x \le 3$; i) $-9 \le -2x + 5 \le 3$; j) $-1 < -3x + 2 \le 5$.

TEST I. 1

1) The first three terms of two patterns are respectively 1,4,7 and 3,7,11 find the first six terms in each pattern.

2) Write each number in standard form.
   a) 800,000+60,000+5,000+30.
   b) seven hundred five thousand eight hundred six.

3) Calculate:
   a) 50×40; b) $5 \times 7,000$; c) 37×78; d) $513 \div 9$;
   e) $632 \div 8$; f) $581 \div 7$.

4) Alex has 736 marbles to share among 8 people. How many marbles will each person get?

5) Vlad played video games for 329 minutes. His brother played 7 times less than Vlad. How long did his brother play?

6) Alex, Matthew, and Vlad worked together cutting grass for their neighbours. At the end of the day, they had earned exactly 9 of each of the 6 kinds of coins.
   a) How much money did they earn?
   b) The children shared the money equally. How much money did each child get?

7) Diana had $10.47. She spent $4.69. How much money does she have left?

8) At school, Alex bought chocolate ice-cream for $1.19, a hot dog for $1.39, and a can of juice for $0.5. On the way home, he found a quarter on a bench. At the end of the day he had $2.45 in his

wallet. How much money did Alex start with?

9) Matthew walks 80 m every minute. Alex walks 85 m every minute. How far will each person walk in 5 minutes? 20 minutes? 1 hour?

10) Vlad got from his father $10. In the first day he bought a can of juice for $0.69 and a cup of vanilla ice-cream for $0.99. In the second day his mother gave him $2 and he bought a ball for $2.99. How much money does he has left?

TEST I. 2

Find out the value of a such that:

1) $18 + a = 49$; 2) $a - 31 = 49$; 3) $7 \cdot 4 + a = 5 \cdot 8$;
4) $a - 3 \cdot 7 = 5 \cdot 7$; 5) $a + 3 \cdot 7 = 5 \cdot 8$; 6) $4 \cdot 5 - a = 3 \cdot 6$;
7) $3 \cdot 7 + 2 \cdot 5 + a = 5 \cdot 9$; 8) $a - 3 \cdot 7 = 3 \cdot 9 + 4 \cdot 7 - 3 \cdot 4$;
9) $5 \cdot 9 + 4 \cdot 8 - a = 1 \cdot 9 + 9 \cdot 3$;
10) $4 \cdot 7 + 5 \cdot 5 + a = 5 \cdot 8 + 7 \cdot 7 + 8 \cdot 0$.

TEST I. 3

1) Find a from this equation
$(100 - 1) + (101 - 2) + (102 - 3) + (103 - 4) + (104 - 5) = 500 - a$.

2) I multiplied my age by 4, added 24, divided by 2, and then subtracted twice my age. What number did I finally get?

3) Calculate $2^7 - 2^6 - 2^5 - 2^4$.

4) The average of six numbers is 12. Subtract 6 from one of these numbers. What would the average be after I subtract 6 from one of these numbers?

5) The length of each side of a hexagon is a whole number. What number(s) can not be the perimeter of this hexagon: 5, 36, 77, or 78?

6) The sum of the squares of the first 20 positive integers is 2870. What is the sum of the squares of the first 19 positive integers?

7) Calculate $(2+4+6+8+10) \div (10+8+6+4+2)$.

8) How many edges does a cube have?

9) Alex has $2.50. He bought 13 gummy bears at 7¢ each. How many gumballs at 14¢ each can he buy with the money he has left?

10) Diana has just begun to read a 270-page book. If she reads 30 pages everyday, how many days will it take her to read the whole book?

TEST I. 4

1) Calculate: $455 + \{455 + [455 + (455 - 455 \div 5)]\} - 7 \cdot 13 \cdot 16$.

2) Solve the equations:
   i) $x \div 16 = 9$ and reminder 9;   ii) $320 \div x = 80$;
   iii) $5x + 8 = 53$;   iv) $3 \cdot (x + 6) = 138$;
   v) $(5x + 40) \div 5 = 17$.

3) If $ab - ac = 138$ and $a = 23$, then calculate $b - c$.

4) If $ab + ac = 432$ and $b + c = 12$, then calculate a.

5) If $a + b + c + d = 30$ and $a + d = 10$, then calculate $7a + 9b + 9c + 7d$.

6) If $x = 10$ and $a + b = 8$, then calculate $3x + 5a + 5b$.

7) Knowing that the difference of two numbers is 152, and one of them is 5 times bigger that the other, find the numbers.

8) The sum of two numbers is 265, and one of them is 4 times bigger that the other. Find the numbers.

9) The sum of two numbers is 513 . Dividing one to the other one we get the quotient 50 and the reminder 3 . Find the numbers.

10) Vlad buys 5 kg of grapes and pays $2 per kg and 4 times more tomatoes paying $3 per kg. How much money does he get back if he gives to the seller a 100 dollar bill?

TEST I. 5

1) Find out what a is:
  a) $a \times 0.45 = 2.70$ **b)** $a \times 68 = 612$
  c) $2.3 \times a = 1.84$ **d)** $a \times 96 = 864$.

2) Find the missing number a in $(1.5 + 1 + a + 2.5) \div 2 = 9$.

3) Find the letter x in the following equation:
$$5 \times [200.31 + (x + 15.45)] = 2005$$

4) Calculate $255 \div (24.5 - 9.5) - (120 \div 6 - 159 \times 0) \div 2$

5) The sum of three numbers is 140. One of the numbers is 45 and the other one is 70. Find out what the third number is.

6) In a farm there are pigs and hens. All together they have 40 eads and 100 legs. How many pigs and hens are in the farm?

7) The sum of two numbers a and b is 345, and a is 15 less than b. Find the numbers.

8) Calculate $44.5 + 5 \times 32 \div 8 + 5 \times [40 + 8 \times (200 \div 5 - 72 \div 2)]$

9) The sum of three numbers is 120. One of them is 30, and the second one is 5 times more than the third number. Find the numbers.

10) Calculate $3 \times 1000 - \left[(5 \times 10 \times 2 - 8 \times 10)\right] \times 25 \div 10$.

TEST I. 6

1) Find out what a is:
a) $a \times 45 = 270$,   b) $a \times 63 = 567$,
c) $a \times 33 = 264$,   d) $a \times 86 = 774$.

2) Find the missing number a in $(2 + 1 + a + 3) \div 2 = 9$.

3) Find the letter x in the following equation:
$2 \times \left[500 + (x + 500)\right] = 2006$.

4) Compute $255 \div (25 - 10) - (240 \div 12 - 144 \times 0) \div 2$.

5) The sum of three numbers is 161. One of the numbers is 35 and the second one is 71. Find out what is the third number.

6) A six digit number has the last digit 6. If the digit 6 is moved in front of the number, then the new number is four times bigger than the initial number. Find the number.

7) Compute $5 \times 1100 - \left[(25 \times 10 \times 2 - 9 \times 10)\right] \times 15 \div 10$.

8) The sum of two numbers a and b is 224, and a is 46 bigger than b. Find the numbers.

9) Compute $33 + 5 \times 64 \div 16 + 5 \times \left[44 + 8 \times (220 \div 5 - 93 \div 3)\right]$.

10) The sum of three numbers is 212. One of them is 30, and the second one is 6 times more than the third one. Find the numbers.

TEST I. 7

1) **a)** $2^3 \cdot 2^5 \cdot 2$; **b)** $3^4 \cdot 3 \cdot 3^2 \cdot 3^3$; **c)** $a^2 \cdot a^3 \cdot a^4 \cdot a^6$;
   d) $7 \cdot 7 \cdot 7^5 \cdot 7^7$; **e)** $\left(5^{10} \div 5^3\right) \div 5^2$; **f)** $2^{10} \div 2^8$; **g)** $\left(3^2\right)^4$;
   i) $\left[\left(5^{12} \div 5^{11} \cdot 5\right)^2\right]^3$; **2) a)** $2^3 \cdot 2^5$; **b)** $3^4 \div 3^2$;
   c) $\left(2^3\right)^4$; **d)** $\left(5^2 \cdot 7^2\right)^4$.

3) Compare the numbers:
   a) 2^{10} & 8^3; **b)** 32^{20} & 16^{25}; **c)** 2^{15} & 3^{30};
   d) 3^{100} & 2^{150} **e)** 5^{30} & 3^{45}.

4) **a)** $\left(2^5 \cdot 3^7 \cdot 5^6\right)^2 \div \left(2^{10} \cdot 9^7 \cdot 5^{11}\right)$; **b)** $\left(2^{3^4} \div 2^{4^3}\right) \div \left(2^3\right)^4$;
   c) $32^{10} \div 4^{25}$; **d)** $\left[\left(5 - 2^2\right)^{10} + \left(17 - 2^4\right)^{13}\right]^5$;
   e) $\left(25^2 \div 5^4\right)^{2010}$; **f)** $\left\{\left[\left(3^7 + 3^8 + 3^9 + 3^{10}\right) \div 3^7 - 1\right] \div 3 - 1\right\} \div 3$;
   g) $\left[2^3 \cdot 2^8 + 5^{70} \div 5^{10} - \left(3^2\right)^{25}\right] \div \left[4^2 \cdot 2^6 \cdot 2^1 + \left(5^5\right)^{2^2} - \left(3^2\right)^{5^2}\right]$.

5) Solve the equations:
   a) $2^x \cdot 2^3 = 2^7$; **b)** $x \cdot 2^{16} = 2^{19}$; **c)** $3^9 \div x = 3^7$; **d)** $\left(2^3\right)^x = 2^{18}$.

6) **a)** $\left(5^3 - 10^2\right)^2 = 5^x$; **b)** $\left(2^{10}\right)^4 \cdot 2^{5^2} \cdot 2^{12^0} = 2^x$;
   c) $\left[\left(5^{2^3} \cdot 5^{3^2}\right) \div 25^2\right]^5 = 5^x$.

7) Compare the numbers $a = 2^5 \cdot 3^7 \cdot 5^{26}$ and $b = 2^{30} \cdot 5^5 \cdot 7^7$.

8) Compare the numbers
   $a = 4^{n+2} \cdot 9^{n+1} - 4^{n+1} \cdot 3^{2n+3}$ and $b = 3^{2n+3} \cdot 4^{2n+3} - 2^{2n+1} \cdot 6^{2n+3}$.

9) **a)** Find the natural number n if
   $8^{(n-4)(n+4)} - 8 = 7 \cdot \left(8 + 8^2 + 8^3 + \dots 8^{2008}\right)$
   b) Prove that $25^{n+1} \cdot 3^{n+1} - 3^{n+2} \cdot 5^{2n+1} + 5^n \cdot 15^n$ is divisible by 31 for any $n \in N^*$.

10) If $a = 2^{100} + (2^{50})^2 + (2^{25})^4 + (2^5)^{20}$ and

$b = 2^{99} + (2^{33})^3 + (2^{11})^9 + (2^3)^{33} + 2^{99} \cdot x$, find x such that $a = b$.

TEST I. 8

1) Calculate: **a)** $1 - 2^3 + 3^2$; **b)** $(5^2 + 5 + 1) \div (3^3 + 2^2)$.

2) Calculate $\left[10^3 \div 5^3 + (2^2)^3 \div 2^{2^2} - 3^3\right] \cdot 2^{1^2} + 1^{2010} + 2010^0$.

3) Calculate

$22 \cdot 44 - \left\{14 \cdot 13 - 5 \cdot \left[2^3 \cdot 5 \div (3^2 \cdot 5 \div 15 + 5) + 2 \cdot (13^4 \div 13^3 - 12)\right]\right\} \cdot 5$.

4) Calculate, taking to common factor: **a)** $4^2 \cdot 15 - 4^2 \cdot 14$,

b) $11^2 \cdot 20 - 11^2 \cdot 18 - 11^2$; **c)** $30 \cdot 11^2 - 11^2 \cdot 28$;

d) $(2^{2007} + 2^{2007}) \div 2^{2006}$; **e)** $(3^{2007} + 3^{2005}) \div 3^{2005}$;

f) $2^7 - 2^6 - 2^5 - 2^4 + 2^3$.

5) Calculate

$\left\{\left[(7^2 - 14) + 2^5 \div 4\right] \div (3^3 + 2^4) + 3^2\right\} \cdot 11 - \left[(4 + 3^2) \cdot 5 - 55\right]$.

6) Calculate $\left\{\left[(5^4 - 5^2) \div 5^2 + (2^5 + 4^3) \cdot 2^2\right] - 3^4\right\} - 2010^0$.

7) Find x from the equality

$15 \cdot (3 \cdot x - 7 \cdot 2 + 3^2) + (2 \cdot x + 5^2 \cdot 2 - 2^2 \cdot 3^2) \cdot 5 = 5 \cdot (2 \cdot 4)^2 \div 2$

8) Find x: $\left\{\left[(x - 4) \cdot 5 + 4^5 \div 2^{3^2} - 3^{10} \div (27)^3\right] \cdot 5 - 4^5 \div (2^3)^3\right\} \div 3 = 1$.

9) Calculate:

a) $1400 + 350 \div 5 + 15 \div 3 \cdot \left[265 - (50 \div 25 + 2) \cdot 65\right] \cdot 150$;

b) $7^{100} \div \left[7^{40} \cdot 7^{58} - (7^{10} \cdot 7^{15})^5 \div 7^{27} + (8^{27} \div 8^{26} - 1^{120})^{97} \cdot 7\right]$;

c) Find the number k such that:

$64 + 2 \cdot 64 + 3 \cdot 64 + \ldots + 49 \cdot 64 = k^2$.

10) For any n a natural number we consider the numbers a, b and c:
$$a = \left[(3+4)^2 - (3^2 + 4^2)\right] \div \left(4^{n+1} \div 2^{2n}\right) - 2^2 \cdot 5 \div (3^2 - 2^2) - 1;$$
$$b = (25^n \div 5^{2n} + 1)^n; \quad c = \left[(2^n + 2^n) \div 2\right]^2.$$
Compare a, b, c. and find the number n such that $2ab = c$.

TEST I. 9

1) Show that the sum of three consecutive natural numbers is divisible by three.

2) Calculate:
   a) $[(2 + 4 + 6 + \ldots + 100) + (1 + 3 + 5 + \ldots + 99)] \div 505$;
   b) $2 + 4 + 6 + \ldots + 100 - (1 + 3 + 5 + \ldots + 99)$.

3) Consider the numbers:
   $a = 10 \cdot \{5 + 10 \cdot [362 - 10 \cdot (2^3 \cdot 3 + 24 \div 2^2)]\} + 2^2$;
   $b = 2^3 + 2 \cdot [(3 \cdot 5^2 - 2^2 \cdot 18 \div 3^2 + 3 \cdot 11) \div 10^2].$
   Prove that $a = 5^4 \cdot b + 4$.

4) Find $x \in \mathbb{N}$ from the equality $(x - 2)^{2x-4} \cdot 2^3 = 8^{17}$.

5) Consider $A = 2^{2n+1} \cdot 3^{n+5}$, $B = 16^n \cdot 9^{n+4}$; $a = 4^n$,
   and $b = 3^{n+4}$, $n \in \mathbb{N}^*$.
   a) Write A in terms of a and b.
   b) Show that $B = (a \cdot b)^2$.

6) Find the numeric value of each number.
   $x = 3^{1999} - (4 \cdot 9^{999} - 729^{333})$, $y = (8^{651} - 1024^{195}) \div (2^3 - 1)$,
   y^x, x^y, and show that the number $3^5 - 3^3$ is a perfect cube.

7) If $\overline{ac} \cdot b = \overline{bc} \cdot a + 1$, find the number $\overline{abc}$.

8) Compute the value of the expression
   $E = 1 \cdot (a - b) + 2 \cdot (b - c) + 3 \cdot (c - 4 \div 3)$ for $a + b + c = 4$.

9) Find the digit a such that the number $\left(\overline{a2} \cdot \overline{a0} + 1\right) \div \overline{aa}$ is a natural number.

10) Show that the number $a = 75 \cdot 5^{2n-1} - 8^n \div 2^{3n}$, for $n \in \mathbb{N}$ is not divisible by 5.

TEST I. 10

1) Calculate $804 - 4\{502 - [58 + 4(31 \cdot 9 - 234 \div 3)] \div 2\}$.

2) At a number we add 46, the result is multiplied by 9 and we obtain 477. What is the number?

3) If
$$a = 8 \cdot 6 - 5 \cdot 4 - 3 \cdot 2;$$
$$b = 8 \cdot 7 + 6 \cdot 7 + 4 \cdot 3 + 2 \cdot 1;$$
$$c = 3 \cdot 4 + 5 \cdot 6 + 6 \cdot 8 + 8 \cdot 10 + 11 \cdot 10;$$
Calculate $c \cdot (b - a)$.

4) Find the right place for a pair of round brackets
$7 \cdot 8 + 4 \cdot 11 - 50 = 14$.

Compute:

5) $9 \cdot (77 - 70) - [3 + (46 - 39) \cdot 3]$;

6) $102 - 10 - [3 \cdot (29 - 20) - (99 - 96) \cdot 8]$;

7) $8 \cdot (14 - 6) - [(16 - 9) + 14] + 2$;

8) $9 \cdot (20 - 11) - [2 + 6(17 - 8)]$;

9) $12.5 + 6 \times 32 \div 8 + 5 \times [40 + 8 \times (600 \div 15 - 144 \div 4)]$;

10) $6 + 5 \times 64 \div 16 + 4 \times [40 + 8 \times (200 \div 5 - 33)]$.

TEST I. 11

1) Compute:

a) $-2 \cdot \{13 - 2 \cdot [14 - 5(-13 + 20 - 6)]\}$;

b) $\left|2 \cdot (-5) - (-4)\right| - 6 + 1\left|-(-5)^2\right|$;

c) $\{[(-12) \div 4 - (8 - 9 + 3 - 5) - (-2)^3 \div 4] \cdot 3 - 2 \cdot 10\} \div (3 - 17)$;

d) $-\left[3 \cdot (-2)^3 + 56 \div (-2)^2 - (-2^5) \div 2^4\right]$;

e) $3.75 \div 0.2\overline{7} \cdot 0.\overline{1} + 0.5$; **f)** $\left[0.\overline{7} + 7.\overline{7}\right] \div (0.7 + 0.7)$;

g) $5.6 \cdot (4.5 \div 0.9 + 1.25) \div 1.4 - (2.127 + 5.873) \div 4$;

h) $0.1 \cdot \{2.5 - [0.4 \cdot (28.2 \cdot 1.5 - 28.1) - 0.08] \div 2.8\}$.

2) Solve the equations and inequations:

a) $x + 25.17 = 32.15$; **b)** $37.5 - x = 17.03$;

c) $3.1 \cdot x = 1.24$; **d)** $x - 1.2 < 3$;

e) $[7.23 - (0.5x - 2.7)] \cdot 1.5 = 2.25$; **f)** $0.4x + 1.4 \le 9.6$;

g) $(2.4x + 5) \cdot 1.9 = 27.74$.

3) Show that the next numbers are perfect squares:

a) 100; **b)** 9^7; **c)** $2 \cdot (1 + 2 + 3 + \ldots + 1000) - 1000$;

d) $(2^7 \div 2^3) \cdot 2^4$.

4) Consider the numbers $a = c^2 + cd$ and $b = c^3 - 5cd$, where $c = -4$ and $d = 3$.

a) Find a and b;

b) Solve the equation $a(x + c) + b(x + d) = x(a + b + c + d)$;

c) Calculate $S = c + 2c + 3c + \ldots + 100c$.

5) Prove that the next numbers are perfect squares:

a) 121; **b)** 4^5; **c)** $2 \cdot (1 + 2 + 3 + \ldots + 2010) - 2010$; **d)** $\left(2^{2^5} \div 2^{5^2}\right) \cdot 2^{3^2}$

6) Consider the numbers $\overline{ab}$, with $a \ne 0$ şi $b \ne 0$.

a) Prove that $\overline{ab} + \overline{ba} = 11 \cdot (a + b)$;

b) Find all the numbers $\overline{ab}$ if $\overline{ab} + \overline{ba} = 99$.

7) If x and y are digits such that $\overline{23xy} + \overline{xy23} + \overline{xyxy} = 11413$, then find $x + y$.

8) a) Show that the number $10^{40} + 455$ is divisible by 5;
   b) Find the sum of the digits of the number $10^{40} + 455$;
   c) Show that $15^{15} - 5^5$ is divisible by 10;
   d) Prove that $7^n - 3$ is divisible by 2, $n \in N^*$.

9) Let $N = 3^{n+2} \cdot 2^{n+3} + 3^{n+1} \cdot 2^{n+4} + 5$; $n \in N^*$ be a number. Show that the number is divisible by 720.

10) If a and b are digits, $a \neq 0$. Show that the number $\left(\overline{a1b} + \overline{a2b} + ... + \overline{a9b}\right) \div \left(\overline{a5b}\right)$ is a perfect square.

TEST I. 12

1) Prove the equality $\overline{xx2} + \overline{x4} = \overline{xx1} + \overline{x2} + 3$, where x is a digit between zero and nine.

2) Determine the non-zero digit number a from the equality,
$a = 0.\overline{1a} + 0.\overline{2a} + 0.\overline{3a} + ... + 0.\overline{9a}$.

3) Determine the non-zero digit a such that:
$2.\overline{1a} + 2.\overline{a1} = a.\overline{12} + 1.\overline{a2}$.

4) Show that the number $N = \underbrace{111...1}_{n\ times}\underbrace{222...2}_{n\ times}$ is a product of two consecutive numbers.

5) If a, b, c, d are non-zero digits such that $\overline{ab} + \overline{bc} + \overline{ca} = \overline{dd}$, show that $\overline{abc} + \overline{bca} + \overline{cab} = \overline{ddd}$.

6) Show that $a = \left(\overline{x1x} + \overline{x2x} + ... + \overline{x9x}\right) \div \overline{x5x}$ is a perfect square.

7) Prove that the number $a = 10^3 \cdot \left(\overline{x.x} + \overline{x.0x} + \overline{x.00x}\right)$ is divisible by 61 for any non-zero digit x a digit number.

8) Find the number $\overline{abc}$ if $\overline{abc0} + \overline{abc} = 2057$.

9) Find the digits a, b, c such that
$$\overline{abc2} + 3 \cdot \overline{abc} = \overline{4abc} + 1234.$$

10) Find the non-zero digits a and b, such that $\overline{abb}^{\,\overline{ab}} = \overline{ab}^{\,20 \cdot a + 2 \cdot b}$.

TEST I. 13

1) Prove that the number $A = 2 + 4 + 6 + ... + 4002 + 2002$ is a perfect square.

2) Prove that the number $A = 3^{2n+3} \times 4^{2n+3} - 2^{2n+1} \times 6^{2n+3}$ is a perfect square, $n \in N$.

3) Prove that the number $n = 2^{2013} - 2^{2012} - 2^{2011} - 2^{2010}$ is a perfect square.

4) Prove that the number $A = 2^{\overline{n0}} + 3^{\overline{n1}} + 5^{\overline{n2}} + 6^{\overline{n3}}$ is not a perfect square, where n is a number verifying the equality
$$2^{6n+10} + 4^{3n+5} + 8^{2n+3} = 5 \cdot 2^{3(n+3)+15}.$$

5) Prove that the number $n = 2^0 + 2^1 + 2^2 + 2^3 + ... + 2^{98}$ is divisible by 7.

6) Show that the number $n = 3^1 + 3^2 + ... + 3^{2010}$ is divisible by 363.

7) Prove that the number
$$N = 2013^{2013} + 2014^{2014} + 2015^{2015} + 2016^{2016} + 2017^{2017}$$
is not a perfect square.

8) Prove that the number $a = 11^{2012} + 22^{2012} + 33^{2012} + ... + 99^{2012}$ is not a perfect square.

9) Prove that the number $a = 4^{n+1} \cdot 9^{n+2} + 6^{2n+1} \cdot 9 - 2^{2n} \cdot 9^{n+1}$ is divizibile by 369, for any n a natural number.

10) Compare the numbers x, y, and z, where n is a natural number.
$x = \{1^2 + 2^3 \cdot [3^4 - 3 \cdot (2^5 - 5)]\} \cdot 2^n$;
$y = [(1 + 2 + 3 + ... + 10) \div 11 - 2]^{n-1}$,
$z = [(1 + 2^2 \div 4) - (3^2 - 2^3)]^3 \cdot (1002 \div 3 - 331)$;
Find the last digit of the number $b = x \cdot y \cdot z$.

TEST I. 14

1) Calculate:
a) $1234 \cdot 56 + 45 \cdot 1234 - 1234$;
b) $c^a + c^b + c^c$ where
$$a = \{[10 \cdot 1000 - (300 - 100) \cdot 50]^{2008} + 1\}^{2009},$$
$$b = \left(2^2 \cdot 2^3 \cdot 2^4\right)^3 - 8 \cdot \left(4^3 \cdot 4^4 \cdot 4^5\right),$$
$$c = \{10^3 + 10^2 \cdot [10 + (5^2 - 5^3) \div 5]\} \div 20^2 + 1.$$

2) Compare the numbers:
a) 9^{60} and 27^{40}; b) $7 \cdot 7^2 \cdot 7^3 \cdot 7^4 \cdot ... \cdot 7^{50}$ and 5^{1275}.

3) If $a = 2^{100} + (2^{50})^2 + (2^{25})^4 + (2^5)^{20}$ and
$b = 2^{99} + (2^{33})^3 + (2^{11})^9 + (2^3)^{33} + 2^{99} \cdot x$ find x such that $a = b$.

4) Consider the numbers
$x = \left[2^{30^2} \cdot (2^6)^{100} \cdot 2 + (32^8)^{50} \div 2^{500}\right]^2 + 2^{3004}$ and
$y = 5 \cdot (9^{1001} - 9^{1000} - 3^{2001})$.
a) Show that x and y are perfect squares;

b) Show that x is divisible by 10;

c) Compare the numbers x and y.

5) Prove that the number $x = \overline{ab0} + \overline{ba0} + \overline{ab} + \overline{ba}$ is divisible by 11 for any digits a and b.

6) Prove that the number $x = \overline{ab0} + \overline{ba0}$ is divisible by 110 for any digits a and b.

7) Prove that the number $x = \overline{ab0} + \overline{ba0} + \overline{a0b} + \overline{b0a}$ is divisible by 211 for any digits a and b.

8) Find the number $\overline{xyz}$ knowing that $\overline{xyz0} + \overline{xyz} = 2002$.

9) Prove that the number $\underbrace{\overline{abcabcabc...abc}}_{2010\ digets}$, is divisible by 7 for

any digits a and b.

10) Prove that the number $y = \overline{abc} + \overline{acb} + \overline{bac} + \overline{bca} + \overline{cab} + \overline{cba}$ is divisible by 37 for any digits a and b.

TEST I. 15

1) What is sum of the digits of the number
$x = 10^{2006} + 9 \cdot 10^{2002} - 2$?

2) Prove that the number $\underbrace{299...998}_{n-1\ times}\underbrace{200...0029}_{n-2\ times}$ can be written as a

sum of three squares of three consecutive numbers.

3) Show that the number $\overline{ab0ab0}$ is divisible by 1001.

4) Find the digit x holding the equality $\overline{2x31} + \overline{24x2} + \overline{5x3} = \overline{5x16}$.

5) Find the number $\overline{xyz}$ such that $\overline{xyz0} + \overline{xyz} = 2002$.

6) Find x from the equalities:

   a) $15 \cdot \left(3 \cdot x - 7 \cdot 2 + 3^2\right) + \left(2 \cdot x + 5^2 \cdot 2 - 2^2 \cdot 3^2\right) \cdot 5 = 5 \cdot \left(2 \cdot 4\right)^2 \div 2$;

   b) $\left\{\left[(x - 4) \cdot 5 + 4^5 \div 2^{3^2} - 3^{10} \div (27)^3\right] \cdot 5 - 4^5 \div \left(2^3\right)^3\right\} \div 3 = 1$.

7) Find the number $\overline{abc}$ such that $\overline{abc} + \overline{bc} = 196$.

8) Prove that the number $N = \underbrace{111...11}_{n \ times}\underbrace{222...22}_{n \ times}$ can be written as a product of two consecutive numbers.

9) Find the natural number n such that $2011^n - n^{2011} = 2010$.

10) Show that the number $444^{2010} + 9$ is divisible by 5.

Chapter II

RATIONAL NUMBERS AND FRACTIONS

The set of *rational numbers* $\mathbf{Q} = \left\{ \frac{a}{b} \mid a \in \mathbf{Z}, b \in \mathbf{Z}, b \neq 0 \right\}$.

Finite decimal numbers $a.b = \frac{ab}{10}$, $a.bc = \frac{abc}{100}$.

Periodic numbers $a.\overline{b} = \frac{ab - a}{9}$, $a.b\overline{c} = \frac{abc - ab}{90}$,

$a.b\overline{cd} = \frac{abcd - ab}{990}$.

Ratio and proportions $\frac{a}{b}$, $b \neq 0$ is a *ratio*. $\frac{a}{b} = \frac{c}{d}$ is a *proportion*.

The fundamental property of a proportion is $ab = cd$.

Denominator Addition/Subtraction property $\frac{a+b}{b} = \frac{c+d}{d}$ or

$\frac{a-b}{b} = \frac{c-d}{d}$, similarly for nominator *Addition/Subtraction*

property

$\frac{a}{b \pm a} = \frac{c}{d \pm c}$. Also $\frac{a}{b} = \frac{c-a}{d-b}$ or $\frac{a^2}{b^2} = \frac{c^2}{d^2}$.

II. 1. Write the numbers as decimals or periodic numbers:

a) $\frac{3}{4}$; **b)** $\frac{3}{5}$; **c)** $\frac{73}{100}$; **d)** $\frac{5}{6}$; **e)** $\frac{2}{3}$; **f)** $\frac{7}{20}$; **g)** $\frac{34}{35}$; **h)** $\frac{13}{15}$; **i)** $\frac{3}{25}$.

II. 2. Write each number as a fraction:

a) 0.7; **b)** 2.31; **c)** 0.01; **d)** 0.013; **e)** $0.\overline{34}$; **f)** $0.\overline{7}$; **g)** $2.\overline{3}$;
h) $1.\overline{43}$; **i)** $0.9\overline{3}$; **j)** $0.5\overline{6}$; **k)** $2.1\overline{3}$; **l)** $1.41\overline{6}$.

II. 3. Simplify the fractions:

a) $\dfrac{35}{105}$; **b)** $-\dfrac{54}{88}$; **c)** $\dfrac{99}{8181}$; **d)** $-\dfrac{35}{49}$; **e)** $\dfrac{2010}{5\cdot12\cdot67}$; **f)** $\dfrac{3^3\cdot7^3}{3^3\cdot7^4}$;

g) $\dfrac{256}{2^7\cdot5}$; **h)** $\dfrac{5^2-4^2}{2^2\cdot3^2}$; **i)** $\dfrac{3^2\cdot2}{105-69}$; **j)** $\dfrac{972}{963}$.

II. 4. Compare the numbers:

a) $-\dfrac{3}{4}$ and -0.75; **b)** $-\dfrac{3}{4}$ and $-\dfrac{2}{4}$; **c)** $-\dfrac{2}{5}$ and $\dfrac{1}{7}$;

d) 0.47 and $\dfrac{47}{100}$; **e)** $1.\overline{4}$ and $1\dfrac{4}{9}$; **f)** -0.35 and 0.34;

g) $-\dfrac{3}{8}$ and $-\dfrac{1}{3}$; **h)** $-\dfrac{2}{3}$ and $-\dfrac{7}{8}$; **i)** $\dfrac{2}{3}$ and $\dfrac{3}{4}$.

II. 5. Write the numbers in an increasing order:

a) $-2,\ -\dfrac{4}{3},\ -\dfrac{1}{3},\ 0,\ \dfrac{2}{3},\ \dfrac{4}{3},\ 2$;

b) $-1\dfrac{1}{5},\ -\dfrac{4}{5},\ -\dfrac{2}{5},\ 0,\ \dfrac{2}{5},\ 1\dfrac{1}{5},\ 1\dfrac{3}{5}$;

c) $-2\dfrac{1}{4},\ -\dfrac{1}{4},\ \dfrac{1}{2},\ \dfrac{3}{4},\ \dfrac{1}{4}$; **d)** $\dfrac{1}{4},\ \dfrac{5}{7},\ \dfrac{3}{3},\ \dfrac{8}{2},\ \dfrac{9}{5},\ \dfrac{2}{9}$;

e) $\dfrac{3}{5},\ 5\dfrac{3}{2},\ \dfrac{4}{1},\ 2\dfrac{5}{7},\ \dfrac{4}{7},\ 1\dfrac{9}{3}$.

II. 6. Calculate:

a) $\dfrac{3}{8}-\dfrac{5}{8}$; **b)** $-\dfrac{4}{9}+\dfrac{7}{9}$; **c)** $-\dfrac{4}{7}+\left(-\dfrac{1}{7}\right)$; **d)** $-\dfrac{4}{7}+\dfrac{5}{7}$; **e)** $\dfrac{2}{3}+\left(-\dfrac{2}{3}\right)$;

f) $0.9+0.7$; **g)** $0.46+0.75$; **h)** $0.36+0.9$; **i)** $-0.6+0.2$;
j) $0.3+(-0.14)$; **k)** $-0.4+(-0.3)$; **l)** $2.6+(-5.82)$;
m) $-0.9+6.15$; **n)** $-0.29+(-0.9)$; **o)** $6.5+(-0.019)$. .

II. 7. Calculate:

a) $-\dfrac{4}{7}+\dfrac{1}{2}$; **b)** $\dfrac{4}{7}+\left(-\dfrac{1}{2}\right)$; **c)** $-\dfrac{5}{7}+\dfrac{4}{8}$; **d)** $-\dfrac{3}{5}+\left(-\dfrac{1}{2}\right)$;

e) $\dfrac{2}{5} + \left(-\dfrac{2}{3} \right)$; **f)** $-\dfrac{3}{5} - \dfrac{1}{3}$; **g)** $-\dfrac{2}{5} + \dfrac{1}{2}$; **h)** $-\dfrac{1}{2} + \dfrac{1}{4}$;

i) $\dfrac{1}{2} + \left(-\dfrac{3}{6} \right) + \left(-\dfrac{1}{4} \right) + \left(-\dfrac{2}{3} \right)$; **j)** $\dfrac{1}{2} - \dfrac{3}{4} + \dfrac{5}{6} - 1\dfrac{3}{6}$;

k) $1\dfrac{3}{4} + \dfrac{4}{3} - 1\dfrac{1}{2} - 0.75$; **l)** $0.45 + 2.3 - 4.43 - 10$;

m) $2.31 - 0.35$; **n)** $0.42 - (-0.5)$; **o)** $-1.5 - (-1.4)$;

p) $0.25 - (+0.26)$.

II. 8. Calculate:

a) $\dfrac{5}{7} - \left(+\dfrac{2}{7} \right)$; **b)** $\dfrac{3}{8} - \left(-\dfrac{5}{8} \right)$; **c)** $-\dfrac{2}{7} + \left(-\dfrac{1}{2} \right)$; **d)** $\dfrac{1}{5} - \left(-\dfrac{1}{9} \right)$;

e) $-\dfrac{1}{8} - \left(-\dfrac{1}{3} \right)$; **f)** $-\dfrac{1}{6} - \left(-1\dfrac{1}{2} \right)$; **g)** $-\dfrac{3}{4} - \left(+\dfrac{2}{5} \right)$;

h) $-1\dfrac{1}{3} + \left(-1\dfrac{1}{4} \right) - \left(-1\dfrac{5}{6} \right)$; **i)** $\left(-1\dfrac{1}{20} \right) + 1\dfrac{2}{15} - \left(-\dfrac{5}{6} \right)$;

j) $-\left[-\dfrac{1}{16} + \left(-\dfrac{1}{32} \right) \right]$.

II. 9. Calculate:

a) $\dfrac{2}{5} - \left(\dfrac{2}{6} - \dfrac{1}{5} \right)$; **b)** $-\dfrac{1}{7} - \left(\dfrac{2}{5} - \dfrac{4}{7} \right)$; **c)** $-2\dfrac{1}{2} - \left(-\dfrac{3}{5} + 1\dfrac{1}{4} \right)$;

d) $-\dfrac{5}{12} - \left(-\dfrac{7}{18} - \dfrac{13}{24} \right)$; **e)** $1\dfrac{1}{3} - \left(-\dfrac{5}{6} + 1\dfrac{1}{4} \right)$; **f)** $-\dfrac{1}{6} - \left(-\dfrac{5}{12} - \dfrac{7}{18} \right)$;

g) $-\dfrac{5}{9} - \left[-\dfrac{11}{24} + \left(\dfrac{5}{36} - \dfrac{7}{48} \right) \right]$; **h)** $1\dfrac{5}{16} - \left[\left(\dfrac{11}{36} - \dfrac{7}{48} \right) - \left(-\dfrac{5}{18} + \dfrac{11}{24} \right) \right]$;

i) $-2\dfrac{5}{4} - \left[\left(-\dfrac{3}{4} + 5.6 \right) - \left(-\dfrac{1}{2} + \left(-\dfrac{1}{6} \right) \right) \right]$;

j) $\left[\dfrac{1}{3} - \left(-\dfrac{1}{6} \right) \right] - \{ (-1.2) - [0.4 - (-0.9)] \}$;

k) $2\dfrac{3}{4}+\left\{-1\dfrac{2}{3}+\left[0.8+(-7)+\left(+1\dfrac{1}{2}\right)\right]-3\dfrac{1}{5}\right\}$;

l) $-3+\left(-\dfrac{3}{4}\right)+\left[\left(-1\dfrac{5}{6}\right)+2.5+2\right]-\left[3\dfrac{1}{2}+(-3)\right]$;

m) $-2\dfrac{1}{4}-\left\{\dfrac{3}{8}-\left[1\dfrac{1}{6}-\left(2\dfrac{1}{2}-1\dfrac{2}{3}\right)\right]-1\dfrac{3}{4}\right\}$.

Multiplication, division, order of the operations.

II. 10. Calculate:

a) $3\cdot\dfrac{2}{3}$; **b)** $-\dfrac{4}{9}\cdot9$; **c)** $-5\cdot\left(-\dfrac{1}{5}\right)$; **d)** $-0,2\cdot\dfrac{3}{2}$;

e) $-6\cdot(-2.5)\cdot\left(+\dfrac{1}{4}\right)$; **f)** $10\cdot(-0.25)\cdot\left(-\dfrac{3}{5}\right)$;

g) $\dfrac{10}{3}\cdot\left(-\dfrac{7}{52}\right)\cdot\dfrac{26}{35}$; **h)** $0.12\cdot(-0.7)\cdot(-2.5)$;

i) $-3\cdot\dfrac{4}{3}\cdot\left(-\dfrac{7}{44}\right)\cdot\dfrac{11}{35}\cdot\left(-\dfrac{25}{75}\right)$.

II. 11. Calculate:

a) $\dfrac{2}{9}\div\dfrac{1}{3}$; **b)** $\dfrac{3}{25}\div1\dfrac{1}{4}$; **c)** $-\dfrac{8}{7}\div\left(-3\dfrac{1}{2}\right)$; **d)** $-\dfrac{24}{35}\div\left(-1\dfrac{5}{7}\right)$;

e) $-\dfrac{33}{4}\div\left(-2\dfrac{1}{6}\right)$; **f)** $-3\dfrac{3}{7}\div\dfrac{48}{14}$; **g)** $14.4\div1.2$; **h)** $-1.19\div0.17$;

i) $17.5\div(-0.25)$; **j)** $4.59\div\left(-\dfrac{9}{10}\right)$; **k)** $6.5\div\left(-\dfrac{5}{2}\right)$;

l) $-9.6\div\dfrac{3}{10}$; **m)** $\dfrac{8}{3}\cdot\left(-\dfrac{5}{6}\right)\div\left(-\dfrac{15}{18}\right)$; **n)** $-\dfrac{5}{6}\div\dfrac{2}{6}\cdot(-0.2)$.

II. 12. Calculate:

a) $\dfrac{5}{7}+\dfrac{2}{7}\cdot\dfrac{1}{3}$; b) $\dfrac{3}{5}-\dfrac{3}{5}\cdot\dfrac{1}{2}$; c) $\dfrac{5}{3}\cdot\dfrac{3}{8}-\dfrac{3}{4}\cdot 2$; d) $\dfrac{4}{5}+\dfrac{1}{5}\cdot\left(1-\dfrac{3}{2}\right)$;

e) $\left(\dfrac{1}{5}+\dfrac{1}{6}\cdot\dfrac{1}{5}\right)\cdot 6$; f) $\left(-1\dfrac{1}{4}+2\dfrac{1}{2}\right)\cdot(-2)+1$; g) $\dfrac{12}{20}+\dfrac{6}{5}\div\dfrac{6}{10}+\dfrac{1}{2}$;

h) $\dfrac{6}{4}-\dfrac{1}{3}\div\left(\dfrac{1}{5}+\dfrac{1}{15}\right)$; i) $0.5\div\dfrac{3}{5}+\dfrac{-6}{4}\cdot\dfrac{-1}{-2}$;

j) $\dfrac{7}{12}\cdot\dfrac{3}{14}-\dfrac{7}{12}\cdot\dfrac{3}{14}$; k) $-\dfrac{4}{10}\cdot\dfrac{4}{9}-\dfrac{2}{5}\cdot\dfrac{14}{9}$;

l) $-\dfrac{5}{12}\cdot\dfrac{5}{27}-\dfrac{7}{12}\cdot\left(-\dfrac{7}{27}\right)$; m) $-\dfrac{3}{7}\cdot\dfrac{4}{5}-\dfrac{3}{7}\cdot\left(-\dfrac{5}{12}\right)$;

n) $\dfrac{11}{12}+\dfrac{2}{3}\div\left(-\dfrac{36}{27}\right)$; o) $\left(-\dfrac{8}{10}+\dfrac{5}{6}\right)\div\dfrac{4}{10}$;

p) $\left(-\dfrac{44}{33}\right)\cdot\left(\dfrac{5}{21}-\dfrac{3}{14}\right)$; q) $\left(\dfrac{4}{18}-\dfrac{7}{15}\right)\div\left(-\dfrac{22}{75}\right)$;

r) $\dfrac{39}{21}\cdot\left(-\dfrac{7}{15}\right)-\dfrac{3}{7}\div\left(-\dfrac{9}{35}\right)$; s) $\left(-3\dfrac{2}{10}\right)\div\left(-1\dfrac{2}{3}+0.4\right)\cdot\left(-1\dfrac{3}{16}\right)$;

t) $10\cdot\left(2\dfrac{1}{7}-1\dfrac{1}{2}\right)\cdot[-0.1\bar{5}]$.

II. 13. Calculate:

a) $-\left(2+\dfrac{7}{9}\right)\cdot\left[-\dfrac{1}{3}+\dfrac{1}{7}\cdot\left(\dfrac{2}{5}-\dfrac{1}{6}\right)\right]$; b) $\dfrac{11}{4}\div\left[-\dfrac{1}{3}-\dfrac{1}{2}\cdot\left(\dfrac{3}{4}-\dfrac{1}{2}\right)\right]$;

c) $\dfrac{3}{2}\div\left[\dfrac{1}{2}+\left(\dfrac{5}{12}-\dfrac{2}{3}\right)\div\left(-\dfrac{1}{4}\right)\right]$;

d) $2\dfrac{13}{15}\div\left[-1\dfrac{1}{6}+1.16\div\left(3\dfrac{4}{15}-1\dfrac{1}{3}\right)\right]$;

e) $\left(\dfrac{1}{4}\cdot\dfrac{6}{2}+\dfrac{3}{4}\right)\div\left[\dfrac{1}{2}+\dfrac{3}{5}\cdot\left(-\dfrac{5}{4}\right)\right]$; f) $8\cdot\left[\dfrac{1}{4}+\left(\dfrac{7}{36}-\dfrac{1}{24}\right)\div\dfrac{11}{18}\right]$;

g) $2\dfrac{3}{4}-0.75\cdot\dfrac{11}{5}\div\left(2+0.5\div 2\dfrac{1}{2}\right)$;

h) $\left(1.91+0.09+\dfrac{3}{4}\right)\div\left(1\dfrac{2}{5}+2.84+0.16\right).$

II. 14. Calculate:

a) $\left(\dfrac{8}{5}\cdot\dfrac{5}{8}+1-\dfrac{6}{10}\cdot\dfrac{5}{6}\right)\div\left(\dfrac{4}{5}+\dfrac{18}{50}\cdot\dfrac{15}{12}-\dfrac{7}{10}\right);$

b) $\left[-3\dfrac{1}{4}-\dfrac{3}{4}\cdot\left(-\dfrac{20}{9}\right)\right]\div\left(-\dfrac{38}{2}\right)+\left(\dfrac{5}{12}-\dfrac{7}{30}+\dfrac{2}{5}\right)\div\left(-\dfrac{1}{4}-\dfrac{4}{5}\right);$

c) $\dfrac{4}{6}-1\dfrac{2}{3}\cdot\left(2\dfrac{9}{20}\cdot2\dfrac{13}{21}-7\dfrac{7}{15}\right)\div\left(1\dfrac{1}{35}\div1\dfrac{13}{14}-1\dfrac{7}{10}\right);$

d) $\left[-\dfrac{9}{14}\cdot2\dfrac{1}{3}+\left(\dfrac{4}{10}-\dfrac{3}{4}\cdot\dfrac{8}{9}\right):1\dfrac{3}{5}\right]\div\left(-2\dfrac{1}{12}\right);$

e) $\left(\dfrac{10}{5}\div\dfrac{3}{2}-\dfrac{3}{2}\div\dfrac{6}{2}\right)\cdot\left(\dfrac{1}{2}+2\right)-\dfrac{5}{4}\cdot\left(3\dfrac{3}{4}-2\dfrac{3}{4}\right);$

f) $\left[\left(3\dfrac{1}{9}\div3\dfrac{4}{9}+\dfrac{6}{62}\right)\div\dfrac{1}{4}-3\cdot\dfrac{2}{7}\right]\div\left[\left(6\dfrac{3}{5}\div1\dfrac{3}{8}-3\dfrac{4}{5}\right)\div\dfrac{1}{2}\right];$

g) $\dfrac{1}{4}+\dfrac{1}{4}\div\left\{\dfrac{1}{4}\cdot\left[\dfrac{1}{4}+\dfrac{1}{4}\right]\right\};$ **h)** $2\dfrac{1}{3}+\left[\left(1\dfrac{1}{2}+\dfrac{4}{9}\cdot2\dfrac{1}{4}\right)\div\dfrac{1}{9}+2\dfrac{3}{4}\right]\div\dfrac{1}{4};$

i) $\left[\left(1-1\dfrac{4}{5}\right)\div\left(2\dfrac{1}{2}-3\dfrac{1}{3}\cdot\dfrac{3}{5}+\dfrac{3}{4}\right)+2\div\left(1\dfrac{3}{4}-\dfrac{1}{4}\right)\right]\div\dfrac{2}{9}.$

II. 15. Simplify the expressions:

a) $34-8\cdot\left[\dfrac{\left(15-9\dfrac{1}{3}\right)\div5\dfrac{2}{3}}{\left(19\dfrac{2}{3}-11\dfrac{7}{9}\right)\cdot\dfrac{9}{71}}\cdot\left[\left(\dfrac{79}{100}\div\dfrac{1}{4}\right)\div0.79+\dfrac{1}{4}\right]\right];$

b) $1+4\cdot\left[\dfrac{(1-0.1)\cdot1\dfrac{2}{3}}{0.625+\dfrac{3}{8}+0.0625\div\dfrac{1}{16}}+2.04\div\dfrac{3}{25}\right]\div\dfrac{71}{4};$

c) $10 - 4 \cdot \left[\dfrac{0.0625 + \dfrac{2}{5} + 2.04 - \dfrac{1}{25}}{\left(2^3 \cdot 3^2 \cdot 5^4\right) \div \left(5^3 \cdot 3^2 \cdot 2^2\right) \cdot \dfrac{0.1}{3} + \dfrac{33.\overline{3}}{5^2 \cdot 2^2}} \right];$

d) $\dfrac{\left[\left(3\dfrac{2}{5} + 5\dfrac{4}{15}\right) \cdot 0.25 \cdot \dfrac{7}{13} - \left(\dfrac{15}{2} - 0.5\right) \div 7\right] \cdot 6}{\left(5.01 - 4\dfrac{7}{50}\right) \div 0.03} + \dfrac{28}{29};$

e) $\dfrac{2\dfrac{3}{20} - 1\dfrac{1}{2}\left(4\dfrac{1}{6} + \dfrac{0.003}{0.25} \div \dfrac{0,004}{0.2} - \dfrac{1}{2} \cdot 9\dfrac{1}{3}\right) + \left(\dfrac{2}{5}\right)^2 \cdot 1.25}{\left(2.625 + \dfrac{3}{8} + 0.0625 + \dfrac{15}{16}\right) - \left(0.03125 + \dfrac{31}{32} + 1.25 + \dfrac{3}{4}\right)} + \dfrac{4}{5};$

f) $\dfrac{\left(5\dfrac{3}{5} - 5\dfrac{4}{15}\right) \cdot 0.25 \div \dfrac{1}{19} + \left(\dfrac{15}{2} - 0.5\right) \div 7}{\left(\sqrt{26.4196} - 4\dfrac{7}{50}\right) \div 0.125 - 6\dfrac{2}{3}};$

g) $\dfrac{\left(\dfrac{7}{15} + \dfrac{14}{45} + \dfrac{2}{9}\right) \cdot 10\dfrac{1}{3} - 1\dfrac{1}{11} \cdot \left(2\dfrac{2}{3} - 1\dfrac{3}{4}\right)}{\left(\dfrac{3}{7} - \dfrac{1}{4}\right) \div \dfrac{3}{28} - 1} \cdot 2 + 2;$

h) $\dfrac{\left(6\dfrac{5}{9} - 3\dfrac{1}{4}\right) \cdot 2\dfrac{2}{17} + 40.5 \cdot \dfrac{2}{9} \div 9}{4 \div 6.25 - 1 \div 5 + \dfrac{1}{7} \cdot 1.96 + 0.28} \div 24;$

i) $\dfrac{9 - 4.7 \div \left(5 - 0.8 \div 2\dfrac{4}{6}\right)}{\left(5\dfrac{3}{9} - 3\dfrac{3}{4}\right) \div 1\dfrac{7}{12} + 2} \cdot \dfrac{3}{8};$

j) $50.05 - \left(1.5 + 3\dfrac{2}{3}\right) \cdot 7\dfrac{1}{2} \div 3.15 \cdot \dfrac{63}{31} - \left(1 - 2\dfrac{1}{5} \div 7\right) \cdot 1\dfrac{11}{24} - \dfrac{1}{20}.$

II. 16. Simplify the following expressions:

a) $\dfrac{\left(4\dfrac{3}{10} \div 3\dfrac{7}{12} - 0.2\right)\cdot 55}{2\dfrac{7}{12} - 0.25\cdot 9\dfrac{1}{3}\cdot\dfrac{5}{7}}\cdot 0.1 + \dfrac{3.04}{2} + 14\dfrac{12}{25}$;

b) $\dfrac{\left(8\dfrac{4}{45} - 7\dfrac{1}{15}\right)\cdot 30}{1\dfrac{1}{3}} + \dfrac{4\dfrac{1}{4} \div 0.85 + 1 \div 0.5}{(5.56 - 4.06)\div 3}$;

c) $\dfrac{\left(1\dfrac{16}{75} + 2.46\right)\div(55.1\div 5)}{1\dfrac{2}{3}\div 1\dfrac{8}{9}\cdot\left(\dfrac{2}{15} + 0.15\right)} + \dfrac{9.72 - 6\dfrac{13}{25}}{40.5\cdot\dfrac{2}{9}\div 9}$;

d) $\dfrac{\left(15 - 9\dfrac{1}{3}\right)\div 2\dfrac{5}{9}}{\left(20\dfrac{2}{3} - 12\dfrac{7}{9}\right)\cdot\dfrac{9}{71}}\cdot\left(0.71 - \dfrac{1}{4}\right)\div\left(0.71 + \dfrac{1}{4}\right)$;

e) $2 + 0.125\cdot\dfrac{5\dfrac{8}{3}\cdot\left(4.2 - 3\dfrac{7}{11} + \dfrac{9}{55}\right)\cdot\dfrac{3}{23}}{4.8\div 5\dfrac{7}{10}\cdot 1\dfrac{3}{16} - \left(3\dfrac{1}{7} - 2.8\right)\cdot 1\dfrac{1}{6}} - \dfrac{20}{11}$;

f) $\dfrac{\left(5.225 - \dfrac{5}{9} - 3\dfrac{5}{6}\right)\cdot\dfrac{36}{43} + 1.3}{\left(2\dfrac{23}{50} + 1\dfrac{16}{75}\right)\div(55.1\div 5) - 0.09 + \dfrac{227}{300}}$;

g) $\dfrac{15}{48}\div 0.125 + 1.456\div\dfrac{91}{250} + 4\dfrac{1}{2}\cdot\dfrac{1}{3} + \dfrac{2.652 \div 1.3 - 1\dfrac{17}{30} + \dfrac{3}{50}}{\left(1.34 + 1\dfrac{1}{10}\div\dfrac{11}{15}\right)\div 5\dfrac{13}{40}}$;

h)
$$\left(\frac{2\frac{1}{7}}{2\frac{1}{7}-1\frac{2}{7}}+\frac{1}{2}\right)\div\frac{\left(1.\overline{7}-\frac{1}{9}\right)\cdot\left(2.\overline{3}-\frac{1}{2}\right)}{2.\overline{3}-\frac{1}{3}}\cdot\frac{\sqrt{\frac{1.\overline{7}-1}{1.\overline{7}+1}}+\frac{18}{25}}{3-7\cdot0.(142857)};$$

i) Find out A, B, C and find out x from this equation: $\dfrac{B}{C}=\dfrac{A}{x}$

$$A=\frac{1+1\frac{1}{2}\cdot\left(0.5+2\frac{1}{3}\right)+0.6\div\frac{1}{5}}{2\cdot\sqrt{0.0121}};$$

$$B=\sqrt{\frac{0.01+0.001+0.0001+1\div1000}{10^{-4}\cdot\left(1-\frac{1}{2}\right)\cdot\left(1-\frac{1}{3}\right)\div\left(1-\frac{1}{4}\right)}};\quad C=\frac{1^2+2^2+3^2+4^2}{4^2-3^2+2^2-1^2}.$$

Integer exponents. Properties of the powers. The order of the operations and the use of brackets.

II. 17. Calculate:

a) $\left(+\frac{3}{5}\right)^3$; **b)** $\left(-\frac{2}{3}\right)^3$; **c)** $\left(-\frac{3}{5}\right)^2$; **d)** $\left(-2\frac{1}{3}\right)^4$; **e)** $\left(-\frac{11}{13}\right)^0$;

f) $\left(-\frac{3}{22}\right)^1$; **g)** $\left(+\frac{137}{219}\right)^0$; **h)** $\left(\frac{23}{11}\right)^1$; **i)** $(-1.1)^3$; **j)** $[-0.\overline{4}]^2$;

k) $-(1.3)^2$; **l)** $-(-0.5+0.2)^3$; **m)** $(-1-0.6)^3$; **n)** $\left(\frac{1}{6}-\frac{5}{4}\right)^3$.

II. 18. Calculate:

a) 3^{-2}; **b)** 5^{-3}; **c)** 4^{-2}; **d)** $(-6)^{-3}$; **e)** $(-5)^{-1}$;

f) $\left(-\frac{1}{4}\right)^{-1}$; **g)** $\left(-\frac{2}{5}\right)^{-1}$; **h)** $\left(-\frac{3}{4}\right)^{-2}$; **i)** $\left(-2\frac{1}{3}\right)^{-3}$;

j) $\left(-0.\overline{6}\right)^{-1}$; **k)** $(-0.2)^{-3}$; **l)** $(-1.5)^{-2}$; **m)** $[-0.\overline{5}]^{-3}$.

II. 19. Use positive powers:

a) $32, \ 16, \ 8, \ 4, \ 2, \ 1, \ \dfrac{1}{2}, \ \dfrac{1}{4}, \ \dfrac{1}{8}, \ \dfrac{1}{16}, \ \dfrac{1}{32}$;

b) $\dfrac{1}{625}, \ \dfrac{1}{125}, \ \dfrac{1}{25}, \ \dfrac{1}{5}, \ 1, \ 5, \ 25, \ 125, \ 625$;

c) $\dfrac{1}{10000}, \ \dfrac{1}{1000}, \ \dfrac{1}{100}, \ \dfrac{1}{10}, \ 1, \ 10, \ 100, \ 1000, \ 10000.$

II. 20. Use negative powers for the numbers:

a) $\dfrac{1}{10^2}$; b) $\dfrac{1}{7^3}$; c) $\dfrac{1}{10000}$; d) $\dfrac{1}{27}$; e) $\dfrac{1}{16}$; f) $\dfrac{1}{121}$; g) $\dfrac{1}{729}$;

h) $\dfrac{1}{216}$; i) $\dfrac{1}{3125}$.

II. 21. Using the properties of the powers, calculate:

a) $2^4 \cdot 2^5$; b) $7^4 \cdot 7^2$; c) $(-8)^5 \cdot 8^3$; d) $(-3)^3 \cdot (-3)^2$; e) $4^5 \cdot 4^{-2}$;

f) $7^4 \cdot 7^{-5}$; g) $(-3)^{-2} \cdot (-3)^{-3}$; h) $\left(-\dfrac{2}{5}\right)^5 \cdot \left(-\dfrac{2}{5}\right)^3$;

i) $\left(-\dfrac{2}{5}\right)^2 \cdot \left(-\dfrac{2}{5}\right)^3$; j) $\left(-\dfrac{3}{7}\right)^3 \cdot \left(-\dfrac{3}{7}\right)$;

k) $\left(-\dfrac{1}{2}\right)^{-2} \cdot \left(+\dfrac{1}{2}\right)^{-1}$; l) $(-3)^3 \cdot \left(-\dfrac{1}{3}\right)^{-2}$;

m) $\left(-\dfrac{2}{3}\right)^2 \cdot \left(-\dfrac{3}{2}\right)^{-3}$; n) $(-3)^{-4} \cdot (-3)^3 \cdot (-3)^2$;

o) $\left(-\dfrac{3}{7}\right)^4 \cdot \left(+\dfrac{3}{7}\right)^3 \cdot \left(+\dfrac{3}{7}\right)^{-4} \cdot \left(-\dfrac{3}{7}\right)^0$;

p) $\left(-\dfrac{5}{11}\right)^{-3} \cdot \left(-\dfrac{5}{11}\right)^4 \cdot \left(-\dfrac{5}{11}\right)^5$; q) $\left(-\dfrac{2}{3}\right)^7 \cdot \left(-\dfrac{3}{2}\right)^{-3} \cdot \left(-\dfrac{2}{3}\right)^{-6}$;

r) $(-2.3)^3 \cdot (-2.3)^2 \cdot (+2.3)^{-4}$; s) $(-2.7)^5 \cdot (-2.7)^{-7} \cdot (-2.7)^2.$

II. 22. Calculate:

a) $5^6 \div 5^2$; b) $2^4 : 2^{-2}$; c) $(-2)^{-5} \div (-2)^{-3}$; d) $4^{-4} \div 4^{-1}$;

e) $\left(\dfrac{2}{5}\right)^6 \div \left(\dfrac{2}{5}\right)^2$; f) $\left(-\dfrac{5}{6}\right)^9 \div \left(-\dfrac{5}{6}\right)^8$; g) $\left[-\left(\dfrac{3}{4}\right)^{12}\right] \div \left(-\dfrac{3}{4}\right)^9$;

h) $\left(-\dfrac{2}{3}\right)^6 \div \left(-\dfrac{3}{2}\right)^{-5}$; i) $\left(+\dfrac{3}{5}\right)^{-5} \div \left(-\dfrac{5}{3}\right)^4$;

j) $\left[\left(-\dfrac{4}{5}\right)^{13} \div \left(-\dfrac{4}{5}\right)^8\right] \div \left(+\dfrac{5}{4}\right)^{-4}$; k) $(-1.9)^8 \div (+1.9)^5$;

l) $(-0.\overline{4})^{21} \div (-0.\overline{4})^{19}$; m) $(-3.8)^{-5} \div (-3.8)^3$;

n) $(-1.3)^2 \div (1.3)^{-5}$; o) $\left[(-0.8)^3 \div (-0.8)^{-7}\right] \div (-0.8)^6$.

II. 23. Calculate:

a) $-(6^2)^8$; b) $\left[(-4)^2\right]^4$; c) $(-3^3)^4$; d) $(-2^4)^3$; e) $\left[\left(-\dfrac{2}{5}\right)^3\right]^2$;

f) $\left[-\left(\dfrac{3}{4}\right)^2\right]^3$; g) $\left[-\left(\dfrac{-3}{4}\right)^3\right]^2$; h) $\left(-\dfrac{2}{3}\right)^{2^3}$; i) $\left\{\left[\left(-\dfrac{3}{4}\right)^3\right]^2\right\}^5$;

j) $\left[(-0.5)^2\right]^2$; k) $\left[(-1.8)^3\right]^2$; l) $\left[-(0.\overline{3})^2\right]^4$; m) $(-0.5)^{3^4}$;

n) $\left\{\left[(-1.5)\right]^2\right\}^3$; o) $\left[(-2)^{-3}\right]^{-2}$; p) $\left[(-4)^{-3}\right]^{-1}$; q) $\left[\left(-\dfrac{2}{5}\right)^{-3}\right]^4$;

r) $\left[\left(-\dfrac{3}{4}\right)^{-1}\right]^{-4}$; s) $\left[\left(-\dfrac{3}{5}\right)^2\right]^{-3}$; t) $\left\{\left[\left(-\dfrac{2}{3}\right)^{-4}\right]^{-3}\right\}^2$;

u) $\left\{\left[\left(-\dfrac{1}{2}\right)^0\right]^3\right\}^{-4}$; v) $\left\{\left[(-3.28)^{-2}\right]^3\right\}^0$.

II. 24. Using the properties of the powers, calculate:

a) $\left(-\dfrac{2}{3}\right)^3 \cdot \left(-\dfrac{3}{2}\right)^2 \cdot \left(-\dfrac{3}{2}\right)^2$;   **b)** $\left(\dfrac{1}{0.2}\right)^{-3} \cdot (-0.2)^3 \div \left(-\dfrac{1}{10}\right)^3$;

c) $(0.3)^{25} \cdot (0.3)^{12} \div \left\{\left[(0.3)^3\right]^9 \div \left[\left(\dfrac{3}{10}\right)^7\right]^2\right\}$;

d) $\left[\left(-\dfrac{1}{2}\right)^4 \cdot \left(-\dfrac{1}{2}\right)^3 \div \left(-\dfrac{1}{2}\right)^6\right]^2 \div \left(-\dfrac{1}{2}\right)^3$.

II. 25. Calculate:

a) $(-3)^5 \cdot (-27)^3 \cdot (-3)^{-10}$;   **b)** $2^{-3} \cdot (-2^7) \cdot 4^3$;   **c)** $10^{-6} \cdot 5^5 \cdot (-10)^3$;

d) $\left[(-3)^2\right]^3 \cdot \dfrac{1}{27}$;   **e)** $4^{-3} \div (-2)^7 \cdot 2^2$;   **f)** $\left[3^{12} \div (-9^2)\right] \div 0.\overline{3}$;

g) $-3^{11} \cdot (-2)^{-7} \div 6^{-7}$;   **h)** $\left|5^{-3} \div (-5)^{-6}\right| \div 25^2$;

i) $2^{-4} \cdot \dfrac{1}{4^{-20}} \cdot 16^{-4} \div \dfrac{1}{2^{12}}$;   **j)** $\left(\dfrac{1}{27}\right)^3 \cdot \left[(-3)^2\right]^3 \cdot 3^3 \div \left(\dfrac{1}{3^{-1}}\right)^6$;

k) $36^5 \cdot \left(-\dfrac{1}{6}\right)^6 \div \left[(-6)^{-1}\right]^2 \cdot \dfrac{1}{216}$;

l) $\left(\dfrac{1}{3}\right)^{-3} \cdot \dfrac{1}{3^8} \cdot 9^5 \cdot 9^{-6}$;   **m)** $\dfrac{2^3 \cdot 4^{-2} \cdot (-2)^4 \cdot 8^{-3}}{2^{-2} \cdot (-2)^3}$;

n) $(-3)^{-2} \cdot \left(-\dfrac{1}{3}\right)^{-3} \cdot (-3)^4 \div \left(\dfrac{1}{3}\right)^{-5}$;   **o)** $\left(\dfrac{1}{2}-\dfrac{2}{3}\right)^{-5} \cdot \left(\dfrac{2}{3}-\dfrac{1}{2}\right)^4 \div \dfrac{1}{27^{-1}}$;

p) $\left[\left(1\dfrac{1}{2}\right)^{-4} \div \left(0.25 + 1\dfrac{1}{4}\right)\right] \div \left(\dfrac{4}{9}\right)^{-2}$;

q) $\left[\left(-\dfrac{1}{2}\right)^{-3} \cdot \left(2-\dfrac{5}{2}\right)^{-2}\right]^2 \div \left(4\dfrac{1}{2}-3\right)^{-5}$.

II. 26. Calculate:

a) $\left(\dfrac{2}{3}\right)^2 \cdot \left(\dfrac{5}{7}\right)^2 \cdot \left(\dfrac{3}{2}\right)^2$;   **b)** $\left(\dfrac{1}{0.2}\right)^{-3} \cdot (-0.2)^3 \div \left(-\dfrac{1}{10}\right)^3$;

c) $\left(\dfrac{1}{2}\right)^4 \cdot (1)^4 \cdot \left(\dfrac{1}{0,5}\right)^4$;   **d)** $\left[\left(\dfrac{1}{2}\right)^5 \cdot \left(\dfrac{1}{2}\right)^3 \div \left(\dfrac{1}{2}\right)^6\right]^2 \div \left(\dfrac{1}{2}\right)^3$;

e) $\left(\dfrac{1}{27}\right)^4 \cdot \left(3^2\right)^3 \cdot 9 \div \left(\dfrac{1}{3}\right)^6$;   **f)** $36^3 \cdot \left(\dfrac{1}{6}\right)^6 \div \dfrac{1}{216} \cdot \left(6^1\right)^2$;

g) $\left(\dfrac{1}{3}\right)^3 \cdot \dfrac{1}{3^8} \cdot 9^5 \cdot 9^6$;   **h)** $\dfrac{4^3 \cdot 2^2 \cdot 2^4 \cdot 8^3}{\left(2^{-1}\right)^2 \cdot 2^3}$;

i) $5^2 \cdot \left(\dfrac{1}{5}\right)^3 \cdot 5^4 \div \left(\dfrac{1}{5}\right)^5$;   **j)** $\left[\left(\dfrac{1}{2}\right)^3 \cdot \left(2 - \dfrac{3}{2}\right)^2\right]^2 \div \left(6\dfrac{1}{2} - 6\right)^7$;

k) $\left(\dfrac{2}{3} - \dfrac{1}{2}\right)^6 \cdot \left(\dfrac{2}{3} - \dfrac{1}{2}\right)^4 \div \dfrac{1}{36}$.

II. 27. Calculate:

a) $\dfrac{5}{6^3} \cdot 6^4 - 1 \div 5^{-2}$;   **b)** $20^{-2} \cdot (-10)^2 - 5^2 \cdot (-4)^{-2}$;

c) $3 \cdot (-3)^{-3} + 3^{-2} - 9^{-2}$;   **d)** $(-2)^{-2} \cdot 3^1 + (-5)^2 \cdot 2^{-2} + (-4)^{-1} \cdot 3^2$;

e) $(-2.12)^0 \cdot (0.5)^{-2} - 10^2 \cdot (0.3)^2 + 3 \cdot (0.5)^{-2}$;

f) $\left[\left(-\dfrac{3}{7}\right)^6 \div \left(\dfrac{4}{7} - 1\right)^4 - \dfrac{1}{7}\right] \cdot (-7)^2$;

g) $\dfrac{15}{(-2)^3} + \left[(-1.25)^2 - \left(-1\dfrac{3}{6}\right)^2\right]^2 \div \left(-2\dfrac{2}{3}\right)$;

h) $\left(5 - \dfrac{3}{8}\right)^2 \cdot \left[(-2.5)^2 - \left(-2\dfrac{1}{2}\right)^2\right]^{10}$;

i) $\left(-\dfrac{1}{2}\right)^{-3} \cdot 0.25 + \left(\dfrac{1}{3}\right)^{-2} \cdot 3^{-1} + 0.3^{-1} \cdot \dfrac{1}{5}$;

j) $\left[\left(-2\dfrac{3}{4}\right)^3 \div (-1-1.75)^3\right]^2 \div \left(2+\dfrac{3}{4}\right)^2$;

k) $\left[(19-3^3)\div(-2)^2 +45-(-8)^2\right]^0 \cdot (-1.5)^3$;

l) $3^{-1} \div \left[2^{-1} \div (-2)^2\right] + (-0.1)^3 \cdot (0.1)^{-2}$;

m) $0^5 \cdot \left(-\dfrac{1}{5}\right)^{15} + \left\{\left[\left(-\dfrac{2}{3}\right)^3\right]^{-2}\right\}^0 + \left[\left(-\dfrac{1}{2}\right)\cdot 4\right]^{-2}$;

n) $\left(-\dfrac{3}{4}\right)^5 \div \left(-\dfrac{3}{4}\right)^3 - \left(-2+\dfrac{1}{2}\right)^6 \div \left(2-\dfrac{1}{2}\right)^4 - \left(-\dfrac{3}{4}\right)^2$;

o) $\left[0.5^2 \cdot \left(1+\dfrac{1}{2}\right) - \left(\dfrac{1}{3}\right)^2\right] \div \dfrac{19}{8} - \left(1.0\overline{5} - 1\dfrac{3}{4}\right) \div \left(-1\dfrac{1}{5}\right)^{-2}$.

II. 28. Calculate:

a) $\left(\dfrac{7}{3}\right)^{-3} \cdot \left[\left(\dfrac{3}{7}\right)^{-2}\right]^{-3} \div \left[\left(\dfrac{3}{7}\right)^{-4}\right]^{-2}$; **b)** $\left[\left(-\dfrac{2}{3}\right)^6 \cdot \left(-\dfrac{2}{3}\right)^{-3} \cdot \left(-\dfrac{2}{3}\right)^2\right]^{-2}$;

c) $\left(-\dfrac{2}{3}\right)^6 \cdot \left(-\dfrac{2}{3}\right)^{-4} + \left(2-\dfrac{2}{3}\right)^2 \div \left(2-\dfrac{2}{3}\right)^2 - \left(-\dfrac{3}{2}\right)^{-2}$;

d) $\left(-\dfrac{3}{2}\right)^{-6} \div \left(-\dfrac{2}{3}\right)^4 + \left(1-\dfrac{1}{3}\right)^2 \div \left(2-\dfrac{4}{3}\right)^2 + \left(-\dfrac{2}{3}\right)^3 \cdot \left(1-\dfrac{1}{3}\right)$;

e) $\dfrac{1}{5^2} \cdot \left(-\dfrac{5}{3}\right)^3 + (-3)^5 \cdot (-3)^{-6} + \left(-\dfrac{2}{3}\right)^2$;

f) $(-0.4)^{-2} \div \left[(-2)^5 \cdot (-2)^{-6}\right]^3 - (-7)^3 \cdot 7^{-6} \div 7^{-5}$;

g) $(-0.2)^{-3} \cdot \left[5^9 \cdot (-5)^{-6}\right]^{-1} - \left[-0.\overline{3} + (-3)^{-2}\right] \div 7^0$;

h) $\left\{\left[\left(\dfrac{6}{5}\right)^2\right]^{-3} \div \left[\left(\dfrac{6}{5}\right)^2 \div \dfrac{125^{-2}}{216^{-3}}\right] + \left(\dfrac{125}{48}\right)^2 \div \left(\dfrac{32}{15}\right)^{-3} \dfrac{2^{-7}\cdot(-5)^{-2}}{(-3)^{-5}\cdot 5^{-1}}\right\} \cdot \left(\dfrac{5}{2}\right)^{-2}$;

i) $\left\{\left[\left(-\dfrac{9}{8}\right)^{-5} \div \left(\dfrac{27}{16}\right)^{-2} \cdot \dfrac{2^{-7}}{(-9)^{-3}}\right] \cdot \dfrac{48^{-1}}{4^{-2}} - \left[\left(\dfrac{9}{16}\right)^{-12} \cdot \left(\dfrac{9}{8}\right)^{16} \div \dfrac{4^{-2}}{9^{-1}}\right] \cdot \dfrac{4^{-2}}{9^2}\right\} \div 2^{-2}$.

Linear equations in the set Q

II. 29. Solve the equations:

a) $x + \dfrac{1}{4} = \dfrac{1}{3}$; b) $3x + 0.3 = \dfrac{1}{2}$; c) $-x + \dfrac{3}{4} = 0,5$; d) $-2x - \dfrac{2}{4} = \dfrac{1}{3}$;

e) $\dfrac{1}{5}x = -3$; f) $0.7 \cdot x = 7$; g) $-0.7x = -\dfrac{4}{5}$; h) $\dfrac{x}{2} = -3$;

i) $\dfrac{a}{5} = -4$; j) $\dfrac{y}{2} = -3.5$; k) $\dfrac{z}{0.6} = 5$; l) $-\dfrac{t}{6} = 0.5$;

m) $\dfrac{x}{3} = 0.\overline{6}$; n) $\dfrac{5x}{9} = 0.\overline{3}$.

II. 30. Solve the equations:

a) $6x - 2 = 7$; b) $6x - 7 = 1$; c) $5x + 2 = 0$; d) $3x - 3 = 8$;

e) $2x - 1 = 8$; f) $\dfrac{14}{3}x - 0.\overline{3} = 0$; g) $\dfrac{2}{5}x + \dfrac{1}{3} = 0$; h) $0.\overline{7}z - \dfrac{6}{7} = 5$;

i) $0.7x - 0.2 = 7$; j) $1.\overline{5}y - \dfrac{15}{8} = 0$; k) $0.4a + 0.4 = 0$;

l) $\dfrac{4}{5}x + \dfrac{2}{3} = 0$; m) $2.4u - \dfrac{2}{7} = 1$; n) $-1.9x + \dfrac{1}{7} = 8$;

o) $-\dfrac{3}{5}y - \dfrac{3}{15} = 1$; p) $21 - \dfrac{19}{3}x = 2$; q) $2 - \dfrac{x}{3} = 5$;

r) $7 + \dfrac{x}{3} = -6$; s) $3 = \dfrac{3}{2}x + 6$; t) $21\dfrac{1}{2}b - \dfrac{2}{5} = -9$.

II. 31. Solve the equations:

a) $2x - 19x = 17$; b) $-5x + (-2+x) = 30$; c) $2x + 4 = x + 26$;

d) $3x - 7 = 2x + 23$; e) $-2 = -10x + 16x$; f) $\dfrac{3}{4} = -2x + 8x$;

g) $\dfrac{1}{4}x + \dfrac{2}{3}x = 11$; h) $\dfrac{4}{5}x + \dfrac{1}{4}x = 42$; i) $0,4x + 0,9x = 6,5$;

j) $\dfrac{4}{9}y - \dfrac{5}{9} = -\dfrac{7}{27}$; k) $\dfrac{4}{3}x + \dfrac{1}{5}x = 2.3$; l) $4 = 4.3x + \dfrac{1}{11}$;

m) $-\dfrac{2}{5}y+\dfrac{1}{5}y-2=8$; **n)** $7x-8=\dfrac{1}{3}x-0.5$;

o) $2(x+7)-4(x-4)=7$; **p)** $4(x+9)-5(x+12)=8$;

q) $4-3x-3(8-x)=2(3-x)$; **r)** $3x-(1-2x)=5x-3$;

s) $\dfrac{x}{4}+\dfrac{x}{3}=7$.

II. 32. Calculate:

a) $-\dfrac{8}{3}x+\dfrac{1}{6}x=\dfrac{15}{2}$; **b)** $10x+\dfrac{5}{4}-3x=\dfrac{3}{8}$; **c)** $\dfrac{3}{4}x-\dfrac{2}{5}-\dfrac{5}{8}x=-\dfrac{1}{10}$;

d) $\dfrac{3}{5}x+\dfrac{3}{10}-\dfrac{1}{2}x=-\dfrac{1}{5}$; **e)** $\dfrac{3}{8}x-\dfrac{1}{2}-\dfrac{1}{4}x=\dfrac{3}{16}$;

f) $\dfrac{3}{4}x-\dfrac{3}{5}-\dfrac{1}{8}x=-\dfrac{1}{10}$; **g)** $-\dfrac{2}{3}+\dfrac{3}{5}x-\dfrac{1}{2}x=-\dfrac{3}{5}$;

h) $-\dfrac{1}{9}-\dfrac{2}{3}x+\dfrac{1}{18}+\dfrac{1}{2}x=\dfrac{5}{6}$.

II. 33. Find x in the equations:

a) $3x-\dfrac{1}{4}=x+\dfrac{7}{4}$; **b)** $x+\dfrac{2}{5}=-x-\dfrac{8}{5}$; **c)** $4x-\dfrac{5}{3}=2x-\dfrac{2}{3}$;

d) $2\left(x-\dfrac{1}{2}\right)=3\left(x-\dfrac{1}{3}\right)-2$; **e)** $4\left(x-\dfrac{1}{4}\right)=5\left(x-\dfrac{1}{5}\right)-2$;

f) $4\left(2x-\dfrac{1}{4}\right)=8\left(4x-\dfrac{1}{8}\right)-12$; **g)** $\dfrac{2y}{5}+\dfrac{6y}{5}+\dfrac{12y}{5}+4y=40$.

II. 34. Solve the equations:

a) $\dfrac{x}{2}-\dfrac{x-3}{3}=\dfrac{x}{4}+\dfrac{1}{6}$; **b)** $\dfrac{x}{3}-\dfrac{4-x}{2}=8\dfrac{5}{6}-\dfrac{x}{4}$;

c) $\dfrac{-2x+9}{3}-\dfrac{3x-3}{4}=\dfrac{x}{2}-x-1$;

d) $\dfrac{-x-7}{2}-\dfrac{3x-5}{4}=\dfrac{x-2}{3}-2x+1$;

e) $\dfrac{2x+1}{2}-\dfrac{4x+1}{3}=\dfrac{2x+9}{4}-\dfrac{25}{6}$;

f) $\dfrac{1}{5}\left\{\dfrac{1}{5}\left[\dfrac{1}{5}\left(\dfrac{1}{5}x-5\right)-5\right]-5\right\}-5=0$;

g) $2x+0.\overline{6}(x-3)=4-0.\overline{3}(5x-12)$;

h) $0.8\left(x-0.\overline{3}\right)+1.\overline{16}(7x-5)=4.\overline{8}$;

i) $0.\overline{3}\cdot\left\{0.\overline{3}\cdot\left|0.\overline{3}\cdot\left(0.\overline{3}\cdot x+0.\overline{3}\right)\right|\right\}-0.\overline{3}=0$;

j) $\left[(x\div5-42)\cdot2-30\cdot14\right]\div6+29=79$.

II. 35. Solve în **Q** the ecuations:

a) $\dfrac{x-2}{3}+\dfrac{x-3}{4}=\dfrac{x-4}{6}+1$; **b)** $\dfrac{3x-1}{4}-\dfrac{x+1}{3}=x-1$;

c) $\dfrac{3x+1}{5}-\dfrac{4x-1}{2}=\dfrac{x+5}{10}-\dfrac{1}{2}$; **d)** $\dfrac{2x+3}{5}+\dfrac{5x-4}{3}-2=\dfrac{x}{15}$;

e) $\dfrac{a+1}{3}+\dfrac{a+3}{9}=0.\overline{3}$; **f)** $\dfrac{3+b}{7}+\dfrac{3b-1}{3}=b+\dfrac{1}{3}$;

g) $\dfrac{2(3x+5)}{15}-\dfrac{2(4x-1)}{3}=-\dfrac{2(3x+1)}{5}-\dfrac{2(7x-1)}{3}$;

h) $\dfrac{3x+1}{3}+\dfrac{3}{4}=\dfrac{x}{6}-\dfrac{3-x}{12}$; **i)** $\dfrac{3(x-1)}{2}-\dfrac{2(3-x)}{3}-\dfrac{1}{6}=\dfrac{x+1}{4}$;

j) $\dfrac{3(x-1)}{4}-\dfrac{5-x}{2}-\dfrac{3-2x}{6}=1-\dfrac{x}{12}$; **k)** $\dfrac{x}{3}-\dfrac{2-x}{5}=\dfrac{4(1-x)}{15}$;

l) $\dfrac{11x}{3}-\dfrac{3x+4}{5}=3x-2$; **m)** $\dfrac{4(2x-3)}{5}=4-\dfrac{3(x-3)}{4}-x$;

n) $\dfrac{x}{2}-\dfrac{x}{12}+3=\dfrac{2x}{9}-\dfrac{3-x}{6}$; **o)** $\dfrac{3x}{2}-\dfrac{5}{4}=2x-\dfrac{2-2x}{5}-\dfrac{9}{20}$;

p) $\dfrac{5x+7}{2}-\dfrac{4x+7}{3}=3x-\dfrac{1}{6}$; **q)** $\dfrac{5x+1}{3}-\dfrac{4-x}{2}=\dfrac{5x-8}{2}-1$;

r) $\dfrac{5-x}{4}-\dfrac{1}{8}+5x=4-\dfrac{3x}{2}-\dfrac{5}{4}$; **s)** $\dfrac{21x}{2}-\dfrac{5-x}{2}+\dfrac{3x}{8}=\dfrac{5x}{8}-\dfrac{3x}{2}$;

t) $\dfrac{t+5}{3}-\dfrac{t-9}{5}=3+\dfrac{3t-1}{15}$; **u)** $\dfrac{4a-1}{9}-\dfrac{5a-1}{6}-3a=\dfrac{1}{3}$;

v) $\dfrac{3(x+8)}{4}-\dfrac{2(x-3)}{3}=\dfrac{3(x+5)}{3}-\dfrac{x}{4}$;

w) $\dfrac{x-7}{4} + \dfrac{1}{3} - x + \dfrac{2x-1}{3} = \dfrac{x}{6}$;

x) $\dfrac{4(1-x)}{7} - \dfrac{2x-3}{3} - \dfrac{4x}{21} = 2(x-4)$;

y) $\dfrac{x+3}{3} - \dfrac{3(x-2)}{4} = 1 + \dfrac{x-12}{2}$;

z) $\dfrac{2(5x-7)}{5} - \dfrac{3x-4}{3} = \dfrac{x+2}{10} - 5x$.

II. 36. Solve the equations:

a) $3(a-2) - \dfrac{1}{2} = 5\left(\dfrac{a}{2} - 1\right) + 2a$; **b)** $\dfrac{7}{3}(x-5) = -2\left(\dfrac{5}{3} - \dfrac{7}{3}x\right)$;

c) $5(x-4) = 6\left(\dfrac{x}{3} - 1\right) + 1$; **d)** $4(2.5x - 1) + \dfrac{1}{3} = -2(3.5 - 3x) + \dfrac{1}{2}$;

e) $-0.\overline{5}(3x-2) - 0.5(4-2x) = 4(1-0.25x)$;

f) $0.4(3x-2) + 1 + 0.3 \cdot (5x-10) = 0$;

g) $0.5 \cdot (2x+1) - 0.4 \cdot (4x-1) = 2.3(x-4)$;

h) $0.5b + 12 = 0.\overline{3}(b+12)$;

i) $4 \cdot (4x - 0.2\overline{3}) - 3(4x - 0.2\overline{3}) = 7x - 7$;

j) $(0.5\overline{3} - 5x) - 5(3 - 0.1\overline{3}x) = 11$.

II. 37. Solve the equations:

a) $a + 80 = \dfrac{9}{2}(10a + 8)$; **b)** $x + \dfrac{9x}{50} + 24\left(\dfrac{x+3}{100}\right) = 4.22$;

c) $0.80 = \dfrac{10+a}{30+a}$; **d)** $\dfrac{x-3}{2} - \dfrac{2(1-3x)}{0.2} + 1 = \dfrac{3(1-0.5x)}{2} - 7$;

e) $\dfrac{1}{2}\left\{\dfrac{1}{2}\left[\dfrac{1}{2}\left(\dfrac{1}{2}x - 1\right) - 1\right] - 1\right\} - 1 = 0$;

f) $x + \dfrac{1}{2}\left[\dfrac{2(x+1)}{3} + \dfrac{4(x+2)}{5}\right] = \dfrac{x-1}{2} + \dfrac{2}{3}\left[\dfrac{3(x-1)}{4} - \dfrac{6(x-1)}{5}\right]$;

g) $\dfrac{1}{2}\left\{\dfrac{1}{2}\left[\dfrac{1}{2}\left(\dfrac{1}{2}x - 1\dfrac{1}{2}\right) - 1\dfrac{1}{2}\right] - 1\dfrac{1}{2}\right\} - 1\dfrac{1}{2} = 0$;

h) $\dfrac{1}{2}\left\{\dfrac{1}{2}\left[\dfrac{1}{2}\left(\dfrac{1}{2}x-3\dfrac{1}{2}\right)-3\dfrac{1}{2}\right]-3\dfrac{1}{2}\right\}-3\dfrac{1}{2}=0$;

i) $\dfrac{1}{2}\left\{\dfrac{1}{2}\left[\dfrac{1}{2}\left(\dfrac{1}{2}x+2\right)+2\right]+2\right\}+2=0$;

j) $[(2x-4.7)\cdot 3-1.9+9]\div 5-3.9=0.1$.

II. 38. Solve the equations:

a) $9(5x-3)-7(3x+5)=5(3x+1)-4(2x+8)-18$;

b) $(x+5)^2+(x-6)(x+6)=2(x-3)+5$;

c) $(5x+7)(5x-7)-(5x-1)^2=-70$;

d) $3(4x-11)+5(7-2x)=16-3(3x-10)$;

e) $(3x+8)(x-1)-3(x+1)^2=-13$.

II. 39. Solve the equations:

a) $\dfrac{(x+2)(x-1)}{2}-\dfrac{(x+1)(x+3)}{6}=\dfrac{(x+5)(x-1)}{3}$;

b) $\dfrac{(x-2)(2x-1)}{3}+\dfrac{(x+2)(x+1)}{4}=\dfrac{11x(x+3)}{12}-\dfrac{x-7}{3}$;

c) $\dfrac{2}{3}+\dfrac{(x+2)(x-5)}{6}+\dfrac{x^2-5x+4}{2}=\dfrac{(x+3)(4x-1)}{6}$;

d) $\dfrac{(x+4)(x+1)}{6}+\dfrac{(1-2x)(3+x)}{4}=\dfrac{1-2x-4x^2}{12}$;

e) $\dfrac{(x-1)^2-3(x-1)+4}{x+1}-5(x+3)=-4x-14$;

f) $\dfrac{2x^2+2x+1}{(x+1)(x+2)}+\dfrac{2x^2+2x+3}{(x+1)(x+3)}=\dfrac{2x^2+2}{(x+2)(x+3)}+2$.

II. 40. Solve the equations:

a) $(3x+1)(-4x+3)=0$; b) $\left(\dfrac{4}{7}x+3\right)\left(\dfrac{1}{3}x+\dfrac{1}{2}\right)=0$;

c) $\left(-\dfrac{5}{8}x-\dfrac{3}{7}\right)(0.\overline{3}x-0.\overline{4})=0$; d) $\left(\dfrac{1}{13}x-\dfrac{7}{6}\right)\cdot\left(-\dfrac{8}{9}x+\dfrac{4}{7}\right)=0$;

e) $\left(-\dfrac{1}{7}x+\dfrac{7}{8}\right)\cdot\left(-\dfrac{5}{9}x-\dfrac{4}{5}\right)=0$.

II. 41. Solve the equations:

a) $\dfrac{2}{x}+\dfrac{3}{4}=\dfrac{5}{x}+\dfrac{1}{2}+1,\ \ x\in\mathbf{Q}\setminus\{0\}$;

b) $\dfrac{3x+3}{2x+5}=\dfrac{3x+1}{2x+1},\ \ x\in\mathbf{Q}-\left\{-\dfrac{5}{2};-\dfrac{1}{2}\right\}$;

c) $\dfrac{4}{x+2}=\dfrac{5}{3-x},\ \ x\in\mathbf{Q}-\{-2;\ 3\}$;

d) $\dfrac{2}{x+2}+\dfrac{3}{x+3}=\dfrac{5}{x},\ x\in\mathbf{Q}-\{-3;\ -2;\ 0\}$;

e) $\dfrac{3}{x-2}+\dfrac{1}{x-3}=\dfrac{4}{x-1},\ x\in\mathbf{Q}-\{1;\ 2;\ 3\}$;

f) $\dfrac{5x+2}{3x-1}=\dfrac{4}{7},\ \ x\in\mathbf{Q}-\left\{\dfrac{1}{3}\right\}$;

g) $\dfrac{5-3x}{2-3x}=5,\ \ x\in\mathbf{Q}-\left\{\dfrac{2}{3}\right\}$; **h)** $\dfrac{3}{4}-\dfrac{7x+2}{3x}=\dfrac{1}{6x},\ \ x\in\mathbf{Q}\setminus\{0\}$;

i) $\dfrac{2x+3}{2x-3}=\dfrac{x+5}{x-4},\ x\in\mathbf{Q}-\left\{\dfrac{3}{2};\ 4\right\}$; **j)** $\dfrac{2}{5\cdot x}+\dfrac{5}{4\cdot x}=\dfrac{3}{8},\ \ x\in\mathbf{Q}\setminus\{0\}$;

k) $\dfrac{2x+1}{3x+1}=\dfrac{2x-1}{3x-1},\ x\in\mathbf{Q}-\left\{-\dfrac{1}{3};\ \dfrac{1}{3}\right\}$;

l) $\dfrac{3x+2}{3x-2}=\dfrac{x-1}{x+4},\ \ x\in\mathbf{Q}-\left\{-4;\ \dfrac{2}{3}\right\}$.

II. 42. Give a correct answer:

a) Knowing that $a=\dfrac{2}{5}$ and $b+c=-\dfrac{25}{2}$ what is the value of $ab+ac$?

b) Knowing that $ab+ac=-4\dfrac{1}{6}$ and $b+c=\dfrac{36}{5}$ what is the value of a?

c) Knowing that $ab=\dfrac{2^2}{3^{-3}}$ and $ac=\dfrac{4}{3^{-4}}$ and $b+c=108$ what is the value of a?

d) Knowing that $ab = \dfrac{3^2}{4^{-3}}$ and $ac = \dfrac{12}{5^{-1}}$ what is the value of $a(b-c)$?

e) Knowing that $a = b + \dfrac{3^{-3}}{5^{-2}}$ and $c = -\dfrac{25}{27}$ what is the value of $ac - bc$?

f) Knowing that $ab = -\dfrac{4}{25}$ and $ac = \dfrac{16}{5}$ what is the value of $\dfrac{b}{c}$ and $\dfrac{c}{b}$?

II. 43. a) Find the value of a such that the equation $ax + 5 = 13 + 3x$ has the solution $x = 5$.

b) Find the value of a such that the equation $ax - 6 = 2x - 5$ has the solution $x = 2$.

c) Find the value of m such that the equation $2x + mx = m + 3$ has the solution $x = 3$.

d) Find the value of m such that the equation $3mx + 1 = x - 2m$ has the solution $x = 4$.

e) Find the value of m such that the equation
$3m(x+3) + 1 - 5m = -(2x-4) - 12m$ has the solution $x = 6$.

II. 44. Solve the equations, a and m are real parameters:
a) $2mx - 1 = x - 2m$; **b)** $2x + a = ax + 3$; **c)** $ax - 5 = 4x + 7$;
d) $ax + 2x = 9 + 3ax$; **e)** $ax + 4 = 3a - 3ax$;
f) $(2+x)(a+1) = 2(a-2x) + 3a$; **g)** $2ax + 1 + x = (a+4)x + 2x$;
h) $2a(2-x) = (1-a)x + 3(1-3x)$;
i) $2(mx - 3m) + 1 + x = (m+4)x - (2x-1)$.

II. 45. Solve the equations:

a) $\dfrac{1}{x(x+1)} + \dfrac{1}{(x+1)(x+2)} + ... + \dfrac{1}{(x+2008)(x+2009)} = \dfrac{2009}{2010}$

b) $\dfrac{1}{1\cdot2} + \dfrac{2}{1\cdot2\cdot3} + \dfrac{3}{1\cdot2\cdot3\cdot4} + ... + \dfrac{x}{1\cdot2\cdot...\cdot(x+1)} = \dfrac{1\cdot2\cdot3\cdot...\cdot2010-1}{1\cdot2\cdot3\cdot...\cdot2010};$

c) $1 + \dfrac{x+1}{2} + \dfrac{x+2}{3} + ... + \dfrac{x+n-1}{n} = n, \quad n \in \mathbf{N}, n \geq 2.$

Problems solved using equations.

II. 46. A 12-foot board is cut into two pieces so that one piece is 4 feet longer than the other. How long is each piece?

II. 47. A 36-foot rope is cut into two pieces so that one piece is twice as long as the other. How long is each piece?

II. 48. A clothing store sells suits at $140 and $210. The store owners observes that they sold 40 suits at $6,650. How many suits of each type did the owner sell?

II. 49. Find five consecutive positive integers if the product of the first three numbers is 216 less than the product of the last three numbers.

II. 50. Two gears have a total of 96 teeth and one gear has 18 less teeth than the other. How many teeth are on each gear?

II. 51. A number is 27.8 less than another number. If their sum is 66.7, find the numbers.

II. 52. A number is three times less than another number. If their sum is $\dfrac{34}{9}$, find the numbers.

II. 53. If multiplying a number by 5.8 and subtracting 4 from the number, we obtain the same result. Find the number.

II. 54. The length of a rectangular field is 4 yeards more than twice the width of the field. If the semiperimeter is 610 yards, what are the dimensions of the field?

II. 55. The side of a square is x cm and the dimensions of a rectangle are $(x + 3)$ cm and $(x + 5)$ cm. Find the sides of the square if the area of the rectangle is 83cm² larger than the area of the square.

II. 56. The length of a room is 2 feet less than twice its width. If the perimeter is 62 feet, what are the dimensions?

II. 57. A triangle ABC has $\angle B = 2x$ and $\angle C = 5x$. Find x such that the triangle ABC is isosceles.

II. 58. Adding the same number to the numerator and the denomerator of the fraction $\dfrac{4}{7}$ we obtain $\dfrac{19}{31}$. Find the number.

II. 59. The perimeter of a rectangle is $4\dfrac{1}{2}$ cm and the length is 6.2 larger than the double of the width. Find the dimensions of the rectangle.

II. 60. One of the angles of a triangle is 72°, and another angle is three times larger than the other angle. Find the angles of the triangle.

II. 61. 3 kg of apples and 2 kg of oranges cost $5.05. If 1 kg of ranges costs $0.30 cents more than 1 kg of apples, find the prices of the apples and oranges.

II. 62. 10 math textbooks and 15 physics textbooks cost $2,000. The price of one math textbook and a physics texbook is $163.
Find the price of each textbook.

II. 63. The sum of two numbers is 150. If the first number is divided by the second number, we obtain the quotient 4 and the remainder 15. Find the numbers.

II. 64. The difference between the length and width of a rectangle is 13 cm. Find the dimensions of the rectangle if the semiperimeter is 63cm.

Ratios, proportions, procents, equal proportions.

II. 65. Find the ratio of the length and width of a rectangle and the ratio of the width and length of the same rectangle knowing that:
a) $l = 17$cm, $w = 10$cm; **b)** $l = 7$dm, $w = 51$cm;
c) $l = 3$m, $w = 25$dm; **d)** $l = 4$m, $w = 360$cm.

II. 66. Find the ratio for:

a) the sum and difference of $\dfrac{2}{5}$ and 1,4;

b) the product and difference of 13 and 7,5.

II. 67. A circle has the circumference 16π cm and the radius 8 cm. Find the ratio of:
a) the circumference and the radius of the circle;
b) the area and the radius of the circle.

II. 68. The ratio of the lengths of the sides of two equilateral triangles is $\dfrac{3}{4}$. Find the ratio of:

a) the perimeters;
b) the areas.

II. 69. Simplify the ratios:

a) $\dfrac{24}{36}$; **b)** $\dfrac{72}{42}$; **c)** $\dfrac{25}{45}$; **d)** $\dfrac{6a}{18a}$; **e)** $\dfrac{32b}{64b}$; **f)** $\dfrac{15a^3}{25a}$; **g)** $\dfrac{10\,\text{min}}{1\text{h}}$;

h) $\dfrac{15\text{cm}}{4\text{dm}}$; **i)** $\dfrac{15\text{m}}{4\text{hm}}$; **j)** $\dfrac{25l}{4\text{dl}}$; **k)** $\dfrac{15\text{mm}}{4\text{cm}}$; **l)** $\dfrac{150\text{s}}{3\text{min}}$.

II. 70. Calculate the ratio:

a) $\dfrac{7a+3b}{5a-7b}$ if $\dfrac{a}{b}=\dfrac{4}{5}$; b) $\dfrac{2y-3x}{9x+5y}$ if $\dfrac{x}{y}=5\dfrac{1}{3}$;

c) $\dfrac{3x+4y}{7x+31y}$ if $\dfrac{x}{y}=\dfrac{11}{23}$; d) $\dfrac{m}{n}$ if $\dfrac{5m-2n}{3m-4n}=\dfrac{4}{7}$;

e) $\dfrac{2x+7y}{3x-9y}$ if $\dfrac{3x}{5y}=\dfrac{4}{15}$.

II. 71. Find the unknown term from the following proportions:

a) $\dfrac{x}{5}=\dfrac{12}{3}$; b) $\dfrac{x}{3.25}=\dfrac{0.54}{0.6}$; c) $\dfrac{14}{x}=\dfrac{35}{22}$; d) $\dfrac{4}{a}=\dfrac{19}{38}$;

e) $\dfrac{38}{z}=\dfrac{72}{18}$; f) $\dfrac{\frac{5}{35}}{z}=\dfrac{\frac{45}{20}}{2.5}$; g) $\dfrac{a}{18}=\dfrac{5}{36}$; h) $\dfrac{n}{85}=\dfrac{0.5}{17}$;

i) $\dfrac{21}{x}=\dfrac{3.9}{1.3}$; j) $\dfrac{x}{4}=\dfrac{5}{2.8}$; k) $3=\dfrac{6}{x}$;

l) $\dfrac{x}{2^5 \div 2^3}=\dfrac{13^2}{3^0+3^1+3^2}$; m) $\dfrac{0.32}{3.2}=\dfrac{5.5}{2x}$; n) $\dfrac{x}{0.2\overline{3}}=\dfrac{0.1\overline{4}}{0.\overline{4}}$.

II. 72. Find x and y from the proportions:

a) $\dfrac{x}{y}=\dfrac{3}{7}$ and $x+y=50$; b) $\dfrac{x}{y}=\dfrac{13}{7}$ and $x-y=7$;

c) $\dfrac{x}{y}=\dfrac{2}{9}$ and $x-y=10$; d) $\dfrac{x}{y}=\dfrac{11}{27}$ and $x-2y=41$;

e) $\dfrac{3x}{2y}=\dfrac{5}{9}$ and $3x-4y=41$; f) $\dfrac{2x}{y}=\dfrac{3}{17}$ and $x+3y=5$.

II. 73. Find the numbers a, c, d, k, x, z from the proportions:

a) $\dfrac{28}{a+2}=7$; b) $\dfrac{2x-4}{7}=8$; c) $4=\dfrac{c-2}{5}$; d) $\dfrac{b+6}{2}=\dfrac{3b}{5}$;

e) $\dfrac{6z}{5}=7z-4$; f) $\dfrac{t+3}{6}=\dfrac{t}{10}$; g) $\dfrac{5x}{3}=\dfrac{2x-1}{4}$; h) $\dfrac{5x+1}{18}=\dfrac{6x-1}{21}$;

i) $\dfrac{2x-1}{14} = \dfrac{3x+2}{28}$; j) $\dfrac{2x-5}{2} = \dfrac{3x-7}{3}$; k) $\dfrac{4x+1}{8} = \dfrac{3x+3}{7}$;

l) $\dfrac{6k+112}{24} = \dfrac{4k+8}{12}$; m) $\dfrac{3x-4}{5} = \dfrac{10-3x}{20}$.

II. 74. Find the value of the ratio $\dfrac{x}{y}$:

a) $\dfrac{2x+3y}{5y} = \dfrac{3}{4}$; b) $\dfrac{5x-2y}{4y} = \dfrac{4}{3}$; c) $\dfrac{5x-y}{6} = \dfrac{4x-y}{4}$;

d) $\dfrac{2x-y}{-5y} = \dfrac{3}{7}$; e) $\dfrac{7x-2y}{2y} = \dfrac{4}{5}$; f) $\dfrac{3x-2y}{0.6} = \dfrac{5x-y}{0.4}$;

g) $\dfrac{3x+2y}{5y-7x} = \dfrac{3}{8}$.

II. 75. Write each percent as a fraction:
a) 75%; b) 25%; c) 12%; d) 0.4%; e) 50%; f) 0.7%;
g) 2.3%; h) 1.6%.

II. 76. Write each percent as a decimal number:
a) 5%; b) 3%; c) 34%; d) 17%; e) 0.33%; f) 4.15%; g) 0.25%;
h) 1.45%.

II. 77. Find the number such that:
a) 25% from 400; b) 8% from 75; c) $5\dfrac{1}{2}$% from 300; d) $\dfrac{1}{2}$% from 108; e) 2.5% from 90; f) 120% from 65; g) 100% from 99; h) 200% from 20.3; i) $4\dfrac{1}{3}$% from 250.

II. 78. Find the percent:
a) 15 from 90; b) 24 from30; c) 300 from400; d) 16 from25;
e) 12 from25; f) 24 from74; g) 450 from 9; h) 180 from 36;
i) 12 from 15; j) 36 from54; k) 160 from 250.

II. 79. Find the number x knowing that:
a) 15 is 20% from x; b) 16 is 60% from x; c) 4.5 is 20% from x;

d) 16 is 15% from x; **e)** 65% from x is 130; **f)** 3.5% from x is 21;
g) 14% from x is 70; **h)** 300% from x is 190;
i) 250% from x is 180.

II. 80. Find the numbers x, y, z if $\dfrac{x}{3} = \dfrac{y}{4} = \dfrac{z}{5}$ and

$x + z = 48$.

II. 81. Find the numbers x, y, z, t if $\dfrac{x}{2} = \dfrac{y}{3} = \dfrac{z}{5} = \dfrac{t}{7}$

and $8x + 5y - 3z - 2t = 4$

II. 82. Find the numbers x, y, z if $\dfrac{x}{3} = \dfrac{y}{2} = \dfrac{z}{4}$ and

$x \cdot y \cdot z = 192$.

II. 83. Let $\dfrac{x}{2} = \dfrac{y}{3}$ and $\dfrac{y}{4} = \dfrac{z}{5}, x, y \in Q_+^*$ be two proportions.

Find the numbers x, y, z if $x^2 + y^2 + z^2 = 433$.

II. 84. Find the numbers x, y, z if $\dfrac{x}{2} = \dfrac{y}{3} = \dfrac{4}{z}$ and

$y = \dfrac{x + z}{2}$.

II. 85. Find the numbers x, y, z if $\dfrac{x}{5} = \dfrac{y}{3}$ and $\dfrac{y}{2} = \dfrac{z}{4}$ and

$x + 2y + 3z = 116$.

II. 86. Find the numbers x, y, z if $\dfrac{5}{x} = \dfrac{y}{3} = \dfrac{z}{4}$.

II. 87. Find the numbers x, y, z if

$x + y + z = 4$, $\dfrac{x}{2} = \dfrac{y}{3}$ and $\dfrac{y}{6} = \dfrac{z}{8}$.

II. 88. Find the numbers a, b, c if $\dfrac{a}{3} = \dfrac{b}{4} = \dfrac{c}{6}$ and

$a \cdot b \cdot c = 576, \ a, b, c \in \mathbf{N}$.

II. 89. Find the numbers $a, b, c, d \in \mathbf{Z}$ if $\dfrac{a}{b} = \dfrac{2}{3}$,

$\dfrac{b}{c} = \dfrac{3}{4}, \dfrac{c}{d} = \dfrac{4}{5}$, and $abcd = 1920$.

II. 90. If $\dfrac{x}{a} = \dfrac{y}{b} = \dfrac{z}{c} = \dfrac{2}{3}$ calculate:

a) $\dfrac{x + y + z}{a + b + c}$; **b)** $\dfrac{x + 3y + z}{a + 3b + c}$; **c)** $\dfrac{x^2 + y^2 + z^2}{a^2 + b^2 + c^2}$.

II. 91. If $\dfrac{a}{3} = \dfrac{b}{4} = \dfrac{c}{5}$ calculate:

a) $\dfrac{2a - 3b}{5c + a}$; **b)** $\dfrac{a + 2b + 3c}{3a + 2b + c}$; **c)** $\dfrac{2a - 3b + 4c}{4a - 3b + 2c}$.

II. 92. If a, b, c are the sides of a triangle, show that the triangle is equilateral in both cases:

a) $\dfrac{(m+1)a + b + c}{a} = \dfrac{a + (m+1)b + c}{b} = \dfrac{a + b + (m+1)c}{c}$, $m \in \mathbf{R}$

b) $\dfrac{(m+1)a + b + c}{a + (m+1)b + c} = \dfrac{a + (m+1)b + c}{a + b + (m+1)c} = \dfrac{a + b + (m+1)c}{(m+1)a + b + c}$, $m \in \mathbf{R}$.

II. 93. Consider the numbers: $a = 2^{n+1} \cdot 5^n + 1$, $b = 2^n \cdot 5^{n+1} + 1$, and $c = 2^{n+3} \cdot 5^n + 7$, $d = 2^{n+1} \cdot 5^{n+3} - 1$, where n is a natural number.

a) Show that a, b, c, d are not prime numbers.

b) Prove that $\dfrac{7d + c}{b - a}$ is a natural number.

c) Can c be a perfect square?

II. 94. Calculate the sums:

a) $S = 13 + 12 \cdot 13 + 12 \cdot 13^2 + 12 \cdot 13^3 + \ldots + 12 \cdot 13^{100}$;

b) $S = 3^{100} - 2 \cdot 3^{99} - 2 \cdot 3^{98} - \ldots - 2 \cdot 3^2 - 2 \cdot 3 - 3$.

II. 95. Show that the number $\dfrac{\overline{abc} + \overline{bcd} + \overline{cda} + \overline{dab}}{a+b+c+d}$ is a natural number.

TEST II. 1

1) Alex has $\dfrac{1}{20}$ of a dollar in his pocket. What coins might he have?

2) Use base ten blocks to represent each fraction. Then write each fraction as a decimal. **a)** $\dfrac{3}{5}$; **b)** $\dfrac{3}{2}$; **c)** $2\dfrac{1}{5}$; **d)** $5\dfrac{4}{5}$.

3) Represent each fraction on a hundredths grid. Then write each number as a decimal. a) $\dfrac{1}{4}$; **b)** $\dfrac{3}{10}$; **c)** $\dfrac{2}{50}$; **d)** $\dfrac{2}{25}$.

4) Write each improper fraction as a decimal number.
 a) $7\dfrac{1}{2}$; **b)** $6\dfrac{2}{5}$; **c)** $3\dfrac{11}{25}$; **d)** $11\dfrac{1}{100}$.

5) Copy and complete. Replace each □ with <, > or = to make the statement true.
 a) $\dfrac{75}{100}$ □ $\dfrac{3}{4}$; **b)** 0.08 □ $\dfrac{8}{10}$; **c)** 0.6 □ $\dfrac{60}{100}$; **d)** $\dfrac{8}{5}$ □ $1\dfrac{3}{10}$.

6) Write two equivalent fractions for each decimal.
 a) 0.9; **b)** 0.40; **c)** 0.75, **d)** 0.5.

7) Write each division statement as an improper fraction and as a mixed number.
   a) $77 \div 8$; b) $84 \div 9$; c) $45 \div 7$; d) $79 \div 6$.

8) Vlad made 7 pizzas for his party. He invited 5 friends. How much pizza did Vlad think each person would eat?

9) Matthew's yard has a rectangular shape. It is 28.54 m wide and 31 m long. What is the area of the yard?

10) A snail traveled 0.94 m each hour. How far would the snail travel in 6 hours?

TEST II. 2

Calculate:

a) $\dfrac{2}{5} \div \dfrac{1}{3}$; b) $\dfrac{6}{25} \div 1\dfrac{1}{4}$; c) $\dfrac{7}{34} \div 2\dfrac{5}{8}$; d) $3\dfrac{3}{5} \div \dfrac{3}{10}$; e) $14.4 \div 1.2$;

f) $1.19 \div 0.17$; g) $2.99 \div \dfrac{23}{10}$; h) $4.5 \div \left(-\dfrac{5}{2}\right)$; i) $-3.6 \div \dfrac{3}{10}$;

j) $\dfrac{2}{3} \cdot \left(\dfrac{5}{6}\right) \div \left(\dfrac{15}{18}\right)$.

TEST II. 3

Find the solution for each of the following equations:

1) $2x - 3 = 5$; 2) $2x - 3 = 8+3x$; 3) $3x + 4 = 10$;
4) $3x + 7 = 8 - 2x$; 5) $2 - x = 6 + x$; 6) $4 - 2x = x + 7$;
7) $2(x-1) = 3x + 4$; 8) $3 - 4x = 5(3x+2)$; 9) $4 \cdot (x-2) = 6$;
10) $3(2-x) = 2x + 1$.

TEST II. 4

1) Calculate the arithmetic average of the numbers x and y, if:
   $x = [(-7+9) \cdot (-3) + (-2)^3 \div (-4)]^2 \div 8$ and

$$y = \left[0.\overline{3} - 1\frac{1}{2}\right] \cdot \left(-\frac{1}{4}\right)^{-1} \cdot 3.$$

2) Solve the equation: $\dfrac{5}{6}(x+3) + 2\dfrac{3}{4} = \dfrac{3x+2}{3} - \dfrac{3-x}{2}.$

3) Find the number x, y, z if they are in proportion with the numbers 9, 10, and 12 and their sum is 217.

4) Simplify the expressions:
   a) $\left(2^3 \cdot 6 - 2^2 \cdot 5\right) \div 7 - \left(1002 \div 3 - 331\right);$
   b) $\left[\left(7^2 - 3^2\right) \div 5 - 2^3\right] \cdot 2007 + 16 \div \left(7 \cdot 3^2 - 55\right).$

5) Calculate $a = \left[\left(0,\overline{3} + \dfrac{1}{4}\right) \cdot \left(1 + \dfrac{5}{7}\right)\right] \div \left[\dfrac{2}{3} \div \left(\dfrac{1}{2} + \dfrac{1}{2} \cdot 0,\overline{3}\right)\right].$

6) Find the positive integer solution of the following equations:
   a) $(x+1) \cdot (x+2) = 72$; b) $3x + 2 = 5(x-2)$;
   c) $\left(2^2 + 2^4 + 2^6\right) \cdot x = 2^3 + 2^5 + 2^7.$

7) Calculate:
   a) $\left[2.4^2 - 0.76 + 0.2 \cdot \left(5.4 + 0.144 \div 1.2^2\right)\right];$
   b) $10 \cdot \left\{21^2 \div 441 + 2 \cdot \left[\left(2^{70} \cdot 3^{30}\right) \div \left(2^{69} \cdot 3^{30}\right) - 15^0\right]\right\};$
   c) $\left\{20.5 \div \left[25 \cdot \left(1 - \dfrac{1}{5}\right) + \left(2\dfrac{1}{3} - 0.75\right) \cdot \dfrac{6}{19}\right]\right\}^2.$

8) Solve the equation: $\left[\left(1 + 1.\overline{6}\right) - \dfrac{5}{3} + 1\dfrac{1}{3}\right] \div 1.\overline{16} \cdot \left(\dfrac{1}{3}\right)^2 - x = \dfrac{2}{3}.$

9) Calculae the areas for a rectangle with the semiperimeter by 80cm and the length $\dfrac{3}{10}$ from the perimeter.

10) Calculate $a = \dfrac{1.\overline{1}}{1} + \dfrac{2.\overline{2}}{2} + \dfrac{3.\overline{3}}{3} + ... + \dfrac{9.\overline{9}}{9}$;

$b = \dfrac{1}{1.\overline{1}} + \dfrac{2}{2.\overline{2}} + \dfrac{3}{3.\overline{3}} + ... + \dfrac{9}{9.\overline{9}}$ and

$c = \dfrac{1.\overline{1}}{1.\overline{1} \cdot 2.\overline{2}} + \dfrac{1.\overline{1}}{2.\overline{2} \cdot 3..\overline{3}} + ... + \dfrac{1.\overline{1}}{8.\overline{8} \cdot 9.\overline{9}}$.

TEST II. 5

1) Find x in each of the following proportion $\dfrac{x}{1.5} = \dfrac{0.5}{0.2}$.

2) Find x in each of the following proportion:

a) $\dfrac{3}{9} = \dfrac{2}{x}$; **b)** $\dfrac{4}{9} = \dfrac{x}{5}$; **c)** $\dfrac{6}{15} = \dfrac{0.3}{x}$; **d)** $\dfrac{5.4 - 1.8}{x} = \dfrac{7.5 - 4.5}{2.4 - 1.4}$.

3) Find the rational numbers x, y such as $x + y = 10$ and $\dfrac{x}{0.2} = \dfrac{y}{0.05}$.

4) Find the rational numbers x, y, and z such as

$x + y + z = 24$ and $\dfrac{x}{3} = \dfrac{y}{4} = \dfrac{z}{5}$.

5) Find the rational numbers x, y, z such as $z - y = 4$ and $\dfrac{x}{5} = \dfrac{y}{2} = \dfrac{z}{6}$.

6) Find the rational numbers x, y, z such that $2x + 3y - 5z = -20$

and $\dfrac{x}{5} = \dfrac{y}{2} = \dfrac{z}{4}$.

7) Find the sides of a triangle if the perimeter is 24 and the sides are proportional with the numbers 3, 4, 5.

8) Find the rational numbers x, y, z such as $\dfrac{x}{12} = \dfrac{y}{15} = \dfrac{z}{30}$ and

$xy\,z = 1600.$

9) Consider the proportions $\dfrac{x}{22} = \dfrac{y}{28} = \dfrac{3z}{40}$. Find x, y, and z if

$\dfrac{x}{11} + \dfrac{y}{6} + \dfrac{z}{17} = 13$.

TEST II. 6

1) Calculate:
   a) $[(9-4.5) \div 0.03] \div [(4.5-3.65) \cdot 4 - 0.4]$;
   b) $[(4.3-4.15) \div 1.5] \div [(1.88+2.12) \cdot 0.0125]$;

2) Calculate:
   a) $\{[2+(3+3\cdot5) \div 2] + [1-(7-3\cdot2) \div 2] \cdot 10\} \div 5$;
   b) $\{[3+(1+1\cdot5) \div 2] + [1-(4-0\cdot2) \div 2] \cdot 10\} \div 5$.

3) If $a = \left(-\dfrac{2}{3}\right)^{-3} \cdot \left(+\dfrac{2}{3}\right)^{4} + \left(\dfrac{7}{6}\right)^{2} \cdot \left(-\dfrac{18}{98}\right) \div \dfrac{3}{4}$ and

   $b = 0.2\overline{3} \div 0.21 \cdot 0.1^{-2} - 0.3\overline{1}$ then find x from the equation
   $ax + b = 1.\overline{3}$.

4) Calculate: $\left(1-\dfrac{1}{2}\right)^{2}\left(1-\dfrac{1}{3}\right)^{2}\left(1-\dfrac{1}{4}\right)^{2}\ldots\ldots\left(1-\dfrac{1}{16}\right)^{2}$.

5) Find x such that $\dfrac{1}{9}\cdot\left\{\dfrac{1}{7}\cdot\left[\dfrac{1}{5}\cdot\left(\dfrac{1}{3}\cdot(x-2)-4\right)+6\right]+8\right\} = \dfrac{1}{3}$.

6) Calculate $(1-7^{-1})\cdot(1-8^{-1})\cdot(1-9^{-1})\cdots\cdots(1-2007^{-1})$.

7) Calculate $\dfrac{1-\sqrt{2}}{\sqrt{1\cdot2}} + \dfrac{\sqrt{2}-\sqrt{3}}{\sqrt{2\cdot3}} + \ldots + \dfrac{\sqrt{2010}-\sqrt{2009}}{\sqrt{2010\cdot2009}}$.

8) Solve the equation $\dfrac{x}{2} + \dfrac{x}{3} + \ldots + \dfrac{x}{n} = \dfrac{1}{2} + \dfrac{1}{3} + \ldots + \dfrac{1}{n}$, $n \in \mathrm{N}$.

9) Let A, B, C D be collinear points and AD = 12cm., AB = BD and CD = 4cm. Find the length of the segments AB, BC, BD, and AM, where M is the midpoint of BC.

10) In the diagram we have:

A   B   C   D   F   G

AG = 57cm, AB = CD = DF, BC = FG = 12cm
Find the length of the segments AB, BF, CG, and the distance from A to the midpoint of CG.

TEST II. 7

1) Calculate the arithmetic average of the numbers x and y, where:

$x = [(-7+9) \cdot (-3) + (-2)^3 \div (-4)]^2 \div 8$ and

$y = \left(0.\bar{3} - 1\dfrac{1}{2}\right) \cdot \left(-\dfrac{1}{4}\right)^{-1} \cdot 3$.

2) Solve the equation $\dfrac{5}{6}(x+3) + 2\dfrac{3}{4} = \dfrac{3x+2}{3} - \dfrac{3-x}{2}$.

3) Find the numbers x, y, z if they are in proportion with the numbers 9, 10, and 12 and their sum is 217.

4) Calculate

a) $(2^3 \cdot 6 - 2^2 \cdot 5) \div 7 - (1002 \div 3 - 331)$;

b) $[(7^2 - 3^2) \div 5 - 2^3] \cdot 2007 + 16 \div (7 \cdot 3^2 - 55)$.

5) Calculate $a = \left[\left(0.\overline{3} + \dfrac{1}{4}\right) \cdot \left(1 + \dfrac{5}{7}\right)\right] \div \left[\dfrac{2}{3} \div \left(\dfrac{1}{2} + \dfrac{1}{2} \cdot 0.\overline{3}\right)\right]$.

6) Solve the equations:

    a) $(x+1) \cdot (x+2) = 72$;

    b) $3x + 2 = 5(x-2)$;

    c) $(2^2 + 2^4 + 2^6) \cdot x = 2^3 + 2^5 + 2^7$.

7) Calculate:

    a) $\left[2.4^2 - 0.76 + 0.2 \cdot (5.4 + 0.144 \div 1.2^2)\right]$;

    b) $10 \cdot \left\{16^2 \div 256 + 2\left[(2^{60} \cdot 3^{20}) \div (2^{59} \cdot 3^{20}) \cdot 2010^0\right]\right\}$;

    c) $61.5 \div \left[25 \cdot \left(1 - \dfrac{1}{5}\right) + \left(2\dfrac{1}{3} - 0.75\right) \cdot \dfrac{6}{19}\right]$.

8) Solve the equation $\left(1 + 1.\overline{6} - \dfrac{2}{9} + 5\dfrac{1}{3}\right) \div 1.\overline{16} \cdot \left(\dfrac{1}{2}\right)^2 - x = \dfrac{2}{3}$.

9) Calculate the area of a rectangle with the perimeter 80 cm and the length is $\dfrac{3}{5}$ from the semiperimeter.

10) Calculate $a = \dfrac{1.\overline{1}}{1} + \dfrac{2.\overline{2}}{2} + \dfrac{3.\overline{3}}{3} + ... + \dfrac{9.\overline{9}}{9}$;

    $b = \dfrac{1}{1.\overline{1}} + \dfrac{2}{2.\overline{2}} + \dfrac{3}{3.\overline{3}} + ... + \dfrac{9}{9.\overline{9}}$; and

    $c = \dfrac{1.\overline{1}}{1.\overline{1} \cdot 2.\overline{2}} + \dfrac{1.\overline{1}}{2.\overline{2} \cdot 3.\overline{3}} + ... + \dfrac{1.\overline{1}}{8.\overline{8} \cdot 9.\overline{9}}$.

TEST II. 8

1) Calculate:

a) $\left[1.4^2 - 0.06 + 4.37 \cdot \left(3.4 + 0.144 \div 1.2^2\right)\right]^0$;

b) $10 \cdot \left\{21^2 \div 441 + 2 \cdot \left[\left(2^{101} \cdot 3^{207}\right) \div \left(2^{100} \cdot 3^{207}\right) \cdot 168^0\right]\right\}$;

c) $\left\{20.5 \div \left[25 \cdot \left(2 - 1\dfrac{1}{5}\right) + \left(3\dfrac{1}{3} - 1.75\right) \cdot \dfrac{6}{19}\right]\right\}^{-5}$.

2) Find x from the proportion $\dfrac{3^{21} + 3^{22} + 3^{23}}{39} = \dfrac{3^{20}}{x}$.

3) Calculate

$$\left(\dfrac{2^2 \cdot 2^5 \cdot 2^1 - 2^3 \cdot 2^4 \cdot 2^5}{2^6 \cdot 2^4 \cdot 2^2 - 2^{4\cdot} \cdot 2^5 \cdot 2^5} \div \dfrac{-3^3 \cdot 3^4 \cdot 3^5 - 3^7 \cdot 3^5 \cdot 3^3}{-3^3 \cdot 3^2 \cdot 3^4 - 3^3 \cdot 3^7 \cdot 3^2}\right) \div \dfrac{1}{6}$$.

4) Solve the equation $\dfrac{5}{6}(-x+4) + 2\dfrac{3}{4} = \dfrac{3x+2}{3} - \dfrac{3-x}{2} + \dfrac{5}{6}$.

5) Solve the equation $\left[1.\overline{6} - \dfrac{2}{9} + 6\dfrac{1}{3}\right] \div 1.1\overline{6} \cdot 5^{-1} - \dfrac{x}{3} = \dfrac{2}{3}$.

6) Calculate $\left(\dfrac{1}{0.\overline{2}} + \dfrac{1}{0.2} + \dfrac{1}{0.3\overline{63}}\right) \div \left(2\dfrac{3}{4} + 9\dfrac{1}{2}\right)$.

7) Calculate:

a) $\left[\left(3^6 \cdot 3^2 \cdot 3^4\right)^{10} \div 3^{118} + 1\right]^3 \div 1000$;

b) $[4^{17} \div 2^{18} \cdot (-8)^{12}]^2 \div (-8)^{20} \cdot 2^{-42}$.

8) Calculate $a = \left[\left(0.\overline{3} + \dfrac{2}{3}\right) \cdot \left(1 + \dfrac{2}{7}\right)\right] \div \left[\dfrac{2}{7} \div \left(\dfrac{5}{6} + \dfrac{1}{2} \cdot 0.\overline{3}\right)\right] \cdot \dfrac{2}{9}$.

9) a) Calculate $11 \cdot [-1 + 343 \div (441 \cdot 21 \div 27)]$;

b) Solve the equation $5x - 4 = 2x + 8$;

c) Find positive integers x such that $2x - 1 < 7$.

10) Solve the equation $\dfrac{2x-9}{3} - \dfrac{x-3}{4} = \dfrac{x-1}{2} - x + \dfrac{1}{12}$.

TEST II. 9

1) Mary and her mother are 40 year old together. Her mother is three times older than Mary.
a) How old is Mary?
b) When does the mother's age become twice the age of Mary's?

2) Calculate:

a) $(2+4+6+...+84)\cdot\left(\dfrac{2}{7\cdot9}+\dfrac{2}{9\cdot11}+\dfrac{2}{11\cdot13}+...+\dfrac{2}{19\cdot21}\right)$;

b) $a = \dfrac{4}{1}+\dfrac{7}{2}+\dfrac{10}{3}+\dfrac{13}{4}+...+\dfrac{301}{100}-\left(1+\dfrac{1}{2}+\dfrac{1}{3}+\dfrac{1}{4}+...+\dfrac{1}{100}\right)$.

3) a) Prove that $\dfrac{1}{n(n+1)} = \dfrac{1}{n} - \dfrac{1}{n+1}$, for any natural number;

b) Calculate the sum $S = \dfrac{1}{2}+\dfrac{1}{6}+\dfrac{1}{12}+\dfrac{1}{20}+...+\dfrac{1}{600}$;

c) Show that $S = \dfrac{1}{1^2}+\dfrac{1}{2^2}+\dfrac{1}{3^2}+...+\dfrac{1}{20^2} < \dfrac{39}{20}$;

d) Find the natural number n such that
$$1+\dfrac{1}{1+2}+\dfrac{1}{1+2+3}+...+\dfrac{1}{1+2+...+n} = \dfrac{200}{101}.$$

4) a) Calculate $\dfrac{a}{b}$ if $a = \left(1+1.\overline{6}-\dfrac{2}{9}+5\dfrac{1}{3}\right)\div1.\overline{16}$ and

$$b = \left(1-\dfrac{1}{2}\right)\cdot\left(1-\dfrac{1}{3}\right)\cdot\left(1-\dfrac{1}{4}\right)\cdot...\cdot\left(1-\dfrac{1}{20}\right).$$

b) Show that the natural number
$n = 7 + 7^2 + 7^3 +...+ 7^{2009} + 7^{2010}$ is divisible by 19.

5) Show that the number $N = 10\cdot2^{6n+2} + 3\cdot2^{6n+3}$, $n \in N$ is also a perfect square and a cub.

6) If $\dfrac{a}{4}=\dfrac{b}{6}=\dfrac{c}{8}$ prove that $a+c=2b$, and if $\dfrac{1}{a}+\dfrac{1}{b}+\dfrac{1}{c}=\dfrac{13}{144}$
find $a, b,$ and $c.$

7) If $\dfrac{a}{b}=\dfrac{3}{2}$ and $\dfrac{c}{d}=\dfrac{4}{5}$ find $\dfrac{3ac-bd}{5ac-bd}.$

8) Find $a,$ $b,$ c such that a and b are in direct proportion with 5 and 8, b and c are in inverse proportions with 3 and 2 prove that $3a+2b-c=57.$

9) Calculate: a) $100+250:5+15:3\cdot\left[265-\left(50:25+2\right)\cdot 65\right]\cdot 50$

b) $6^{100}\div\left[6^{40}\cdot 6^{58}-\left(6^{10}\cdot 6^{15}\right)^{5}\div 6^{27}+\left(7^{57}\div 7^{56}-1^{100}\right)^{97}\cdot 6\right]$

c) Find k such that $64+2\cdot 64+3\cdot 64+...+49\cdot 64=k^{2}.$

10) Calculate $a^{2}+3ab-3ac+d^{2}$ knowing that
$a=3,$ $b-c=5,$ and
$d=\left[2^{18}\cdot\left(2^{3}\cdot 3\right)^{202}\right]\div\left(8\cdot 9\cdot 2^{309}\cdot 3^{99}\right)^{2}+3\cdot(2+4+6+...+2010)^{0}.$

TEST II. 10

1) Find the integer x from the equalities:

a) $0.\overline{1}x+0.\overline{2}x+0.\overline{3}x+...+0.\overline{9}x=\dfrac{26x}{11};$

b) $0.\overline{1}x+0.\overline{2}x+0.\overline{3}x+...+0.\overline{9}x=\dfrac{50+x}{11}.$

2) Show that the number $a=2^{2n+2}\cdot 3^{2n+2}+4^{n+3}\cdot 9^{n+1}-36^{n+1}$ is divizible by 1296, where n is a natural non-zero number.

3) Calculate the sums:
$S_{1}=2^{n}\cdot 3^{n+1}\cdot 5+2^{n+1}\cdot 3^{n}\cdot 7+2^{n}\cdot 3^{n},n\in N;$
$S_{2}=3^{n+1}\cdot 5^{n+1}+3^{n+2}\cdot 5^{n}+15^{n},\ n\in N;$

$$S_3 = 3^n \cdot 7^{n+2} \cdot 11^{n+1} + 21^{n+1} \cdot 11^{n+2} - 40 \cdot 3^{n+3} \cdot 77^n, \ n \in N;$$

$$S_4 = 2 \cdot 4^n \cdot 2^n \cdot 3^{4n} + 8^n \cdot 16 \cdot 81^n \cdot 9 + 2^{3n+2} \cdot 27^n \cdot 3^n \cdot 243.$$

4) Prove that the fraction $\dfrac{9^n + 21 \cdot 3^n + 68}{2 \cdot 3^n + 8}$ is a natural number for any n a natural number.

5) Prove that $A = \dfrac{36^n \cdot 5^{n+1} \cdot 7 - 2^{2n+1} \cdot 3^{2n+1} \cdot 5^n}{6^n \cdot 5^{n-1} \cdot 19 + 2^{n+1} \cdot 15^n}$ is a natural number.

6) Simplify the fractions:

a) $\dfrac{2 \cdot 7^{x+2} - 3 \cdot 7^{x+1} + 3 \cdot 7^x}{2^{x+4} \cdot 3^{x+1} - 2^{x+3} \cdot 3^x}$; b) $\dfrac{11^{x+2} - 11^{x+1} + 10 \cdot 11^x}{13^{x+1} - 13^x}$.

7) Consider the number $N = \overline{abc} + \overline{bca} + \overline{cab}$ where $a \neq b \neq c$.
   a) Find N, if a, b, c are consecutive digits such that $a + b + c = 6$;
   b) Show that $\dfrac{\overline{abc} + \overline{bca} + \overline{cab}}{185}$ is divisible by 37.

8) From a 3-digit number, if taking out the digit from the middle we obtain a 2-digit number which is 6 times less than the initial number. Find the initial number.

9) Find the natural number $\overline{abc}$ if $\overline{a.b} + \overline{b.c} + \overline{c.a} = 3.\overline{3}$.

10) Find a and b, $a < b$ such that the number $A = \sqrt{0.\overline{ab2} + 0.\overline{b2a} + 0.\overline{2ab}}$ is an integer number.

11) Find the non-zero digits a, b, c, d such that $a + b + c = 2d$, $a < c < d < b < 9$, and $\overline{a.bcd} + \overline{b.cda} + \overline{c.dab} + \overline{d.abc} = 10$.

Chapter III

ALGEBRAIC EXPRESSIONS AND LINEAR FUNCTIONS

$$a^2 - b^2 = (a-b)(a+b);$$
$$(a+b)^2 = a^2 + 2ab + b^2;$$
$$(a-b)^2 = a^2 - 2ab + b^2;$$
$$a^3 - b^3 = (a-b)(a^2 + ab^2 + b^2);$$
$$a^3 + b^3 = (a+b)(a^2 - ab^2 + b^2);$$
$$(a+b)^3 = a^3 + 3a^2b + 3ab^2 + b^3;$$
$$(a-b)^3 = a^3 - 3a^2b + 3ab^2 - b^3.$$
$$(a+b+c)^2 = a^2 + b^2 + c^2 + 2ab + 2bc + 2ca, \ a,b,c \in \mathbf{R}.$$

III. 1. Collect like terms.

a) $\dfrac{1}{2}a^2b + 3b^2x - 4a^2b - \dfrac{5}{2}b^2x$; **b)** $2m^2n - 5an + 3m^2n - 7an$;

c) $0.6ax - 0.04ax - 0.24ax - 0.14ax$;

d) $\dfrac{1}{2}x^2 - \dfrac{1}{2}x + x^2 + \dfrac{3}{4}x + 3x - 4x^2$;

e) $\dfrac{1}{0.5}bxy - \dfrac{0.4}{4}x^2y - 0.5x^2y - \dfrac{5}{2}bxy$;

f) $\dfrac{1}{0.1}x + 0.2x - \dfrac{0.5}{2}x - 0.15x - 0.2x$;

g) $0.3a - \dfrac{0.5}{2}a + \dfrac{3}{0.2}b - \dfrac{3}{4}a + b$;

h) $\dfrac{2}{3}ax^3 + \dfrac{1}{2}ax^3 - \dfrac{2}{3}ax^3 - \dfrac{3}{2}ax^3$;

i) $\dfrac{2}{5}by - \dfrac{3}{2}by - 3by + 12by$; **j)** $3a^2b - \dfrac{2}{3}a^2b + 3\dfrac{1}{2}ab^2 - \dfrac{1}{6}a^2b.$

III. 2. Collect like terms.

a) $0.22ab^2 + 0.28ab^2 - \dfrac{2}{5}ab^2 + \dfrac{1}{2}ab^2$;

b) $3(m-n)^2 + 2(m-n)^2 - (m-n)^2$;

c) $2(a+b)^3 + 5(a+b)^3 - 3(a+b)^3$;

d) $-2x^2 - (-5x^2 + 2) - 8x^2 + 10 + 4x^2$;

e) $(3x^3 - 4x^2 + x + 1) - (4x^3 - x^2 - x - 2) - (4x^3 - 2x^2 + 2x + 8)$;

f) $6a + \{4b + [6c - 2a - (2a - c)]\} - [8a - (7b + c)]$;

g) $(5x^2 - 3x + 4) - (4x^2 - 3x + 1) - (2x^2 + 4)$;

h) $(2a - b) - (b + c - 2d) + (b + 2c - d) + (2b - a)$;

i) $\left(\dfrac{4}{7}ab + \dfrac{1}{5}bc - \dfrac{2}{3}ac\right) - \left(-\dfrac{4}{7}ab + \dfrac{3}{10}bc - \dfrac{1}{5}ac\right)$;

j) $\left(\dfrac{1}{3}a^2x + \dfrac{1}{4}ax^2 + \dfrac{2}{3}x^2\right) - \left(-\dfrac{5}{6}a^2x + \dfrac{1}{3}ax^2 - \dfrac{3}{8}x^2\right)$.

III. 3. Calculate and collect like terms:

a) $(3a^3) \cdot (2a^3) + (3a^4) \cdot a^2$; **b)** $(4x^3) \cdot (3x^7) - (2x^6) \cdot (-x^4)$;

c) $(3x^3) \cdot (2x^2) \cdot (5x) - (3x) \cdot (2x^5) - (3x^2) \cdot (6x^4)$;

d) $(4x^2) \cdot (3x) \cdot \left(\dfrac{1}{4}x^6\right) - \left(\dfrac{2}{3}x\right) \cdot (-3x^6) \cdot (-2x^2)$;

e) $(10a^2b) \cdot \left(\dfrac{3}{5}a^4b^2\right) - (5ab^2) \cdot (2a^5b) + 2ab \cdot (-a^5b^2)$;

f) $(8a^2b^2) \cdot (0.5b^3) - (-3ab) \cdot (ab^4) + (-2b) \cdot (a^2b^4)$;

g) $(0.2xy^2z^3) \cdot (9x^3yz) - (4xyz) \cdot \left(\dfrac{1}{2}x^3y^2z^3\right) + (12xyz) \cdot \left(\dfrac{1}{4}x^3y^2z^3\right)$.

III. 4. Calculate:

a) $(x+1)^2$; **b)** $(x-1)^2$; **c)** $(x-2)^2$; **d)** $(x+3)^2$; **e)** $(2x+1)^2$;

f) $(3x-1)^2$; **g)** $(2x-2)^2$; **h)** $(4x-2)^2$; **i)** $(5x+3)^2$;

j) $(4x-3)^2$; **k)** $(-3x+1)^2$; **l)** $(-x-3)^2$; **m)** $(6x-1)^2$;
n) $(0.5x+4)^2$; **o)** $(-a^2-1)^2$; **p)** $(5u-2)^2$; **q)** $(a+7)^2$;
r) $(2b-3)^2$; **s)** $(-1-3x)^2$; **t)** $(2x^2-x)^2$; **u)** $(x+5x^2)^2$;
v) $(a^2-2x)^2$; **x)** $(t^2-5y^3)^2$.

III. 5. Calculate:

a) $\left(\dfrac{x}{2}-\dfrac{y}{3}\right)^2$; **b)** $\left(x-\dfrac{1}{3}\right)^2$; **c)** $\left(-\dfrac{b}{4}+\dfrac{4}{3}\right)^2$; **d)** $\left(\dfrac{3a}{2}-\dfrac{2b}{5}\right)^2$;

e) $\left(2r-\dfrac{4}{3}\right)^2$; **f)** $\left(\dfrac{2a}{5}+\dfrac{5}{4}\right)^2$; **g)** $\left(-\dfrac{3a}{8}-\dfrac{2}{3}\right)^2$; **h)** $\left(\dfrac{a^2}{6}+\dfrac{2a}{3}\right)^2$;

i) $\left(\dfrac{3}{2}x^4y+\dfrac{2}{3}x^3y^2\right)^2$; **j)** $\left(\dfrac{2}{7}a^5b+0.3a^3b^2\right)^2$; **k)** $\left(3x^2y-2y\right)^2$;

l) $(2y-0.5x^3y^2)^2$; **m)** $(2x^2y^2-0.2x^3y)^2$; **n)** $(\sqrt{2}+a)^2$;

o) $(\sqrt{3}a+\sqrt{2}b)^2$; **p)** $(x-\sqrt{3})^2$; **q)** $\left(\sqrt{2}x-\dfrac{1}{\sqrt{2}}y\right)^2$;

r) $\left(\dfrac{2}{\sqrt{3}}x+\sqrt{3}\right)^2$; **s)** $\left(\dfrac{3}{\sqrt{2}}x+\sqrt{2}\right)^2$; **t)** $(2y-\sqrt{3})^2$.

III. 6. Calculate:

a) $(x-4)(x+4)$; **b)** $(2x+3)(2x-3)$; **c)** $(3x-1)(3x+1)$;
d) $(3x-5)(3x+5)$; **e)** $(x-0.4)(x+0.4)$; **f)** $(3-2x)(3+2x)$;

g) $(0.2x-1)(0.2x+1)$; **h)** $(3x-4)(3x+4)$; **i)** $\left(\dfrac{x}{3}+\dfrac{1}{2}\right)\left(\dfrac{x}{3}-\dfrac{1}{2}\right)$;

j) $\left(\dfrac{1}{2}a+\dfrac{2}{3}b\right)\left(\dfrac{1}{2}a-\dfrac{2}{3}b\right)$; **k)** $\left(3x^2+\dfrac{1}{3}\right)\left(3x^2-\dfrac{1}{3}\right)$;

l) $\left(\dfrac{y}{3}-\dfrac{3}{2}\right)\left(\dfrac{y}{3}+\dfrac{3}{2}\right)$; **m)** $\left(\dfrac{a}{3}-3a^2\right)\left(\dfrac{a}{3}+3a^2\right)$;

n) $\left(\dfrac{5}{3}x+\dfrac{2}{5}y\right)\left(\dfrac{5}{3}x-\dfrac{2}{5}y\right)$; **o)** $\left(\dfrac{4}{3}x+\dfrac{3}{2}y\right)\left(\dfrac{4}{3}x-\dfrac{3}{2}y\right)$;

p) $(\sqrt{3}-x)(\sqrt{3}+x)$; **q)** $(\sqrt{5}x-1)(\sqrt{5}x+1)$.

III. 7. Calculate:

a) $x^2-(2x+1)^2$; **b)** $3(x+3)-2(x-3)^2$;

c) $8-(2x+5)^2-2x(x-3)$; **d)** $2(x^2+2)^2-3(x^2-2)^2$;

e) $(x^2+2x)^2-(x^2-2x)^2$; **f)** $3(x-3)(x+3)+(x+9)^2-4x^2$;

g) $(x-3)(x+3)-(x-4)^2$;

h) $5x-2(x+3)^2-3(2x-5)(2x+5)+14x^2$;

i) $3x^2+7(2x+3)^2+4(-2x+5)(2x+5)-15x^2$.

III.8. Calculate:

a) $(x+y+z)^2$; **b)** $(x+y+2)^2$; **c)** $(x+y+0.1)^2$; **d)** $(a^2+a+1)^2$;
e) $(x^2+y^2+z^2)^2$; **f)** $(\sqrt{2}+\sqrt{3}+1)^2$; **g)** $(a+b-1)^2$;
h) $(x+2y-1)^2$; **i)** $(x-2y-2z)^2$; **j)** $(\sqrt{2}-2x+1)^2$.

III. 9. Factor:

a) $x(a+b)-y(a+b)$; **b)** $a(x-y)-b(x-y)$;
c) $m(x+y)-n(x+y)$; **d)** $x(a-b)+2(a-b)$;
e) $(xy+yz)+(xa+za)$; **f)** $(ab-ac)+(bd-cd)$;
g) $mx+xy-ny-mn$; **h)** $4x-6-10ax+15a$;
i) $2ax^2+bx^2-2ay^2-by^2$; **j)** $ay+a\sqrt{3}-y\sqrt{2}-\sqrt{6}$;
k) $a^2x+bx-a^2y-by$; **l)** x^3-x^2+x-1;
m) x^3+3x^2+3x+1; **n)** x^3+x^2-2x-2;
o) $(4x-3)(3x-5)+(4x-3)(-2x+1)-3+4x$;
p) $(5x-1)(-5x+2)-(5x+6)(2-5x)$;
q) $(x+2)(x+2\sqrt{3})-(x+2\sqrt{3})$;
r) $(1-2x)(x+\sqrt{3})-(x-3)(x+\sqrt{3})$.

III. 10. Simplify the fractions. Assume that no denominator is equal to zero.

a) $\dfrac{x^3}{y^3} \cdot \dfrac{y^2}{x^2}$;　b) $\dfrac{-5x^3}{6y^2}\left(-\dfrac{18}{x}\right)$;　c) $\left(\dfrac{-8x}{7y}\right)\left(\dfrac{-7}{x}\right)$;　d) $-2x^2 \cdot \dfrac{5}{x^5} \cdot \dfrac{x^2}{4}$;

e) $\dfrac{2x^2}{6y}\left(-\dfrac{9xy}{y^2}\right)\left(\dfrac{-2}{6x^2}\right)$;　f) $\dfrac{mx^2+my}{x^2-3x} \cdot \dfrac{x-3}{x^2+y}$;　g) $\dfrac{ax^2}{ny^3} \div \dfrac{ax^2}{ny^3}$;

h) $\dfrac{x+y^2}{x-y^2} \div \dfrac{ax+ay^2}{bx-by^2}$;　i) $\left(\dfrac{b}{2}-\dfrac{5b}{4}-\dfrac{b}{6}\right)\div\dfrac{b}{12}$;

j) $\left(\dfrac{3}{a}+\dfrac{2}{a}-\dfrac{1}{a}\right)\div\left(\dfrac{5}{a}-\dfrac{1}{a}-\dfrac{4}{3a}\right)$.

III. 11. Perform the operations. Assume that no denominator is equal to zero.

a) $\dfrac{x}{3}+\dfrac{x}{3}-\dfrac{2x}{3}+\dfrac{5x}{3}$;　b) $\dfrac{x}{y}+\dfrac{3x}{y}+\dfrac{4y-3x}{y}$;　c) $\dfrac{x}{2}+\dfrac{3x-1}{3}+\dfrac{2x+5}{4}$;

d) $x+\dfrac{3x}{5}+\dfrac{1-2x}{2}$;　e) $\dfrac{x}{x+2y+4z}+\dfrac{2y}{x+2y+4z}+\dfrac{4z}{x+2y+4z}$;

f) $x-\dfrac{x}{5}+\dfrac{2-x}{15}+\dfrac{1-x}{3}$;　g) $2x-\dfrac{2-3x}{3}+\dfrac{1-2x}{2}$;

h) $3x+\dfrac{2x-1}{2}-\dfrac{1-x}{4}$;　i) $\dfrac{1}{x}+\dfrac{1}{2x}-\dfrac{1}{3x}+\dfrac{1}{4x}$;

j) $1-\dfrac{2}{7x}-\dfrac{5}{14x^2}$;　k) $\dfrac{1}{2(x+1)}-\dfrac{1}{4(x+1)}-\dfrac{1}{6(x+1)}$;

l) $\dfrac{3}{5x-10}+\dfrac{1}{3x-6}-\dfrac{1}{x-2}$.

III. 12. Simplify the fractions. Assume that no denominator is equal to zero.

a) $\dfrac{xy+3y}{x^2+3x}$;　b) $\dfrac{5x-20}{x^2-16}$;　c) $\dfrac{7x^2+21xy}{9x^3-81xy^2}$;　d) $\dfrac{40x^2-40}{5x+5}$;

e) $\dfrac{(2x+3)(x-2)-9(2x+3)}{(x-11)(4x^2-9)}$;　f) $\dfrac{x^6-y^6}{x^9-x^6y^3}$;　g) $\dfrac{x^2-8x+7}{x^2-49}$;

h) $\dfrac{x^3-8y^3}{x^2-4y^2}$; **i)** $\dfrac{x^2-4y^2-2x-4y}{4y^2-x^2-2y-x}$; **j)** $\dfrac{14x^3+14a^3}{7x+7a}$.

III. 13. Simplify the fractions. Assume that no denominator is equal to zero.

a) $\dfrac{72x^2y^3z}{288x^3y^2z^2}$; **b)** $\dfrac{5a^3y^3}{5a^2y^5}$; **c)** $\dfrac{24ax^2}{18abx^4}$; **d)** $\dfrac{(x-4)(x-2)}{(x-2)(x+3)}$;

e) $\dfrac{3x^2-3x}{6x-6}$; **f)** $\dfrac{x^2-4}{(x+2)^2}$; **g)** $\dfrac{x^2+x}{2x+2}$; **h)** $\dfrac{9x^2-1}{3x+1}$; **i)** $\dfrac{5x+1}{25x^2-1}$;

j) $\dfrac{x^2-6x+9}{x^2-9}$; **k)** $\dfrac{a^2-25}{a+5}$; **l)** $\dfrac{36x^2-25}{36x^2+60x+25}$;

m) $\dfrac{4x^2+32x+64}{2x^2-32}$; **n)** $\dfrac{2x^3-5x^2-12x}{4x^3-20x^2+16x}$;

o) $\dfrac{3x^3-4x^2-4x}{4x^3-6x^2-4x}$; **p)** $\dfrac{a^3-a^2+2a-2}{a^3-a^2-2a+2}$; **q)** $\dfrac{x^2+3x-4}{x^2-5x+4}$;

r) $\dfrac{2x^4+x^3-6x^2}{2x^4-7x^3+6x^2}$; **s)** $\dfrac{(x^2-5x+2)(x^2-5x)-3}{(x^2-5x-2)(x^2-5x)+1}$;

t) $\dfrac{(x^2-4x+4)(x^2-4x)-5}{(x^2-4x+2)(x^2-4x)-15}$; **u)** $\dfrac{(2x^2-3x+4)(2x^2-3x)+4}{(2x^2-3x)(2x^2-3x-3)+2}$.

III. 14. Calculate the expressions and simplify the fractions. Assume that no denominator is equal to zero.

a) $\dfrac{2x}{x^2+3x}+\dfrac{6}{x^2+3x}$; **b)** $\dfrac{2x+3}{x^2-4}+\dfrac{x+3}{x^2-4}$; **c)** $\dfrac{x^2(x-2)}{x^3+8}+\dfrac{4x}{x^3+8}$;

d) $\dfrac{3x-2y}{x-3y}-\dfrac{2x-y}{x-3y}-\dfrac{2y}{x-3y}$; **e)** $\dfrac{1}{x^2-1}+\dfrac{1}{x^2-1}\cdot\dfrac{1-x}{x+1}$;

f) $\dfrac{3a}{6a^2x^2}-\dfrac{x}{6a^3x}+\dfrac{a}{3a^3x^2}$; **g)** $\dfrac{3}{x^2-1}-\dfrac{3}{x^2+1}$;

h) $\dfrac{1+3x}{x^2-1}+\dfrac{4}{x-1}+\dfrac{1}{x+1}$; **i)** $\dfrac{x-1}{x^2-3x}+\dfrac{x+2}{x^2-3x}+\dfrac{1}{x^2-3x}$;

j) $\dfrac{2x+17}{(x-2)(x+5)}+\dfrac{1}{(x-2)(x+5)}$; **k)** $\dfrac{x-1}{x+1}+\dfrac{x+1}{x-1}+\dfrac{2}{x^2-1}$;

l) $\dfrac{x^2}{x^2-3x}+\dfrac{9-6x}{x^2-3x}$; **m)** $\dfrac{2}{x-1}-\dfrac{x+5}{x^2-1}+\dfrac{5}{x+1}$;

n) $\dfrac{3}{x^2+2}+\dfrac{1}{x+1}+\dfrac{9}{\left(1-x^2\right)\left(x^2+2\right)}$.

III. 15. Simplify the expressions. Assume that no denominator is equal to zero.

a) $\dfrac{x+1}{x-1}-\dfrac{x+3}{x-2}$; **b)** $3x-1+\dfrac{2x-3x^4}{x^3-1}$; **c)** $\dfrac{3(x-y)}{x(x+y)}-\dfrac{2(x-2y)}{x(x+y)}$;

d) $\dfrac{x-a+2b}{6(a-b)}-\dfrac{2x+4a-b}{2(a-b)}+\dfrac{2x+a+b}{3(a-b)}+\dfrac{2x+10a-2b}{12(a-b)}$;

e) $\dfrac{\left(2x^3-4x^2\right)\div 2x+3x(1-2x)}{x-2}-\dfrac{2-x}{4}$;

f) $\dfrac{x^2+2xy}{2x+y}+\dfrac{2\left(x^2+3xy+y^2\right)}{2x+y}$;

g) $\dfrac{x^2+xy}{x}+\dfrac{\left(2x^4y-6x^3y\right)\div 2x^2y}{x^4-3x^3}-\dfrac{2xy-x^2}{2x}$;

h) $\dfrac{x-y+1}{x-y}+\dfrac{x+y-2}{4(x-y)}-\dfrac{x+y+7}{3(x-y)}+\dfrac{x+y+22}{12(x-y)}$.

Algebraic rational expressions

III. 16. Simplify the expressions. Assume that no denominator is equal to zero.

a) $\left[\dfrac{(x+y)^2+2y^2}{x^3-y^3}-\dfrac{1}{x-y}+\dfrac{x+y}{x^2+xy+y^2}\right]\div\left(\dfrac{1}{y}-\dfrac{1}{x}\right)$;

b) $\left\{\dfrac{x^2-y^2}{(x+y)^2}+\left(\dfrac{x^2-y^2}{(x+y)^2}\right)^3+2\left[\dfrac{x^2-y^2}{(x+y)^2}\right]^2\right\}\div\dfrac{x^2-y^2}{(x+y)^4}$;

c) $\dfrac{x^2-1}{xy}\div\left[\left(\dfrac{x^2-xy}{x^2y+y^3}-\dfrac{2x^2}{y^3-xy^2+x^2y-x^3}\right)\cdot\left(1-\dfrac{y-1}{x}-\dfrac{y}{x^2}\right)\right]$;

d)
$$\left\{\frac{1}{m^2 n}+\left[\frac{1}{n^4}+\frac{1}{m^2}+\frac{2}{m-n^2}\left(\frac{1}{m}-\frac{1}{n^2}\right)\right]\div\frac{\left(m-b^2\right)^2}{-mn^2}\right\}$$
$$\cdot\left(\frac{m}{m+n}+\frac{n}{m-n}+\frac{2an}{m^2-n^2}\right);$$

e)
$$\left(\frac{x}{y}+\frac{y}{x}+2\right)\left(\frac{x+y}{2x}-\frac{y}{x+y}\right)\div\left[\left(x+2y+\frac{y^2}{x}\right)\left(\frac{x}{x+y}+\frac{y}{x-y}\right)\right];$$

f)
$$\left(\frac{1}{2x-y}+\frac{3y}{y^2-4x^2}-\frac{2}{2x+y}\right)\div\left(\frac{4x^2+y^2}{4x^2-y^2}+1\right);$$

g)
$$\left(\frac{1}{a-2b}+\frac{6b}{4b^2-a^2}-\frac{2}{a+2b}\right)\div\left(\frac{a^2+4b^2}{a^2-4b^2}+1\right);$$

h)
$$\left[\frac{x^2}{x^2-y^2}-\frac{x^2 y}{x^2+y^2}\left(\frac{x}{xy+y^2}+\frac{y}{x^2+xy}\right)\right]\div\frac{x}{x-y};$$

i)
$$\left(\frac{x^2+y^2}{xy}-2\right)\div\left(\frac{2x^2+2xy}{x^2+2xy+y^2}-1\right)\cdot\left(\frac{1}{x-y}-\frac{1}{x+y}\right);$$

j)
$$\left(\frac{a+b}{b}-\frac{2b}{b-a}\right)\cdot\frac{b-a}{a^2+b^2}-\left(\frac{a^2+1}{2a-1}-\frac{a}{2}\right)\div\frac{2+a}{1-2a}\,.$$

III. 17. Consider the rational expression
$$E(x)=\frac{x^2+5x+6}{x^2+4x+4}-\frac{1}{x+2}.$$

a) Show that $E(x)$ is a positive integer for any $x\in\mathbf{R}\setminus\{-2\}$;

b) Find $x\in\mathbf{N}$ such that the number $\dfrac{2x+4}{x^2-4}$ is an integer number.

III. 18. Let
$$E(x)=\left(\frac{x+2}{x^2-x}+\frac{x-3}{x^2-1}-\frac{x-1}{x^2+x}\right)\div\frac{x^2+3x+2}{x^3-2x^2+x}\quad\text{be a}$$
rational expression.

a) Determine the domain of $E(x)$;

b) Solve the equation $E(x) = \dfrac{3(x+1)}{x+2}$.

III. 19. Consider the algebraic expression

$$E(x) = \frac{9}{x-5} \cdot \left(\frac{x}{x+5} + \frac{5}{x-5} - \frac{10x}{x^2 - 25} \right)$$

a) Determine the domain of $E(x)$;

b) Find a simple form for $E(x)$;

c) Find the positive integer x such that $E(x)$ is an integer number.

III. 20. Consider the algebraic expression

$$E(x) = \left(\frac{x^2}{x+2} - \frac{x^3}{x^2 + 4x + 4} \right) \div \left(\frac{x}{x+2} - \frac{x^2}{x^2 - 4} \right),$$

a) Determine the domain of $E(x)$;

b) Find a simple form for $E(x)$;

c) Find the positive integers x, such that $E(x)$ is an integer positive number.

d) Find the numeric value of x, such that $\dfrac{1}{x} \cdot E(x) = \dfrac{4}{x+2}$.

III. 21. Consider the rational expression

$$F(x) = \left(\frac{x^2 + 1}{x^2 - 2x + 1} - \frac{x+1}{x-1} \right) \div \frac{-4}{1 - x^2}.$$

a) Determine the domain of $E(x)$;

b) Find a simple form for $E(x)$;

c) Find the positive integer x such that $E(x)$ is an integer number.

III. 22. Consider the algebraic expression

$$E(x) = \left(\frac{1}{x+2} + \frac{2x}{2-x} + \frac{2x^2}{x^2 - 4} \right) \div \frac{3x+2}{3 \cdot (2-x)}.$$

a) Determine the domain of $E(x)$.

b) Show that $E(x) = \dfrac{3}{x+2}$.

c) Find the positive integer x such that $E(x)$ is an integer number.

III. 23. Consider $A = x^2 - x - 12$ and $B = x^2 - 12x + 36$.
a) Find $A - B$;
b) Find the numeric value of x such that $A = B$;
c) Prove that $\dfrac{A}{x-4} - \dfrac{B}{x-6}$ is constant for any $x \in \mathbf{R} \setminus \{4, \ 6\}$.

III. 24. Consider the rational

expression $E(x) = \left(\dfrac{2}{x-5} + \dfrac{x}{x+5} \right) \div \dfrac{x^2 - 3x + 10}{x^2 - 6x + 5}$.

a) Prove that $E(x) = \dfrac{x-1}{x+5}$, for any $x \in \mathbf{R} \setminus \{ -5; \ 1; \ 5 \ \}$;

b) Find the integer x such that $E(x)$ is an integer number.

III. 25. Consider the rational algebraic expression
$$E(x) = \left(\dfrac{x^4 + x^2}{x^6 + x^4 + x^2 + 1} + \dfrac{1}{x^4 + 1} \right) \div \left(\dfrac{1}{x^2 - 1} - \dfrac{2x^2}{x^6 - x^4 + x^2 - 1} \right).$$
a) Determine the domain of $E(x)$;
b) Find a simple form for $E(x)$;

III. 26. Consider the expression
$$E(x) = \left(\dfrac{1}{x-1} - \dfrac{2x}{x^2 - 1} \right) \dfrac{x-1}{x+1} + \dfrac{2-2x}{x+1} + 1\dfrac{3}{4};$$
a) Determine the domain of $E(x)$;

b) Prove that $E(x) = \dfrac{4}{(x+1)^2}$;

c) Find $E(\sqrt{2})$.

III. 27. Consider the expressions

$$E_1(x) = \left(\frac{x}{x+1} - \frac{x}{x-1}\right)(x^2 - 1) \text{ and } E_2(x) = \frac{x^3 - 2x^2 + x}{x^3 - x}.$$

a) Determine the domain of $E_1(x)$ and $E_2(x)$;

b) Find a simple form for $E_1(x)$ and $E_2(x)$;

c) Prove that $\dfrac{E_1(x)}{x+1} + E_2(x)$ is an integer for any x from the domain.

III. 28. Consider the expression

$$E(x) = \left(\frac{1}{1-2x} - \frac{9x^2 - 18x^3}{1-4x+4x^2}\right) \div \left(1 - \frac{5x^2}{1-4x^2}\right).$$

a) Determine the domain of $E(x)$;

b) Find the integer x such that $\dfrac{E(x)}{2x-1}$ is an integer number.

III. 29. Let $E(x) = \left(\dfrac{x}{x+3} - \dfrac{6x}{x^2 - 9} - \dfrac{3}{3-x}\right) \div \dfrac{x-3}{2x-1}$ be an algebraic expression.

a) Determine the domain of $E(x)$;

b) Show that $E(x) = \dfrac{2x-1}{x+3}$;

c) Find the integer x such that $\dfrac{2E(x)}{2x-1}$ is an integer number.

III. 30. Find the domain and simplify the following rational expressions:

a) $E(x) = \left[\left(\dfrac{x}{x+1} - \dfrac{x^2}{x^2 + 2x + 1}\right) \div \left(\dfrac{x}{x^2 - 1} - \dfrac{1}{x+1}\right)\right]\dfrac{x+1}{x}$;

b) $E = \left[\left(\dfrac{x}{x+2} + \dfrac{1}{x^2 - 4}\right)\left(\dfrac{x+1}{x-1} + \dfrac{2x+5}{1-x^2}\right)\right] \div \left(\dfrac{1}{x+1} - \dfrac{1}{2x}\right)$;

c) $E(x) = \dfrac{x-1}{4} \cdot \left[\dfrac{x+3}{x+1} - \dfrac{x+1}{x+4}\left(\dfrac{2x+3}{x-1} - 1 \right) \right]$;

d) $\left(\dfrac{x^2+8}{x^3-8} + \dfrac{x}{x^2+2x+4} - \dfrac{1}{x-2} \right)\left(\dfrac{x^2}{x^2-4} - \dfrac{2}{2-x} \right)$;

e) $\left\{ \left[\left(\dfrac{x}{x+1} - \dfrac{1}{1-x} - \dfrac{2x}{x^2-1} \right)\left(\dfrac{x-1}{x+1} \right) - \dfrac{2x-2}{x+1} + 1 \right] - 1 \right\} \div \dfrac{x+3}{x+1}$;

f) $\left(\dfrac{5}{2x+3} + \dfrac{2}{3-2x} + \dfrac{2x+9}{4x^2-9} \right)\dfrac{(2x+3)^2}{4} - \left(\dfrac{y}{x} - 2 + \dfrac{x}{y} \right) \div$

$\div \left(\dfrac{y}{x^2+xy} - \dfrac{2}{x+y} + \dfrac{x}{y^2+xy} \right)$;

g) $\left[\left(\dfrac{2x^3+4x^2}{8x^2} - \dfrac{x^2+4}{4x} \right) \div \dfrac{x-2}{2x} + 1 \right]^2 +$

$+ \left(\dfrac{2x-1}{4} - \dfrac{x+1}{3} + \dfrac{x^2+8}{12} \right) \div \dfrac{(x+1)^2}{12} + x$;

h) $E(x) = \left(\dfrac{2x+1}{x^2+3x} + \dfrac{2-x}{x^2-3x} - \dfrac{x-7}{x^2-9} \right) \div \dfrac{x+3}{x^3+2x^2-3x}$

also solve the equation $E(x) = x - 1$;

i) $\left[\dfrac{y-1}{(y+1)^2-y} - \dfrac{1-3y+y^2}{y^3-1} - \dfrac{1}{y-1} \right] \div \dfrac{y^2+1}{1-y}$;

j) $\left[\left(\dfrac{x^2-x}{x^2+1} + \dfrac{2x^2}{x^3-x^2+x-1} \right) \div \dfrac{x^2}{x^2-1} \right] \div \dfrac{2x^2+5x+3}{2x^2+3x}$.

III. 31. Solve the equations. Assume that no denominator is equal to zero.

a) $\dfrac{1}{x-3} = 2$; **b)** $\dfrac{2x-1}{x} = \dfrac{4x}{2x-1}$; **c)** $3x + \dfrac{1}{x} = 3x - \dfrac{1}{x}$;

d) $\dfrac{2x-6}{x-3} = 2$; **e)** $\dfrac{3x-3}{x-1} = \dfrac{3x-4}{x-1}$; **f)** $\dfrac{2}{x} - \dfrac{2}{x+1} = 0$;

g) $x + 1 + \dfrac{x-1}{x-2} = 3 + \dfrac{x-1}{x-2}$;

h) $\dfrac{1}{x+1}-\dfrac{1}{2x+2}-\dfrac{1}{3x+3}-\dfrac{1}{6x+6}=0$;

i) $\dfrac{2}{1+x}-\dfrac{1}{x-3}=\dfrac{4}{(x+1)(3-x)}$; **j)** $\dfrac{x^2}{x-5}=\dfrac{25}{x-5}$;

k) $\dfrac{a+b}{x}+\dfrac{a-b}{a}=1$; **l)** $\dfrac{2x}{x-2}-\dfrac{1}{x+2}=\dfrac{2x^2}{x^2-4}$;

m) $\dfrac{3(x+5)}{3x}=\dfrac{4x+1}{4x+2}$; **n)** $\dfrac{x^2-9}{x+5}=x+3$;

o) $\dfrac{3x-2}{x-5}=\dfrac{3x+1}{x+5}$; **p)** $\dfrac{x+\dfrac{1}{2}}{2}-\dfrac{2x-\dfrac{x+3}{4}}{5}=4$;

q) $\dfrac{2x+1}{2x-1}-\dfrac{3(2x-1)}{2x+1}-\dfrac{8x^2}{1-4x^2}=0$;

r) $\dfrac{a^2+x}{b^2-x}-\dfrac{a^2-x}{b^2+x}=\dfrac{4(abx+2a^2-2b^2)}{b^4-x^2}$, $a\neq b$;

s) $\dfrac{x}{8}+\dfrac{1-\dfrac{x+1}{2}}{2-\dfrac{2}{2-\dfrac{2}{2+\dfrac{x}{2}}}}=-\dfrac{1}{4}$; **t)** $\dfrac{5+\sqrt{7}}{x}=\dfrac{9}{5-\sqrt{7}}$.

III. 32. Solve the equations.

a) $2x(3x-2)-3\left[(2-x)(-2x+3)-\dfrac{x-3}{2}\right]=-4$;

b) $\dfrac{1}{(x+1)^2}+\dfrac{4}{x(x+1)^2}=\dfrac{5}{2x(x+1)}$;

c) $\dfrac{2x+26}{5x^2-5}-\dfrac{12}{x^2-1}-\dfrac{2}{1-x}=0$; **d)** $\dfrac{10}{4-x}=\dfrac{4}{1-3x}-\dfrac{5}{x-4}$;

e) $\dfrac{x+a}{a-x}+\dfrac{x-a}{a+x}=\dfrac{8a}{a^2-x^2}$; **f)** $\dfrac{x}{a}+1\div\left(1-\dfrac{a^2}{b^2}\right)=1\div\left(\dfrac{a}{b}-\dfrac{b}{a}\right)$;

g) $\dfrac{x}{a}(3a+1)=\dfrac{3a}{1+a}+\dfrac{(2a+1)x}{a(a+1)^2}+\dfrac{a^2}{(a+1)^3}$;

h) $\dfrac{a^3-1}{a^3+1}=\dfrac{a(x-1)+a^2-x}{a(x+1)-a^2-x}$;

i) $\dfrac{ax-x}{2a+2}+\dfrac{ax}{a^2-1}-\dfrac{a-x}{a-1}=\dfrac{x}{2}+\dfrac{a+x}{a+1}$;

j) $x\div\left[\dfrac{(a-1)^2}{4a}+1\right]=\left[\dfrac{(a-1)^2+a}{3}\right]\div\left[\dfrac{(a+1)^3}{3a}-a-1\right]$;

k) $\dfrac{(x-1)^2+5(x-1)+16}{x+3}-5(x+2)=-4x-7$;

l) $\dfrac{16a^2x^2-48ax+36}{8ax-12}-\dfrac{ax-5}{5}-\dfrac{9ax}{5}=2-x$;

m) $\dfrac{36-x^2}{x+6}-\dfrac{12x^2+36x+27}{3(2x+3)}=\dfrac{2x-1}{5}$.

III. 33. Solve the equations:

a) $-0.7x+3.8=x+0.4$; **b)** $0.25x-1.5=0$;

c) $5x-2\sqrt{3}=\sqrt{3}$; **d)** $\sqrt{3}x-\sqrt{6}=0$;

e) $-\sqrt{3}x=\sqrt{3}-3$; **f)** $2.\overline{1}-2x=\dfrac{1}{4}+\dfrac{5}{12}+\dfrac{7}{18}$;

g) $x-\left|7-\sqrt{5}\right|=\left|\sqrt{5}-7\right|$; **h)** $3x-\sqrt{2}=2x+\sqrt{18}$;

i) $5\left(x+\sqrt{2}\right)+\dfrac{1}{2}=0.25$; **j)** $(x-1)\sqrt{2}=\left(\sqrt{3}-1\right)\cdot x+1-\sqrt{3}$;

k) $\left(\sqrt{5}-2\right)\left(\sqrt{5}+2\right)=\left(\dfrac{\sqrt{2}}{2}-x\right)\left(\sqrt{7}-\sqrt{5}\right)\left(\sqrt{7}+\sqrt{5}\right)$;

l) $\sqrt{2}x+3=5\sqrt{2}(x-1)$; **m)** $3(x-1)=\dfrac{5x-1}{2}$;

n) $-\dfrac{3}{4}x+\dfrac{1-\sqrt{3}}{3}=0$; **o)** $\dfrac{1-0.4x}{0.6}-\dfrac{1-3x}{2}=\dfrac{0.5x-0.4}{0.3}$;

p) $0.\overline{3}z-\sqrt{2}=1.5z+0.\overline{5}$; **q)** $3\sqrt{2}+x=\sqrt{2}\left(x-\sqrt{2}\right)+5$;

r) $\dfrac{0.2(2-3x)}{0.01} - \dfrac{0.25-2x}{0.1} = \dfrac{0.6-x}{0.02} + 7.5$;

s) $\dfrac{1}{2}\left[\dfrac{1}{3}\left(\dfrac{1}{4}\cdot x + 2\right) - 1\right] = 0$; t) $\dfrac{1}{9}\left\{\dfrac{1}{7}\left[\dfrac{1}{5}\left(\dfrac{1}{3}\cdot(x-1)+4\right)+6\right]+8\right\} = 1$;

u) $\dfrac{3x}{2} - \dfrac{x-2}{3} - 1 = x - \dfrac{2-x}{6}$;

v) Find $a \in \mathbf{R}$ such that $x = -2$ is a solution for the equations:
i) $ax + 5 = 6 - a$; ii) $3x + a = 5$; iii) $2a - 3x = 3(a+1) - 1 - 4x$;
iv) $2x + 2a = 3a + x - 5$; w) Find m such that in the equation
$mx + 5 = 11$ has the solution i) 1; ii) 2; iii) 3; iv) -1.

III. 34. Linear equations containing absolute value.

a) $|x+2| = 6$;   b) $|x-3| = 5$;   c) $|3x+2| = 7$;

d) $20 \div |x-1| = 4$; e) $2|x| + 1 = 7$; f) $|x| + \dfrac{1}{2} = \dfrac{3}{4} + |2x|$;

g) $|2x+1| - 6 = -3\dfrac{1}{3}$; h) $3(2|3x+1|+6) - 10 = 14$;

i) $|x^2 - 9| + |2x - 6| = 0$; j) $|x| = m$; k) $|x+1| = a$;

l) $|x| + |x+1| = -1$; m) $|a| \cdot x = a$, $a \in \mathbf{R}$; n) $|x| = -x$;

o) $|2m+1| \cdot x = 2m+1$; q) $\sqrt{m^2} \cdot x = m(m-1)$;

r) $\sqrt{m^2 + 4m + 4} \cdot x = m + 2$; s) $|m| \cdot x = m$;

t) $|m| \cdot x + 2x = 2 + m$;   u) $|a| \cdot x + 2ax = 2$, $a \in \mathbf{R}$;

v) $|x| = x + 2$; w) $\sqrt{x^2 - 4x + 4} = 3x - 1$; x) $|2x - 1| = |2x + 3|$;

y) $|x-2| - 2|2x+3| = 0$; z) $\||x| + |x + 2\|| = x + 5$.

III. 35. Solve each equation and state the domain for solutions.

a) $2x = 3 - mx$; b) $mx - m = 1 - x$; c) $mx - 4 = 3x$;
d) $3(x+1) = m + 2$; e) $a^2 x = b(3a - bx)$;
f) $m^2 x + mx + 3 = 2x + m$; g) $m^2 x + 3 - 3x = m + 2mx$;
h) $m^2 x - 1 = 3mx + 4x$; i) $mx - 4m = m^2 + 4x$;

j) $m^2 x = 2m(1-x)$; **k)** $mx - 1 = x - m^2$; **l)** $2ax + 7 = x + 3a$;

m) $mx + 3(x+m) = x + m - 1$; **n)** $(m+1)x + a = x$, a, $m \in \mathbf{R}$;

o) $am - ax = mx - am$; **p)** $3mx + 2a - 5 = 2(xm + a)$;

q) $1 - (x - m) \cdot a = (a - x) \cdot m$; **r)** $a \cdot (x - a) = m(x - m)$;

s) $4\sqrt{2}x - \sqrt{2}m = 4mx$; **t)** $\dfrac{m}{x-3} = \dfrac{x+2}{x^2-9}$, $m \in \mathbf{R} \setminus \{-3,\ 3\}$;

u) $2a(x - 2a) = m(x - 3m)$.

III. 36. Compute, factor, and solve each equation.

a) $-2x = 4x^2$; **b)** $x^2 - 9 = 0$, $x \in \mathbf{N}$; **c)** $x^2 - 3 = 0$;

d) $3x^2 + 5x = 0$; **e)** $\dfrac{x^2 - 2x}{x} = 0$; **f)** $4x^2 - 25 = 0$;

g) $(x+1)^2 - 16 = 0$; **h)** $4(x+1)^2 - 9(x-1)^2 = 0$;

i) $(x-1)(x-2)(x+3) = 0$; **j)** $(x+2)(x-5) + (x+2)(3x-4) = 0$;

k) $8x + 16 + x^2 = 0$; **l)** $3 - 2x - x^2 = 0$; **m)** $8x^2 - 7x - 1 = 0$;

n) $1 + 4x^2 = 5x$, $x \in \mathbf{N}$; **o)** $2x^2 - 3x - 9 = 0$;

p) $(x+3)(5x+9) - x^2 + 9 = 0$; **q)** $\dfrac{x^2 + 5x + 4}{x^2 + 3x + 2} = 1$;

r) $\dfrac{x^3 - 9x + 3x^2 - 27}{x^2 - 9} = x^2 + 3x$.

III. 37. Solve the inequations:

a) $-x + 2 \geq 0$; **b)** $-3x + 1 < 0$; **c)** $-7x - 1 \geq 0$;

d) $-4x \geq -2x + 3$; **e)** $-6x \geq -8x$;

f) $4(x-1) - \dfrac{1-x}{2} \leq 3(x+2) - \dfrac{12-x}{5} - 2x$;

g) $2(2 - 3x) - \dfrac{1-x}{2} + \dfrac{12-x}{3} \geq 2(2x - 1) - \dfrac{3-x}{6}$;

h) $(x-1)^2 - 5 > (x+4)^2$;

i) $(x+1)(x+2) + 3(1-x) < (x-1)^2$; **j)** $\dfrac{x-1}{3} - (1 - 2x) > \dfrac{1}{4}x - \dfrac{1-2x}{6}$;

k) $\dfrac{3-x}{4} - \dfrac{3}{2}(x-1) > \dfrac{2}{3} - \dfrac{4x-3}{6}$;

l) $(x-1)(2x+3) \le (2x-5)(x+4)$;

m) $\dfrac{x-1}{4} - \dfrac{3-x}{2} \ge \dfrac{1}{3}x - \dfrac{5(x+3)}{6} + \dfrac{3}{4}$;

n) $\dfrac{2x+27}{35} < \dfrac{3x+5}{7} + 1 + \dfrac{1-3x}{5}$;

o) $\dfrac{x+1}{21} - \dfrac{5-x}{3} - 8 \cdot \left(\dfrac{4-x}{7} - \dfrac{2-x}{3} \right) \ge 5x - \dfrac{3x - \dfrac{2}{3}}{7}$;

p) $\dfrac{3-x}{5} - \dfrac{2-x}{2} \ge x + 2$.

III. 38. Solve the systems of equations.

a) $\begin{cases} 2x+3y = -1 \\ 4x-3y = 7 \end{cases}$; **b)** $\begin{cases} 4x+3y = 5 \\ 2x-5y = 9 \end{cases}$; **c)** $\begin{cases} 4x-5y = 1 \\ 7x-9y = 2 \end{cases}$;

d) $\begin{cases} 8a-11b = 5 \\ 3a+4b = 10 \end{cases}$; **e)** $\begin{cases} a-b = 2 \\ a+b = 4 \end{cases}$; **f)** $\begin{cases} 4a-3x = 1 \\ 3a+4x = 7 \end{cases}$;

g) $\begin{cases} 5m-4n = -1 \\ 8m-2n = -6 \end{cases}$; **h)** $\begin{cases} \dfrac{1}{5}x + \dfrac{1}{2}y = -2 \\ \dfrac{1}{4}x - \dfrac{1}{15}y = -\dfrac{71}{60} \end{cases}$; **i)** $\begin{cases} \dfrac{1}{7}x + \dfrac{2}{3}y = 5 \\ \dfrac{3}{7}x - \dfrac{1}{3}y = 1 \end{cases}$;

j) $\begin{cases} \dfrac{1}{3}a + \dfrac{2}{5}b = 6 \\ \dfrac{1}{6}a - \dfrac{3}{10}b = -2 \end{cases}$.

III. 39. Solve the systems of linear equations:

a) $\begin{cases} x+y = -1 \\ 4x+3y = -2 \end{cases}$; **b)** $\begin{cases} \dfrac{x-1}{2} + \dfrac{y+1}{5} = \dfrac{1}{5} \\ \dfrac{x-4}{7} - \dfrac{1-y}{2} = -\dfrac{13}{14} \end{cases}$;

$$\text{c)} \begin{cases} \dfrac{2y-x+5}{2x+y-5}=4\dfrac{1}{2} \\[2mm] \dfrac{x-2}{3}-\dfrac{3-y}{2}+2x-y=1 \end{cases} ;\quad \text{d)} \begin{cases} 1-\dfrac{x+2y}{7}=\dfrac{1}{3}\left[2-2(x-2)-y\right] \\[2mm] \dfrac{x-2}{4}-\dfrac{1+y}{3}+x+5y=-3 \end{cases} ;$$

$$\text{e)} \begin{cases} x\sqrt{2}+y\sqrt{3}=1 \\ 5x\sqrt{2}-2y\sqrt{3}=16 \end{cases} ;\quad \text{f)} \begin{cases} \left(\sqrt{5}+\sqrt{3}\right)x+\left(\sqrt{5}-\sqrt{3}\right)y=4 \\ 2x-y=3\sqrt{5}-\sqrt{3} \end{cases} ;$$

$$\text{g)} \begin{cases} \dfrac{1}{x}-\dfrac{1}{y}=6 \\[2mm] \dfrac{3}{x}+\dfrac{2}{y}=3 \end{cases} ;\quad \text{h)} \begin{cases} \dfrac{1}{x}+10y=-3 \\[2mm] \dfrac{3}{x}+4y=4 \end{cases} ;$$

$$\text{i)} \begin{cases} \dfrac{28}{x}=\dfrac{82}{y} \\[2mm] x\left(\dfrac{28}{x}-2\right)=y\left(\dfrac{82}{y}-2\right)+54 \end{cases} ;\quad \text{j)} \begin{cases} \dfrac{x-1}{5}+\dfrac{y+3}{3}=\dfrac{x}{4}-\dfrac{1}{4} \\[2mm] x+y=-2 \end{cases} .$$

III. 40. Solve the systems:

$$\text{a)} \begin{cases} \dfrac{2}{x}+\dfrac{3}{y}=-\dfrac{1}{6} \\[2mm] \dfrac{3}{x}-\dfrac{2}{y}=\dfrac{5}{6} \end{cases} ;\quad \text{b)} \begin{cases} \dfrac{2x-3y}{4}-\dfrac{2x-y}{3}=-\dfrac{3}{4} \\[2mm] (x-2)^2-(3-y)^2=(x-y)(x+y)-7 \end{cases} ;$$

$$\text{c)} \begin{cases} \dfrac{x+y-9}{x-2y-3}=-2 \\[2mm] \dfrac{x+y}{4}-\dfrac{3x-2y-6}{6}=-\dfrac{7}{12} \end{cases} ;\quad \text{d)} \begin{cases} x+2\cdot\left[y-6\cdot(x-1)\right]=-1 \\ y-3\cdot\left[x-6\cdot(y+1)\right]=-4 \end{cases} ;$$

$$\text{e)} \begin{cases} \dfrac{x}{a}+\dfrac{y}{b}=3 \\[2mm] \dfrac{x}{3a}+\dfrac{y}{4b}=\dfrac{5}{6} \end{cases} ,\; a,b\in R\setminus\{0\};\quad \text{f)} \begin{cases} \dfrac{x}{a+b}+\dfrac{y}{a-b}=2a \\ x-y=4ab \end{cases} ;$$

$$\text{g)} \begin{cases} \dfrac{2}{x+y-1}+\dfrac{3}{x-y+2}=5 \\[2mm] \dfrac{1}{x-y+2}-\dfrac{2}{x+y-1}=-1 \end{cases} ;\quad \text{h)} \begin{cases} \dfrac{3}{x+y}=\dfrac{4}{x+z}=\dfrac{5}{y+z} \\ (x+y)(x+z)(y+z)=60 \end{cases} ;$$

~ 94 ~

i) $\dfrac{x+2y-7}{1}=\dfrac{2y-x+15}{2}=\dfrac{2x+y+19}{3}$.

III. 41. The sum of two by two sides of a triangle is 45 m, 52 m, and 48 m, respectively. Find the length of the sides of the triangle.

III. 42. Two triangles are similar. The sides of the first are equal to 7 cm, 10 cm, 11 cm respectively, and the perimeter of the second triangle is 70 cm. Find the sides of the second triangle.

III. 43. The sum of three natural numbers is 222. If we divide the first number by the second one, we will obtain 10, and if we divide the second number by the third one, we will also obtain 10. Find the numbers.

III. 44. Solve the linear systems of equations:

a) $\begin{cases}\dfrac{x-3}{y+1}=1\\[2mm]\dfrac{x+1}{y-2}=3\end{cases}$;   b) $\begin{cases}\dfrac{x+y}{2}-\dfrac{x-y}{2}=1\\[2mm]\dfrac{x+y}{3}+\dfrac{x-y}{4}=\dfrac{2}{3}\end{cases}$;

c) $\begin{cases}4x+7y+10z=21\\x+2y+2z=5\\3x+5y-2z=6\end{cases}$;   d) $\begin{cases}3x+y-3z=1\\x-3y+4z=2\\2x+2y-z=3\end{cases}$;

e) $\begin{cases}(x+y)^2-z^2=9\\(y+z)^2-x^2=45\\(z+x)^2-y^2=27\end{cases}$;

f) If $\dfrac{3y-2x}{4y-3x}=\dfrac{5}{7}$, calculate $\dfrac{x}{y}$;

g) If $\dfrac{x}{y}=0,\overline{6}$. Find $A=\dfrac{2y}{6x+5y}$.

Linear functions

III. 45. Consider the linear function $f : \mathbf{R} \to \mathbf{R}$, $f(x) = 3x - 6a + 2$, where $a \in \mathbf{N}$. Find a such that the point $A(a, -a + 8)$ lies on the graph of f.

III. 46. Let $f : \mathbf{Z} \to \mathbf{Z}$, $f(x) = ax + b$, $a, b \in \mathbf{R}$ be a linear function. Find the function f such that $f(-5) = -13$ and $f(3) = 3$.

III. 47. Draw the graph of the function $g : \{-1, \ 0, \ 1, \ 2\} \to \mathbf{R}$, $g(x) = 2x + 2$.

III. 48. Draw the graph of the function $f : \mathbf{R} \to \mathbf{R}$, $f(x) = 2x - 3$.

III. 49. Consider the function $f : \mathbf{R} \to \mathbf{R}$, $f(x) = ax + a + 2$, $a \in \mathbf{R}$. Find the value of the number a if the point $A(1, 4)$ lies on the graph of f.

III. 50. Consider the functions $f, g : \mathbf{R} \to \mathbf{R}$, $f(x) = 2x + 1$, $g(x) = -x + 2$. Find the coordinates of the point of intersection of both graphs. Are the points $A(0, 1)$, $B\left(\dfrac{1}{2}, \ 2\right)$, and $C(-1, -1)$ collinear?

III. 51. Find the function $f : \mathbf{R} \to \mathbf{R}$, $f(x) = ax + b$, where $a, b \in \mathbf{R}$, $a, b < 0$, if the area between the graph of f and Ox and Oy axes is 4 and the slope of f is -3.

III. 52. Consider the function $f : \mathbf{R} \to \mathbf{R}$, $f(x) = x - 3$.
a) Find $f(4)$;
b) Draw the graph of $f(x)$;
c) Find the points of intersection of f with the axis;
d) Which of the points $A(5, 2)$ and $B(-2, 5)$ lay on the graph?
e) Find the area of the triangle formed by the graph of $f(x)$ and the axis;

f) Find the distance from the origin to the graph of $f(x)$;

g) Find the measure of the angle between the graph of $f(x)$ and x-axis ;

h) Solve the equation $f(x) + 2 = 1 - x$;

i) Solve the inequation $3 - 2f(x) \geq 3f(x) - 2$.

III. 53. Find the function $f : \mathbf{R} \to \mathbf{R}, f(x) = ax + b$ such that the graph passes through the points $A(1, 2)$ and $B(3, -2)$.

III. 54. Consider the function $f : \mathbf{R} \to \mathbf{R}$, $f(x) = 3x - 9$, find $f(4) + f(1)$. Which of the points $A(-1, -12)$, $B(0, -9)$, and $C(1, 2)$ lay on the graph of f?

III. 55. Consider the functions $f, g : \mathbf{R} \to \mathbf{R}$, $f(x) = 2x - 8$ and $g(x) = x - 7$.

a) Find the coordinates of the points of intersection between G_f (*graph of the function f*) and x-axis and y-axis;

b) Draw the graph for the function f;

c) Find the point of intersection between the graphs of the functions f and g;

d) Find the perimeter and the area of the triangle formed by the graph of f and the x-axis and y-axis;

e) The distance from the origin to the graph of f;

f) The radius of the incircle and circumcircle of the triangle formed by the graph of f and the x-axis and y-axis;

g) Find the linear function $h : \mathbf{R} \to \mathbf{R}, h(x) = ax + b$ if $h(x) + h(2 + 2x) = f(1 + x) + g(3 + x)$.

h) Solve the inequation $f(x + 2) + 2f(3 - 2x) \geq 6$;

i) Solve the system of equations $\begin{cases} f(x) + 3g(1 - 2y) = -38 \\ 3f(x + 3) - 2g(y - 1) = 0 \end{cases}$;

j) Prove that $\dfrac{f(a) + f(b)}{2} = f\left(\dfrac{a + b}{2}\right)$;

k) Which of the points $A(1,3)$, $B(0,4)$, $C(11,14)$ lay on the graph of f?

III. 56. Find the point M on the function:

a) $f : \mathbf{R} \to \mathbf{R}, f(x) = 8x - 7$, with equal coordinates;

b) $g : \mathbf{R} \to \mathbf{R}, g(x) = -3x + 6$, x intercept is 2;

c) $h : \mathbf{R} \to \mathbf{R}, h(x) = \dfrac{3}{7}x + 1$, y intercept is 1;

d) $i : \mathbf{R} \to \mathbf{R}, i(x) = -\dfrac{1}{3}x + 3$, y intercept is two thirds of x intercept.

III. 57. Find $m \in \mathbf{R}$ such that the graph of $f : \mathbf{R} \to \mathbf{R}$, $f(x) = (m - 3)x - m^2 - 6$ passes through the point $P(m + 2, 5m - 6)$.

III. 58. Find $m \in \mathbf{R}$ such that the point $A\left(\sqrt{2},\ 6\right)$ lies on the graph of $f : \mathbf{R} \to \mathbf{R}$, $f(x) = mx\sqrt{3} + 5\sqrt{3} - 3$.

III. 59. Find the linear function $f : \mathbf{R} \to \mathbf{R}$ such that the graph contains the points:

a) $A(2, 0)$ and $B(-1, 3)$;

b) $C(-3, -9)$ and $D(5, 7)$;

c) $E(0, 5)$ and $F(2, 4)$;

d) $G(1, -6)$ and $H(-4, 14)$;

e) $I(-3, -10)$ and $J(4, -3)$;

f) $K(-1, -4)$ and $L(1, 6)$.

III. 60. Check if the following points are collinear:

a) $A(-2, 1)$, $B(1, 7)$, and $C(-1, 1)$;

b) $D(1, 6)$, $E(3, 8)$, and $F(-5, 0)$;

c) $G(2, -5)$, $H(-3, 10)$, and $I(4, 3)$;

d) $J(1, 0)$, $K(2, 7)$, and $L(3, 14)$.

III. 61. Find $m, n \in \mathbf{R}$ such that the points $A(2, -1)$ and $B(3, 1)$ lay on the graph of the linear function $f : \mathbf{R} \to \mathbf{R}$, $f(x) = (m + 2n - 3)x - 3m - n$.

III. 62. Find $m, n \in \mathbf{R}$ such that the points $A(m, -10)$ and $B(0, -4)$ lay on the graph of the linear function $f : \mathbf{R} \to \mathbf{R}$, $f(x) = (3n + 1)x - 2m + 2$.

III. 63. Find $a \in \mathbf{R}$ such that the points $A(1, -2)$, $B(a, -2a)$ and $C(3, 5)$ are collinear.

III. 64. Calculate the area and the perimeter for the triangle between the graphs of the functions $f : \mathbf{R} \to \mathbf{R}$, $f(x) = 2x + 1$, $g : \mathbf{R} \to \mathbf{R}$, $g(x) = -x + 7$, and $h : \mathbf{R} \to \mathbf{R}$, $h(x) = 3$.

III. 65. Consider the function $g : \mathbf{R} \to \mathbf{R}$, $f(x) = -2x + 5$.
a) Solve the equation $3f(x) - 6x = x - 11$;
b) Calculate $f(-x) + f(x)$ and show that it is an integer number.

III. 66. Consider the function $f : \mathbf{R} \to \mathbf{R}$, $f(x) = 3x - 6$.
a) Draw the graph of f;
b) Find the area of the triangle formed by G_f and x-axis and y-axis;
c) Find $a \in \mathbf{R}$ knowing that the point $A(2a, a - 1)$ lies on the graph of f;
d) Calculate the sum $f(1) + f(2) + f(3) + \dots + f(10)$.

III. 67. Consider the function $f : \mathbf{R} \to \mathbf{R}$, $f(x) = x - 3$:
a) Draw the graph of f;
b) Prove that the point $M(-2, -5)$ lies on the graph of f;
c) Find the area of the triangle formed by G_f and x-axis and y-axis;
d) Find the distance from the origin to the graph of f;
e) Calculate $E = f(\sqrt{3}) \cdot f(-\sqrt{3})$;
f) Prove that $\dfrac{1}{f(1)} + \dfrac{1}{f(2)} + \dfrac{1}{f(3)} + \dots + \dfrac{1}{f(2007)} < 502$.

III. 68. Find the linear function $f : \mathbf{R} \to \mathbf{R}$ such that $f(x - 3) = 2x + 1$.

III. 69. Consider the function

$f : \mathbf{R} \rightarrow \mathbf{R}, f(x) = (2a - 2)x + b - 3.$

a) Find a such that the graph of f is parallel to x-axis;
b) Calculate b such that the origin is on the graph of f;
c) Find $a \in \mathbf{N}$ such that the function is decreasing;
d) If $a = 2$ and $b = 1$, find $M \in G_f$ such that the coordinates are equal.

III. 70. Consider the function $f : \mathbf{R} \rightarrow \mathbf{R}$, $f(x) = \frac{3}{2}x - 3$:

a) Draw the graph of f;
b) Find a point on the graph of f having equal coordinates.

III. 71. Consider the function

$f : \mathbf{R} \rightarrow \mathbf{R}$, $f(x) = x + 3$ and $g : \mathbf{R} \rightarrow \mathbf{R}$, $g(x) = 2x + 2.$

a) Draw the graphs of f and g;
b) Find the area of the triangle formed by G_f and G_g and x-axis.

c) Find the area of the triangle formed by G_f and x-axis and y-axis.

III. 72. Consider the function $f : \mathbf{R} \rightarrow \mathbf{R}$, $f(x) = 2x - 4.$

a) Draw the graph of f;
b) Find x-axis and y-axis intercepts;
c) Find the area of the triangle AOB formed by G_f and x-axis and y-axis;
d) Prove that $Q(1, -2)$ is the circumcenter of $\triangle AOB$;
e) Find the altitude of $\triangle AOB$.

III. 73. Consider the linear function

$f : \mathbf{R} \rightarrow \mathbf{R}$, $f(x) = x + 3$ and $g : \mathbf{R} \rightarrow \mathbf{R}$, $g(x) = 2x + 2.$

a) Draw the graphs of f and g;
b) Find the area of the triangle formed by G_f and G_g and x-axis.

III. 74. Consider the function $g : \mathbf{R} \rightarrow \mathbf{R}$, $g(x) = 5 - 2x.$

Find:
a) the value of a such that the point $R(2, a)$ lies on the graph of f;
b) the geometric mean of $g(\sqrt{6})$ and $g(-\sqrt{6})$;

c) *x*-axis and *y*-axis interception.

Integer part (floor functions) function

$$f : \mathbf{R} \to \mathbf{Z}, \quad f(x) = [x], \quad [x] \le x < [x] + 1.$$

III. 75. Solve the equations:

a) $\left[\dfrac{2x+1}{3}\right] = \dfrac{x-4}{2}$; b) $\left[\dfrac{x+1}{2}\right] = \dfrac{x-1}{3}$; c) $3[x-1] = 2x + 4$;

d) $4[x] - 5 = 3x$; e) $\left[3x - \dfrac{2}{5}\right] = 2x + 1$; f) $\left[2x - \dfrac{2}{3}\right] = 4x - 3$;

g) $\left[\dfrac{5x+2}{3}\right] = x - \dfrac{1}{2}$; h) $\left[5x - \dfrac{1}{4}\right] = 6x - 1$; i) $\left[\dfrac{x+1}{3}\right] = \dfrac{x-1}{2}$.

III. 76. Solve the equations:

a) $\left[x + \dfrac{2}{3}\right] + \left[x + \dfrac{3}{5}\right] = 1$; b) $\left[x + \dfrac{1}{2}\right] + \left[x + \dfrac{1}{3}\right] = 1$, where $[a]$ is the integer part (floor) of a.

Chapter IV

POWERS AND RADICALS

$\mathbf{N} = \{0,1,2,...\}$ Natural number, $\mathbf{N}^* = \{1,2,3,...\}$

$\mathbf{Z} = \{-2,-1,0,1,2,...\}$ Integer numbers,

$\mathbf{Q} = \left\{\dfrac{a}{b} \mid a \in \mathbf{Z}, b \in \mathbf{Z}, b \neq 0\right\}$ Rational numbers,

$\mathbf{R} - \mathbf{Q}$ Irrational numbers, $\mathbf{R}$ The set of real numbers.

Properties of the powers $a^n = \underbrace{a \cdot a \cdot ... \cdot a}_{n \ times}$, $n \in \mathbf{N}^*$.

a) $a^0 = 1$,

b) $(ab)^n = a^n b^n$

c) $\left(\dfrac{a}{b}\right)^n = \dfrac{a^n}{b^n}$

d) $a^n a^m = a^{n+m}$

e) $\dfrac{a^n}{a^m} = a^{n-m}$

f) $\left(a^m\right)^n = a^{mn}$, $a,b \in \mathbf{R}$, $b \neq 0$, $m,n \in \mathbf{R}$.

Power function with integer positive (natural) exponent
$f : \mathbf{R} \to [0,\ \infty)$, $f(x) = x^2$

a similar behavior has the graph
$f(x) = x^{2n}$, $n \in \mathbf{N}$.

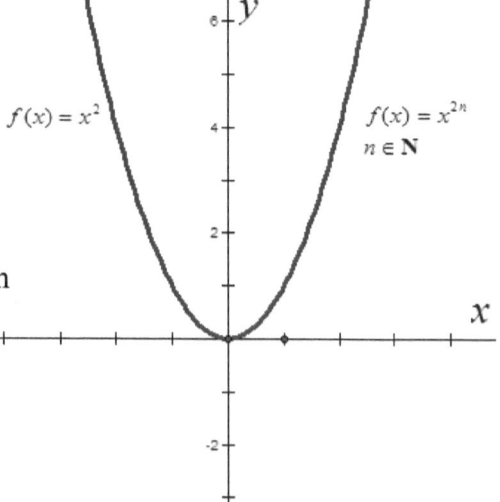

$f(x) = x^2$

$f(x) = x^{2n}$
$n \in \mathbf{N}$

Power function with integer positive (natural) exponent

$f : \mathbf{R} \to \mathbf{R}, \ f(x) = x^3$

A similar behavior has the graph
$f(x) = x^{2n+1}, \ n \in \mathbf{N}.$

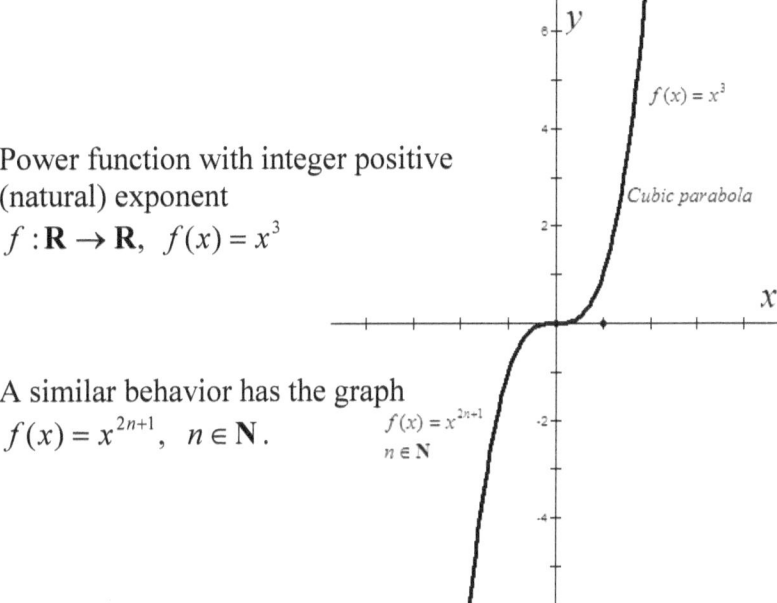

Power function with integer negativ exponent

$f : \mathbf{R} - \{0\} \to [0, \ \infty) \ f(x) = x^{-2}$

A similar behavior has the graph
$f(x) = x^{-n}, \ n \in \mathbf{N}, \ n$ even number.

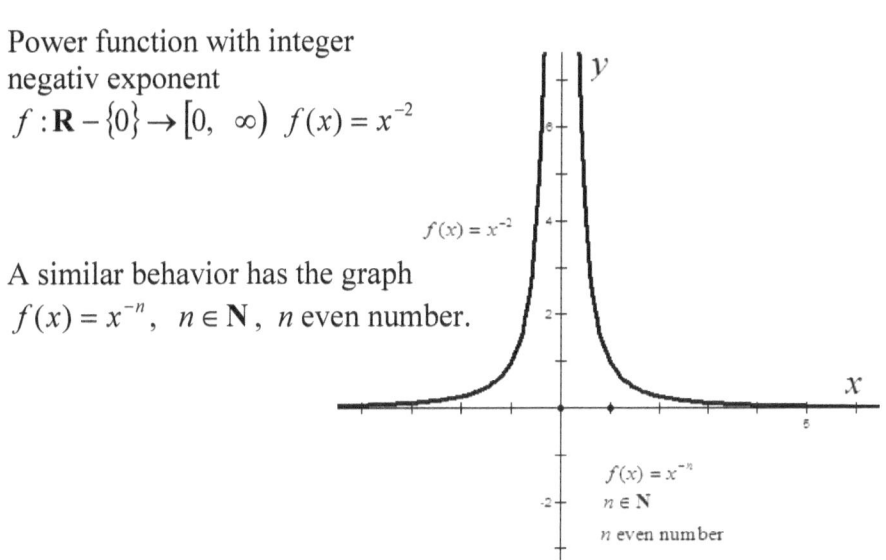

$f : \mathbf{R} - \{0\} \to \mathbf{R} - \{0\}, \; f(x) = x^{-3}$
and a similar behavior has
the graph $f(x) = x^{-n}, \; n \in \mathbf{N}$,
n odd number.

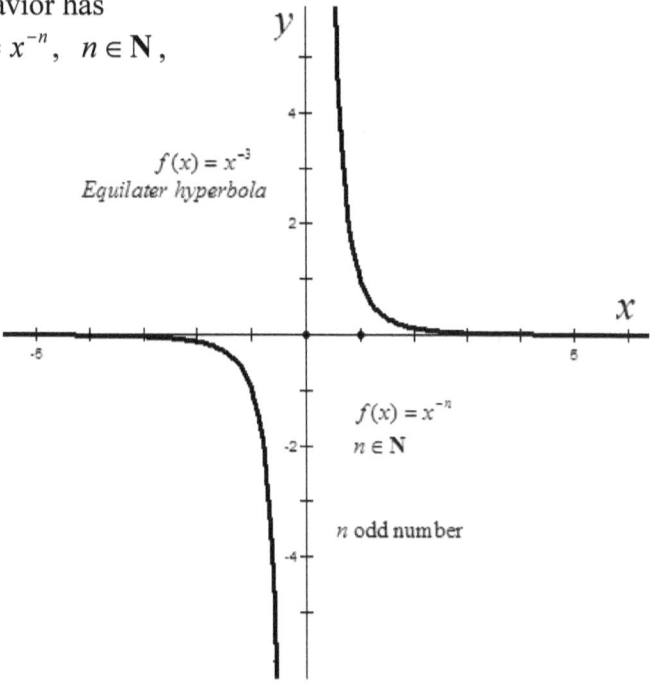

$f(x) = x^{-3}$
Equilater hyperbola

$f(x) = x^{-n}$
$n \in \mathbf{N}$

n odd number

Properties of the radicals $\quad \sqrt[n]{(a)^n} = a, \; n \in \mathbf{N}^*$

a) $\sqrt[n]{ab} = \sqrt[n]{a}\sqrt[n]{b}$,

b) $\sqrt[n]{\dfrac{a}{b}} = \dfrac{\sqrt[n]{a}}{\sqrt[n]{b}}$,

c) $\sqrt[n]{a^{nm}} = a^m$,

d) $\left(\sqrt[n]{a}\right)^m = \sqrt[n]{a^m}$, $\quad \sqrt[n]{a^m} = a^{\frac{m}{n}}$,

e) $\sqrt[n]{\sqrt[m]{a}} = \sqrt[nm]{a}$, $\quad a \geq 0, \, b > 0, \; m, n \in \mathbf{N}$.

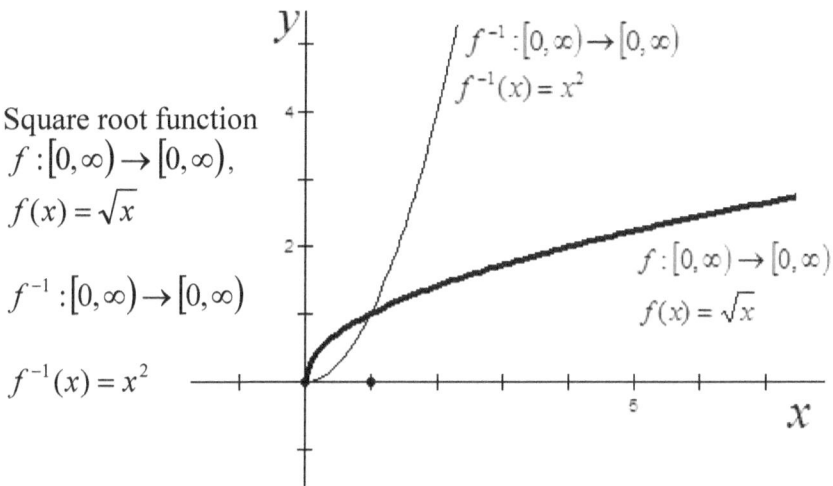

Square root function
$f:[0,\infty)\to[0,\infty)$,
$f(x)=\sqrt{x}$

$f^{-1}:[0,\infty)\to[0,\infty)$

$f^{-1}(x)=x^2$

$f^{-1}:[0,\infty)\to[0,\infty)$
$f^{-1}(x)=x^2$

$f:[0,\infty)\to[0,\infty)$
$f(x)=\sqrt{x}$

IV. 1. Write using powers:

a) $x\cdot x\cdot x\cdot x$; **b)** $y\cdot y\cdot y\cdot y$; **c)** $4\cdot a\cdot a\cdot a$; **d)** $-3\cdot b\cdot b\cdot b$;

e) $-2\cdot x\cdot x\cdot y$; **f)** $3\cdot a\cdot a\cdot b\cdot b\cdot b$; **g)** $(2a)\cdot(2a)\cdot(2a)$;

h) $(a^2b^3)\cdot(a^2b)\cdot(a^2b^2)$; **i)** $a^4\cdot a^5\cdot a^6$;

j) $(-3x^3y^3)(-3x^2y^3)(-3x^7y^3)$; **k)** $a^{12}\cdot a^3\cdot a^6$;

l) $x^2\cdot x^3\cdot x^6$; **m)** $x^n\cdot x$, $n\in\mathbf{N}*$;

n) $x^n\cdot x^n$, $n\in\mathbf{N}*$; **o)** $a^m\cdot a^n\cdot a$, $m,n\in\mathbf{N}*$.

IV. 2. Simplify the expressions and eliminate any negative exponents.

a) $(3y^2)(2y^3)$; **b)** $\left(6x^2y^4\right)\left(\dfrac{1}{2}x^5y\right)$; **c)** $(2y)^3$;

d) $\dfrac{x^9(2x)^4}{x^4}$; **e)** $\dfrac{a^{-4}b^4}{a^{-5}b^5}$; **f)** $b^{-4}\left(\dfrac{1}{3}b^7\right)\left(6b^{-8}\right)$;

g) $(rs)^3(2s)^{-2}(2r)^4$; **h)** $(2u^2t^4)^3(3u^3t)^{-2}$; **i)** $\dfrac{(2y^2)^4}{2y^5}$;

j) $\dfrac{(2x^4)^2(3x^4)}{(x^3)^4}$; **k)** $\dfrac{(x^2y^3)(xy^4)^{-3}}{x^2y^3}$; **l)** $\left(\dfrac{c^4d^3}{c^3d^2}\right)\left(\dfrac{d^2}{c^3}\right)^3$;

m) $\dfrac{(xy^2z)^4}{(x^2y^2z)^3}$; **n)** $\dfrac{\left(x^{-1}\right)^3\left(x^2\right)^3}{x^4\left(x^{-2}\right)^2}$; **o)** $\left(\dfrac{xy^{-2}z^{-3}}{x^2y^2z^4}\right)^{-3}$;

p) $\left(\dfrac{q^{-1}rs^{-2}}{r^{-5}s^2q^{-8}}\right)^{-1}$; **q)** $\left(3ab^2c^2\right)\left(\dfrac{3a^2b}{c^3}\right)^{-2}$; **r)** $\dfrac{x^2\left(y^{-3}\right)^2}{\left(xy^{-1}\right)^2\left(x^3\right)^2}$.

IV. 3. Simplify the expressions and eliminate any negative exponents. Assume that all letters denote positive numbers.

a) $x^{\frac{2}{3}}x^{\frac{1}{5}}$; **b)** $(-2a^{\frac{3}{4}})(5a^{\frac{3}{2}})$; **c)** $(4b)^{\frac{1}{2}}(8b^{\frac{2}{5}})$; **d)** $\left(8x^6\right)^{-\frac{2}{3}}$;

e) $(c^2d^3)^{-\frac{1}{3}}$; **f)** $\left(4x^6y^8\right)^{\frac{3}{2}}$; **g)** $(y^{\frac{3}{4}})^{\frac{2}{3}}$; **h)** $(a^{\frac{8}{3}})^{-\frac{3}{4}}$;

i) $(2x^4y^{-\frac{4}{3}})^3(8^{-1}x^{-9}y^3)^{\frac{2}{3}}$; **j)** $(x^{-5}y^{15}z^{10})^{-\frac{3}{5}}$; **k)** $\left(\dfrac{x^6y^2}{y^4}\right)^{\frac{5}{2}}$;

l) $\left(\dfrac{-2x^{\frac{3}{4}}}{y^{\frac{1}{4}}z^{\frac{1}{2}}}\right)^4$; **m)** $\left(a^{\frac{1}{3}}a^{\frac{4}{5}}a^{-\frac{1}{10}}\right)^{30}$; **n)** $\left(-27a^{\frac{3}{4}}\right)^{\frac{2}{3}}\left(9a^{\frac{3}{2}}\right)^{\frac{1}{2}}$;

o) $\left(8x^{\frac{3}{4}}\right)^{\frac{2}{3}}\left(9x^{\frac{2}{3}}\right)^{\frac{1}{2}}$; **p)** $(3b^3)^{-\frac{2}{3}}(9b^6)^{-\frac{1}{2}}(3b^6)^{\frac{2}{3}}$;

q) $(x^{-2}y^{-6})^{-\frac{1}{2}}(x^2y^3)^{-\frac{1}{6}}$; **r)** $\left(4x^2y^4\right)^{\frac{3}{2}}\left(2^2x^{-2}y^{-4}\right)^{\frac{3}{2}}$; **s)** $(y^{\frac{3}{4}})^{\frac{2}{6}}(y^{\frac{6}{4}})^{\frac{2}{3}}$.

IV. 4. Calculate:

a) $-\sqrt{2}\cdot\left(-\sqrt{7}\right)$; **b)** $\sqrt{8}\div\sqrt{2}$; **c)** $\left(\sqrt{5}\right)^2$; **d)** $\sqrt{13}\cdot\left(\sqrt{3}-\sqrt{2}\right)$;
e) $-2\sqrt{3}+4\sqrt{3}$; **f)** $5\sqrt{2}+8-3\sqrt{2}-10$; **g)** $4\sqrt{2}-\sqrt{18}$;

h) $3\sqrt{45}+4\sqrt{20}$; **i)** $3\sqrt{75}-5\sqrt{27}$;

j) $\left(2\sqrt{18}-3\sqrt{8}\right)+\left(3\sqrt{32}-\sqrt{50}\right)$;

k) $\left(2\sqrt{20}-\sqrt{45}+2\sqrt{18}\right)-\left(2\sqrt{20}+3\sqrt{72}\right)$;

l) $\left(\sqrt{6}-3\sqrt{3}+5\sqrt{2}-\sqrt{8}\right)\cdot2\sqrt{6}$; **m)** $\left(5\sqrt{48}-6\sqrt{27}+4\sqrt{12}\right)\div\sqrt{3}$;

n) $\left(3\sqrt{50}+5\sqrt{200}-3\sqrt{450}\right)\div\sqrt{10}$;

o) $\left(2\sqrt{8}+3\sqrt{5}-7\sqrt{2}\right)\left(\sqrt{72}-5\sqrt{20}-2\sqrt{2}\right)$;

p) $\sqrt{2+\sqrt{3}}\cdot\sqrt{2+\sqrt{2+\sqrt{3}}}\cdot\sqrt{2+\sqrt{2+\sqrt{2+\sqrt{3}}}}\cdot\sqrt{2-\sqrt{2+\sqrt{2+\sqrt{3}}}}$

q) $2\cdot\sqrt{2\sqrt{2\sqrt{2\sqrt{2\sqrt{2\sqrt{2\sqrt{4}}}}}}}$; **r)** $\sqrt{5\sqrt{5\sqrt{5\sqrt{5\sqrt{5\sqrt{5\sqrt{25}}}}}}}$;

s) $\sqrt{33}\cdot\sqrt{6+\sqrt{3}}\cdot\sqrt{3+\sqrt{3+\sqrt{3}}}\cdot\sqrt{3-\sqrt{3+\sqrt{3}}}$.

IV. 5. Calculate:

a) $\sqrt{3}+2\sqrt{3}-4\sqrt{3}$; **b)** $3\sqrt{5}-6\sqrt{5}+5\sqrt{5}$;

c) $3\sqrt{6}+2\sqrt{6}-5\sqrt{6}+3\sqrt{6}$; **d)** $3\sqrt{6}\cdot2\sqrt{3}\cdot(-2\sqrt{2})$;

e) $2\sqrt{75}+\sqrt{50}+2\sqrt{32}-3\sqrt{12}$;

f) $\left(2\sqrt{24}-\sqrt{54}+3\sqrt{96}-\sqrt{150}\right)\cdot\sqrt{6}$); **g)** $-3\sqrt{25}\div\sqrt{5}$;

h) $\sqrt{2+\sqrt{4}}\div(-2)$; **i)** $\left(5\sqrt{32}+3\sqrt{8}-4\sqrt{18}\right)\div\sqrt{2}$);

j) $12+2\sqrt{5}\cdot\left[2\sqrt{500}+2(\sqrt{45}-\sqrt{125})\right]$;

k) $\left(1-\sqrt{2}\right)\left(1+\sqrt{2}\right)+\left(\sqrt{2}-\sqrt{3}\right)\left(\sqrt{2}+\sqrt{3}\right)$; **l)** $\dfrac{2}{\sqrt{2}}+\dfrac{1}{\sqrt{3}}$;

m) $\dfrac{3}{\sqrt{5}-\sqrt{2}}\cdot\dfrac{1}{\sqrt{2}-1}$; **n)** $\sqrt{2}+\left(\dfrac{1}{\sqrt{2}}\right)^{-2}$; **o)** $\dfrac{12}{\sqrt{13}-1}-\sqrt{13}$;

p) $\left(\dfrac{1}{2\sqrt{3}}-\dfrac{3}{\sqrt{75}}+\dfrac{7}{\sqrt{12}}\right)\cdot\sqrt{27}$;

q) $\dfrac{2}{\sqrt{3}+\sqrt{5}}+\dfrac{2}{\sqrt{5}+\sqrt{7}}+\dfrac{2}{\sqrt{7}+\sqrt{9}}$;

r) $\left(\dfrac{\sqrt{3}-\sqrt{2}}{\sqrt{6}}+\dfrac{\sqrt{4}-\sqrt{3}}{\sqrt{12}}+\dfrac{\sqrt{5}-\sqrt{4}}{\sqrt{20}}\right)\div\dfrac{1}{\sqrt{10}}$;

s) $\left(\dfrac{\sqrt{4}-\sqrt{3}}{\sqrt{12}}+\dfrac{\sqrt{5}-\sqrt{4}}{\sqrt{20}}+\dfrac{\sqrt{6}-\sqrt{5}}{\sqrt{30}}\right)\div\dfrac{1}{\sqrt{6}}$;

t) $\dfrac{4-\sqrt{6}}{\sqrt{2}}+\dfrac{\sqrt{6}-1}{\sqrt{3}}-\dfrac{2-\sqrt{6}}{\sqrt{2}}-\dfrac{\sqrt{6}+5}{\sqrt{3}}$;

u) $\dfrac{1}{1+\sqrt{2}}+\dfrac{1}{\sqrt{2}+\sqrt{3}}+....+\dfrac{1}{\sqrt{99}+\sqrt{100}}$.

IV. 6. Give a simple form:

a) $\sqrt{54}$ and $\sqrt[3]{-108}$;

b) $2\times\sqrt[3]{432}+3\times\sqrt[3]{-54}-6\times\sqrt[3]{-128}+5\times\sqrt[3]{16}+7\times\sqrt[3]{-250}$;

c) $4\times\sqrt[3]{-3}-\sqrt[3]{\dfrac{8}{9}}+\sqrt[3]{\dfrac{3}{8}}-\sqrt[3]{7\dfrac{1}{9}}-\sqrt[3]{-0{,}375}+\sqrt[3]{46\dfrac{7}{8}}$.

IV. 7. Rationalize the denominators:

a) $\dfrac{2}{\sqrt{3}}$; **b)** $\dfrac{6}{\sqrt{12}}$; **c)** $\dfrac{\sqrt{27}}{3\sqrt{2}}$; **d)** $\dfrac{1-\sqrt{2}}{1+\sqrt{2}}$; **e)** $\dfrac{6-\sqrt{6}}{\sqrt{6}-1}$;

f) $\dfrac{\sqrt{3}}{2+\sqrt{3}}$; **g)** $\dfrac{37}{7-2\sqrt{3}}$; **h)** $\dfrac{4}{\sqrt{7}+\sqrt{3}}$; **i)** $\dfrac{6}{3+\sqrt{2}-\sqrt{5}}$;

j) $\dfrac{1+3\sqrt{2}-2\sqrt{3}}{\sqrt{2}+\sqrt{3}+\sqrt{6}}$; **k)** $\dfrac{10}{\sqrt[3]{3}+\sqrt[3]{7}}$; **l)** $\dfrac{4}{\sqrt{2\left(3+\sqrt{5}\right)}}$;

m) $\dfrac{2}{\sqrt{3+\sqrt{5}-\sqrt{13+4\sqrt{3}}}}$; **n)** $\dfrac{\sqrt{2}}{\sqrt{3+2\sqrt{2}}}+\dfrac{\sqrt{2}}{\sqrt{3-2\sqrt{2}}}$;

o) $\dfrac{2}{\sqrt{13+30\sqrt{2+\sqrt{9+4\sqrt{2}}}}}$.

IV. 8. Calculate:

a) $\sqrt{\dfrac{1-x}{1+x}}$ for $x = \dfrac{\sqrt{3}}{2}$; **b)** $\dfrac{(a+x)(x+b)+(a-x)(x-b)}{(a+x)(x+b)-(a-x)(x-b)}$ for $x = \sqrt{ab}$.

IV. 9. Simplify:

a) $A = \left(x^{\frac{1}{3}} - 1 \right)\left(x^{\frac{2}{3}} + x^{\frac{1}{3}} + 1 \right)\left(x^{\frac{1}{3}} + 1 \right)\left(x^{\frac{2}{3}} - x^{\frac{1}{3}} + 1 \right)$;

b) $B = \left(x^2 - x^{\frac{3}{2}} y^{\frac{1}{2}} - 2x^{\frac{1}{2}} y^{\frac{1}{4}} + 2y^{\frac{3}{4}} \right) \div \left(x^{\frac{1}{2}} - y^{\frac{1}{2}} \right)$;

c) $D = \dfrac{x-y}{x^{\frac{1}{3}} - y^{\frac{1}{3}}} - \dfrac{x+y}{x^{\frac{1}{3}} + y^{\frac{1}{3}}}$.

IV. 10. Consider $a,\ b,\ c,\ d \in \mathbf{R}$, prove that :

a) $(a+b)^5 - a^5 - b^5 = 5ab(a+b)(a^2 + ab + b^2)$;

b) $a^3 - b^3 = \left(a\dfrac{a^3 - 2b^3}{a^3 + b^3} \right)^3 + \left(b\dfrac{2a^3 - b^3}{a^3 + b^3} \right)^3$, $a + b \neq 0$;

c) $\left(a^3 + b^3 + c^3 + d^3 \right)^2 = 9(bc - ad)(ca - bd)(ab - cd)$, where $a + b + c + d = 0$.

IV. 11. Draw the graph of each function and its parent function:

a) $f_1 : \mathbf{R} \to [0, \infty)$, $f_1(x) = 4x^2$;

b) $f_2 : \mathbf{R} \to \mathbf{R}$, $f_2(x) = x^3 - 8$;

c) $f_3 : \mathbf{R} \to [0, \infty)$, $f_3(x) = (x-2)^2$;

d) $f_4 : \mathbf{R} \to [0, \infty)$, $f_4(x) = (x+1)^4$;

e) $f_5 : \mathbf{R} \to [0, \infty)$, $f_5(x) = |x - 1|^3$;

f) $f_6 : \mathbf{R} \to [0, \infty)$, $f_6(x) = |x + 2|^3$;

g) $f_7 : \mathbf{R} \setminus \{0\} \to \mathbf{R} \setminus \{-1\}$, $f_7(x) = x^{-2} - 1$;

h) $f_8 : \mathbf{R} \setminus \{0\} \to \mathbf{R} \setminus \{-1\}$, $f_8(x) = x^{-3} - 1$;

i) $f_9 : \mathbf{R} \setminus \{1\} \to \mathbf{R} \setminus \{0\}$, $f_9(x) = \dfrac{1}{x+1}$;

j) $f_{10} : \mathbf{R} \setminus \{-1\} \to (0, \infty)$, $f_{10}(x) = \dfrac{1}{(x+1)^2}$.

IV. 12. Find the range of the function

$f(x) = \sqrt{x^2 + x + 1} - \sqrt{x^2 - x + 1}$, $x \in \mathbf{R}$.

IV. 13. Calculate:

a) $2x - 1 + \sqrt{(x-4)^2}$ **b)** $x - 3 + \sqrt{(x-3)^2}$.

IV. 14. Find the value of x such that the radicals are defined:

a) $\sqrt{2x - 1}$; **b)** $\sqrt[3]{x^2 - 3x + 1}$; **c)** $\sqrt{-2 - 6x}$; **d)** $\sqrt[3]{x - 2}$;

e) $\sqrt[6]{1 - 3x}$; **f)** $\sqrt{(x^2 - 4x)^2}$; **g)** $\sqrt[4]{x^2 - 7x + 12}$; **h)** $\sqrt[6]{(4 - x^2)x^2}$;

i) $\sqrt{(x+1)(x-2)(x+4)}$; **j)** $\sqrt{3 - x} - \sqrt{x - 1}$; **k)** $\sqrt{x^2 - 1} + \sqrt[3]{x^2 - 2}$;

l) $\sqrt{2x - 4} + 3 \cdot \sqrt{5 - x} + \sqrt[3]{5 - 4x}$; **m)** $\sqrt[3]{1 + 2x} + \sqrt{6 - 3x}$.

IV. 15. Without using a calculator compare the radicals:

a) $4\sqrt{2}$ and $\sqrt{27}$; **b)** $3\sqrt[4]{5}$ and $\sqrt[4]{216}$; **c)** $21\sqrt{5}$ and $12\sqrt{14}$;

d) $2 \cdot \sqrt{\dfrac{6}{7}}$ and $\sqrt{\dfrac{47}{14}}$; **e)** $-2 \cdot \sqrt{\dfrac{1}{75}}$ and $-\sqrt{\dfrac{24}{450}}$;

f) $-3\sqrt[3]{14}$ and $-2\sqrt[3]{18}$; **g)** $2 \cdot \sqrt[3]{\dfrac{1}{44}}$ and $\sqrt[3]{\dfrac{7}{325}}$.

IV. 16. Let $a = \dfrac{\sqrt{7} - \sqrt{5}}{\sqrt{7} + \sqrt{5}}$ and $b = \dfrac{\sqrt{7} + \sqrt{5}}{\sqrt{7} - \sqrt{5}}$ be the numbers.

Compute $a + b$, $a \times b$ and $a^2 + b^2$.

IV. 17. Compute

$a = \sqrt{\dfrac{2 + \sqrt{3}}{2 - \sqrt{3}}} + \sqrt{\dfrac{2 - \sqrt{3}}{2 + \sqrt{3}}}$ and $b = \left(\sqrt{11 + 4\sqrt{7}} - \sqrt{11 - 4\sqrt{7}} \right)^2$.

IV. 18. Write in an increasing order the numbers (without calculators):

a) $\sqrt[3]{5},\ \sqrt[4]{6},\ \sqrt[12]{628}$; **b)** $\sqrt{4},\ \sqrt[4]{15},\ \sqrt[3]{11}$; **c)** $\sqrt{5},\ \sqrt[3]{6},\ \sqrt[12]{777}$;

d) $2\cdot\sqrt[3]{3},\ \sqrt[6]{16},\ \sqrt[12]{250}$; **e)** $\sqrt[5]{50},\ \sqrt[10]{20},\ \sqrt[15]{448}$;

f) $\sqrt[3]{7},\ \sqrt[4]{17},\ \sqrt[12]{4000}$.

IV. 19. Prove that:

a) $\dfrac{7-4\sqrt{3}}{2-\sqrt{3}}=\sqrt[3]{26-15\sqrt{3}}$; **b)** $\dfrac{11-6\sqrt{2}}{3-\sqrt{2}}=\sqrt[3]{45-29\sqrt{2}}$;

c) $\dfrac{2\cdot\sqrt[3]{2}}{1+\sqrt{3}}=\dfrac{\sqrt[3]{20+12\sqrt{3}}}{2+\sqrt{3}}$;

d) $\sqrt{10+\sqrt{24}+\sqrt{40}+\sqrt{60}}=\sqrt{2}+\sqrt{3}+\sqrt{5}$;

e) $\dfrac{\sqrt{7+4\sqrt{3}}\cdot\sqrt{19-8\sqrt{3}}}{4-\sqrt{3}}=2+\sqrt{3}$;

f) $\dfrac{2+\sqrt{3}}{\sqrt{2}+\sqrt{2+\sqrt{3}}}+\dfrac{2-\sqrt{3}}{\sqrt{2}-\sqrt{2-\sqrt{3}}}=\sqrt{2}$;

g) $\dfrac{\sqrt{2}\cdot\sqrt{\sqrt[4]{8}-\sqrt{\sqrt{2}+1}}}{\sqrt{\sqrt[4]{8}+\sqrt{\sqrt{2}-1}}-\sqrt{\sqrt[4]{8}-\sqrt{\sqrt{2}-1}}}=1$;

h) $\dfrac{\sqrt{5-2\sqrt{6}}\left(5+2\sqrt{6}\right)\!\left(49-20\sqrt{6}\right)}{3\sqrt{3}-3\sqrt{18}+3\sqrt{12}-\sqrt{8}}=1$;

i) $\dfrac{\sqrt{21+8\sqrt{5}}}{4+\sqrt{5}}\cdot\sqrt{9-4\sqrt{5}}+2=\sqrt{5}$;

j) $\sqrt{8+2\sqrt{10+2\sqrt{5}}}-\sqrt{8-2\sqrt{10+2\sqrt{5}}}=\sqrt{20-4\sqrt{5}}$;

k) $5\cdot\sqrt[3]{3\sqrt{128}}-3\cdot\sqrt[3]{9\sqrt{162}}-11\cdot\sqrt[6]{18}+2\cdot\sqrt[3]{75\sqrt{50}}=0$;

l) $\left[\dfrac{\left(1+3^{-1/2}\right)^{1/6}}{\left(3^{1/2}+1\right)^{-1/3}}-\dfrac{\left(3^{1/2}-1\right)^{1/3}}{\left(1-3^{-1/2}\right)^{-1/6}}\right]^{-2}\cdot\dfrac{3^{1/12}\div 3}{\sqrt{3}+\sqrt{2}}=\dfrac{\sqrt[6]{3}}{6}$.

IV. 20. Show that:

a) $\left(4+\sqrt{15}\right)\left(\sqrt{10}-\sqrt{6}\right)\cdot\sqrt{4-\sqrt{15}}=2$;

b) $\dfrac{1-2^{-2}}{2^{\frac{1}{2}}-2^{-\frac{1}{2}}}-\dfrac{2}{2^{\frac{3}{2}}}+\dfrac{2^{-2}-2}{2^{\frac{1}{2}}-2^{-\frac{1}{2}}}=-\dfrac{3\sqrt{2}}{2}$;

c) $\sqrt{\dfrac{\sqrt{2}}{3}+\dfrac{3}{\sqrt{2}}+2}-\dfrac{1}{3}\cdot\dfrac{9\cdot\sqrt[4]{2}-2\sqrt{3}}{\sqrt{6}-\sqrt[4]{8}}=-1$;

d) $\left(\dfrac{\sqrt[4]{8}-1}{\sqrt[4]{2}-1}+\sqrt[4]{2}\right)^{\frac{1}{2}}\left(\dfrac{\sqrt[4]{8}+1}{\sqrt[4]{2}+1}+\sqrt[4]{4}\right)\left(1-\sqrt{2}\right)^{-1}=1$;

e) $\left(\dfrac{\sqrt{3}+1}{1+\sqrt{3}+\sqrt{5}}+\dfrac{\sqrt{3}-1}{1-\sqrt{3}+\sqrt{5}}\right)\cdot\left(\sqrt{5}-\dfrac{2}{\sqrt{5}}+2\right)=2\sqrt{3}$;

f) $\sqrt[3]{\sqrt[5]{\dfrac{32}{5}}-\sqrt[5]{\dfrac{27}{5}}}=\sqrt[5]{\dfrac{1}{25}}+\sqrt[5]{\dfrac{3}{25}}-\sqrt[5]{\dfrac{9}{25}}$;

g) $\dfrac{\sqrt[3]{7+2\sqrt{6}}}{\left(\sqrt{6}+1\right)^{-1/3}}+\dfrac{\sqrt[3]{7-2\sqrt{6}}}{\left(\sqrt{6}-1\right)^{-1/3}}=2\sqrt{6}$;

h) $\sqrt[3]{5+2\sqrt{13}}+\sqrt[3]{5-2\sqrt{13}}=1$; **i)** $\sqrt[3]{6+\sqrt{\dfrac{847}{27}}}+\sqrt[3]{6-\sqrt{\dfrac{847}{27}}}=3$;

j) $\sqrt[3]{45+29\sqrt{2}}-\sqrt[3]{45-29\sqrt{2}}=2\sqrt{2}$; **k)** $\sqrt[3]{9+\sqrt{80}}+\sqrt[3]{9-\sqrt{80}}=3$;

l) $\left(\dfrac{3}{4-\sqrt[3]{25}}+\dfrac{\sqrt[3]{40}}{\sqrt[3]{8}+\sqrt[3]{5}}-\dfrac{10}{\sqrt[3]{25}}\right)\div\left(\sqrt[6]{8}+\sqrt[6]{5}\right)=\sqrt{2}-\sqrt[5]{5}$;

m) $\dfrac{\sqrt{\sqrt[4]{27}+\sqrt{\sqrt{3}-1}}-\sqrt{\sqrt[4]{27}-\sqrt{\sqrt{3}-1}}}{\sqrt{\sqrt[4]{27}-\sqrt{\sqrt{3}+1}}}=\sqrt{2}$;

n) $\dfrac{\sqrt[3]{\sqrt{3}+\sqrt{6}}\cdot\sqrt[6]{9-6\sqrt{2}}-\sqrt[6]{18}}{\sqrt[6]{2}-1}=-\sqrt[3]{3}$;

o) $\dfrac{25 \cdot \sqrt[4]{2} + 2\sqrt{5}}{\sqrt{250} + 5 \cdot \sqrt[4]{8}} - \sqrt{\dfrac{\sqrt{2}}{5} + \dfrac{5}{\sqrt{2}}} + 2 = -1.$

IV. 21. Calculate the sums

$$S_1 = \dfrac{1}{1+\sqrt{2}} + \dfrac{1}{\sqrt{2}+\sqrt{3}} + \dfrac{1}{\sqrt{3}+\sqrt{4}} + \ldots + \dfrac{1}{\sqrt{n-1}+\sqrt{n}} \quad \text{and}$$

$$S_2 = \dfrac{1}{2\sqrt{1}+1\sqrt{2}} + \dfrac{1}{3\sqrt{2}+2\sqrt{3}} + \dfrac{1}{4\sqrt{3}+3\sqrt{4}} + \ldots + \dfrac{1}{(n+1)\sqrt{n}+n\sqrt{n+1}}$$

.

IV. 22. Calculate the irrational expressions:

a) $a = \sqrt{x \times \sqrt[3]{x\sqrt{x}}}$ and $b = \sqrt{\dfrac{\sqrt[3]{a^3}}{a^2}} \times \sqrt[3]{\dfrac{a^2\sqrt{a}}{\sqrt{a^3}}} \times \sqrt[4]{\dfrac{\sqrt{a}}{a\sqrt[3]{a^2}}} \times \sqrt[8]{\dfrac{\sqrt{a^{11}}}{\sqrt{a}}}$;

b) $\sqrt[4]{\dfrac{x}{32}} \cdot \dfrac{\left(\sqrt[8]{x} - \sqrt[8]{2}\right)^2 + \left(\sqrt[8]{x} + \sqrt[8]{2}\right)^2}{\sqrt{x} - \sqrt[4]{2x}} \div \dfrac{\left(\sqrt[4]{x} + \sqrt[4]{2} - \sqrt[8]{2x}\right)\left(\sqrt[4]{x} + \sqrt[4]{2} + \sqrt[8]{2x}\right)}{2 - \sqrt[4]{2x^3}}$

c) $\dfrac{a(a-2) - b(b+2) + \sqrt{ab}(b-a+2)}{a+b-\sqrt{ab}} \div \left(1 + 2\dfrac{a^2 + b^2 + ab}{b^3 - a^3}\right)$, $a \neq b$;

d) $\dfrac{\left(\sqrt{a} + \sqrt{ax} + x + x\sqrt{x}\right)^2 \left(1 - \sqrt{x}\right)^2}{\left(x + x^{-1} - 2\right)a^{-\frac{1}{4}}} - \dfrac{\left(x\sqrt{a}\right)^{\frac{3}{2}}}{\left(ax^{-1} + 4\sqrt{a} + 4x\right)^{-\frac{1}{2}}}$, $x \in R_+$;

e) $\dfrac{\sqrt{a^2 - b + \sqrt{c}} \times \sqrt{a - \sqrt{b + \sqrt{c}}} \times \sqrt{a + \sqrt{b + \sqrt{c}}}}{\sqrt{\dfrac{a^3}{b} - 2a + \dfrac{b}{a} - \dfrac{c}{ab}}}$;

f) $\dfrac{\left(\sqrt[4]{a} + \sqrt[4]{b} - \sqrt[8]{ab}\right)\left(\sqrt[4]{b} + \sqrt[4]{a} + \sqrt[8]{ab}\right)}{\sqrt[4]{a^3 b} - b} \div \dfrac{\left(\sqrt[8]{a} + \sqrt[8]{b}\right)^2 + \left(\sqrt[8]{a} - \sqrt[8]{b}\right)^2}{b^{-\frac{1}{4}}\left(\sqrt{a} - \sqrt{b}\right)}$;

g) $\left(\dfrac{2(a+1) + 2\sqrt{a^2 + 2a}}{3a + 1 - 2\sqrt{2a^2 + a}}\right)^{\frac{1}{2}} - \left(\sqrt{2a+1} - \sqrt{a}\right)^{-1}\sqrt{a+2}$, $a \in R_+$;

h) $E = \left[\dfrac{\left(x^2 + a^2\right)^{\frac{1}{2}} + \left(x^2 - a^2\right)^{\frac{1}{2}}}{\left(x^2 + a^2\right)^{-\frac{1}{2}} - \left(x^2 - a^2\right)^{-\frac{1}{2}}} \right]^{-2}$ for $x = a\left(\dfrac{m^2 + n^2}{2mn}\right)^{\frac{1}{2}}$. $a > 0$,

$n > m > 0$;

i) $\dfrac{\sqrt{\sqrt{3}+2} \cdot \sqrt[4]{7 - 4\sqrt{3}} + \sqrt[3]{\sqrt{x}(x+27) - 9x - 27}}{\sqrt{x} - 2 - \sqrt{2 - \sqrt{3}} \cdot \sqrt[4]{7 + 4\sqrt{3}}}$;

j) $\dfrac{\left(\sqrt[8]{x} + \sqrt[8]{y}\right)^2 + \left(\sqrt[8]{x} - \sqrt[8]{y}\right)^2}{x - \sqrt{xy}} \div \dfrac{\left(\sqrt[4]{x} + \sqrt[8]{xy} + \sqrt[4]{y}\right)\left(\sqrt[4]{x} - \sqrt[8]{xy} + \sqrt[4]{y}\right)}{\sqrt[4]{x^3 y} - y}$.

IV. 23. Simplify the expressions:

a) $\dfrac{\left[a^{\frac{2}{3}} \times (ab)^{\frac{1}{3}} \times b^{\frac{2}{3}} \right]^{\frac{1}{2}}}{\left[a^{-\frac{3}{2}} \times (ab)^{-\frac{1}{2}} \times b^{-\frac{3}{2}} \right]^{-\frac{1}{4}}}$;

b) $\left[\dfrac{\left(1 + x^{-\frac{1}{2}}\right)^{\frac{1}{6}}}{\left(x^{\frac{1}{2}} + 1\right)^{-\frac{1}{3}}} - \dfrac{\left(x^{\frac{1}{2}} - 1\right)^{\frac{1}{3}}}{\left(1 - x^{-\frac{1}{2}}\right)^{-\frac{1}{6}}} \right]^{-2} \dfrac{x^{\frac{1}{12}}}{x^{\frac{1}{2}} + (x-1)^{\frac{1}{2}}}$, $x > 1$;

c) $\dfrac{\dfrac{1}{\sqrt{x-1}} - \sqrt{x+1}}{\dfrac{1}{\sqrt{x+1}} - \dfrac{1}{\sqrt{x-1}}} \div \dfrac{\sqrt{x+1} \cdot \sqrt{x^2 - 1}}{(x-1)\sqrt{x+1} - (x+1)\sqrt{x-1}}$;

d) $E = \dfrac{\sqrt{\left(\dfrac{a^4 + b^4}{a^4 - a^2 b^2} + \dfrac{2b^2}{a^2 - b^2}\right)\left(a^3 - ab^2\right)} - 2b\sqrt{a}}{\sqrt{\dfrac{a}{a-b} - \dfrac{b}{a+b} - \dfrac{2ab}{a^2 - b^2}}(a-b)}$.

IV. 24. Simplify the expression:

a) $\dfrac{\sqrt{x^2-4xy+4y^2}}{\sqrt{x^2+4xy+4y^2}}-\dfrac{8xy}{x^2-4y^2}+\dfrac{2y}{x-2y}$, $0<x<2y$;

b) $E=\left(\dfrac{1+\sqrt{x}}{\sqrt{1+x}}-\dfrac{\sqrt{1+x}}{1+\sqrt{x}}\right)^2-\left(\dfrac{1-\sqrt{x}}{\sqrt{1+x}}-\dfrac{\sqrt{1+x}}{1-\sqrt{x}}\right)^2$, $x\in[0,\infty)$;

c) $\left(\sqrt{\dfrac{1}{x^2}-1}+\dfrac{1}{x}\right)\left(\dfrac{1-x}{\sqrt{1-x^2}+x-1}+\dfrac{\sqrt{1+x}}{\sqrt{1+x}-\sqrt{1-x}}\right)$, $x\in(-1,0)$.

d) $\dfrac{\sqrt{1+\sqrt{1-x^2}}\left((1+x)\sqrt{1+x}-(1-x)\sqrt{1-x}\right)}{2+\sqrt{1-x^2}}$, $x\in(-1,1)$;

e) $\dfrac{\left(\sqrt{\sqrt{\dfrac{x-1}{x+1}}+\sqrt{\dfrac{x+1}{x-1}}-2}\right)\left(2x+\sqrt{x^2-1}\right)}{(x+1)\sqrt{x+1}-(x-1)\sqrt{x-1}}$, $x>1$;

f) $\left[\left(\sqrt{xy}-\dfrac{y^2\sqrt{x}}{\sqrt{xy^2}+\sqrt{y^3}}\right)\div\dfrac{\sqrt[4]{xy}-\sqrt{y}}{x-y}-x\sqrt{y}\right]^2\div$

$\div\sqrt[3]{xy\sqrt{xy}}-\left(\dfrac{x}{\sqrt{x^4-1}}\right)^{-2}$;

g) $\left(\sqrt[3]{x^2}+\dfrac{2\sqrt[3]{x^2}}{\sqrt[3]{x}-2}\right)\dfrac{\sqrt[3]{x^2}-4}{\sqrt[3]{x^3}+2\sqrt[3]{x^2}}-\left(2+\dfrac{\sqrt[3]{x^2}}{\sqrt[3]{x}+2}\right)^{-1}\dfrac{x-8}{\sqrt[3]{x}+2}$,

$x\in R\setminus\left\{-\sqrt[3]{2},0\right\}$;

h) $\left(\dfrac{x^2y^3}{(x+y)^{\frac{5}{2}}}-\dfrac{2x^2y^2}{(x+y)^{\frac{3}{2}}}+\dfrac{x^2y}{(x+y)^{\frac{1}{2}}}\right)\div\left(\dfrac{x^3}{(x+y)^{\frac{5}{2}}}-\dfrac{x^3y}{(x+y)^{\frac{7}{2}}}\right)$;

i) $\left(\sqrt[3]{\dfrac{8x^3+24x^2+18x}{2x-3}}-\sqrt[3]{\dfrac{8x^3-24x^2+18x}{2x+3}}\right)-\left(\dfrac{1}{2}\sqrt[3]{\dfrac{2x}{27}}-\dfrac{1}{6x}\right)^{-1}$,

$x\in\left(\dfrac{3}{2},\ \infty\right)$.

j)
$$\frac{\left(\sqrt[8]{x}+\sqrt[8]{y}\right)^2+\left(\sqrt[8]{x}-\sqrt[8]{y}\right)^2}{x-\sqrt{xy}} \div \frac{\left(\sqrt[4]{x}+\sqrt[8]{xy}+\sqrt[4]{y}\right)\left(\sqrt[4]{x}-\sqrt[8]{xy}+\sqrt[4]{y}\right)}{\sqrt[4]{x^3 y}-y}.$$

IV. 25. Find the value of the expressions for indicated value:

a) $\left[\dfrac{(x+2y)\div 8y^3}{x^2+2xy+2y^2}-\dfrac{(x-2y)\div 8y^3}{x^2-2xy+2y^2}\right]+\left(\dfrac{y^{-2}}{4x^2-8y^2}-\dfrac{1}{4x^2 y^2+8y^4}\right)$

$x=\sqrt[4]{6},\;\; y=\sqrt[8]{2}\;;$

b) $\dfrac{a^{-1}-b^{-1}}{a^{-3}+b^{-3}}\div\dfrac{a^2 b^2}{(a+b)^2-3ab}\left(\dfrac{a^2-b^2}{ab}\right)^{-1}$, $a=1-\sqrt{2},\;\; b=1+\sqrt{2}\;;$

c) $\left(\dfrac{a}{b}+\dfrac{b}{a}+2\right)\left(\dfrac{a+b}{2a}-\dfrac{b}{a+b}\right)\div\left[\left(a+2b+\dfrac{b^2}{a}\right)\cdot\left(\dfrac{a}{a+b}+\dfrac{b}{a-b}\right)\right]$,

$a=0.75,\;\; b=\dfrac{4}{3}\;;$

d) $\sqrt{a^3-b^3+\sqrt{a}}\cdot\dfrac{\sqrt{a^{3/2}+\sqrt{b^3}+\sqrt{a}}\cdot\sqrt{a^{3/2}-\sqrt{b^3}+\sqrt{a}}}{\sqrt{\left(a^3+b^3\right)^2-a\left(4a^2 b^3+1\right)}}$,

$a=4,\;\; b=\sqrt[3]{62}\;;$

e) $\dfrac{\sqrt{2}(x-a)}{2x-a}-\left[\left(\dfrac{\sqrt{x^2}}{\sqrt{2x^2}+\sqrt{ax}}\right)^2+\left(\dfrac{\sqrt{2x}+\sqrt{a}}{2\sqrt{a}}\right)^{-1}\right]^{1/2}$,

$a=0.32, x=0.08;$

f) $\dfrac{\sqrt{x-2\sqrt{x+3}}+4}{x^{1/2}-(x-3)^{1/2}-\sqrt{3x+x^2}+\sqrt{x^2-9}}-\dfrac{1}{\sqrt{x}+\sqrt{x-3}}$, $x=9;$

g) $\left(-4a\sqrt[3]{\dfrac{\sqrt{a^3 x}}{a^3}}\right)^3+\left[-10a\sqrt{x}\cdot\sqrt{(ax^{-1})}\right]^2+\left[-2\sqrt[3]{a\sqrt[4]{\dfrac{x}{a}}}\right]^3$,

$a=3\dfrac{4}{7},\;\; x=0.28\;;$

h) $\left(\dfrac{2-n}{n-1}+4\dfrac{m-1}{m-2}\right)\div\left(n^2\dfrac{m-1}{n-1}+m^2\dfrac{2-n}{m-2}\right)$, $m=\sqrt[4]{400}$, $n=\sqrt{5}$;

i) $\dfrac{\sqrt{x+4\sqrt{x-4}}+\sqrt{x-4\sqrt{x-4}}}{\sqrt{1-\dfrac{8}{x}+\dfrac{16}{x^2}}}$, $x=5$;

j) $\dfrac{\sqrt{3}x^{3/2}-5x^{1/3}+5x^{4/3}-\sqrt{3x}}{\sqrt{3x+10\sqrt{3}x^{5/6}+25x^{2/3}}\,\sqrt{1-2x^{-1}+x^{-2}}}$, $x=-1$.

IV. 26. Simplify the expressions:

a) $\dfrac{\left[(3x+2)^{-1/2}+(3x-2)^{-1/2}\right]^{-1}+\left[(3x+2)^{-1/2}-(3x-2)^{-1/2}\right]^{-1}}{\left[(3x+2)^{-1/2}+(3x-2)^{-1/2}\right]^{-1}-\left[(3x+2)^{-1/2}-(3x-2)^{-1/2}\right]^{-1}}$;

b) $\dfrac{\left(\sqrt{a}+\sqrt{ax}+x+x\sqrt{x}\right)^2\left(1-\sqrt{x}\right)^2}{\left(x+x^{-1}-2\right)a^{-1/4}}-\dfrac{\left(x\sqrt{a}\right)^{3/2}}{\left(ax^{-1}+4\sqrt{a}+4x\right)^{-1/2}}$;

c) $\left(\sqrt{\dfrac{(1-n)\sqrt[3]{1+n}}{n}}\cdot\sqrt[3]{\dfrac{3n^2}{4-8n+4n^2}}\right)^{-1}\div\sqrt[3]{\left(\dfrac{3n\sqrt{n}}{2\sqrt{1-n^2}}\right)^{-1}}$;

d) $\dfrac{2a\left(a+2b+\sqrt{a^2+4ab}\right)}{\left(a+\sqrt{a^2+4ab}\right)\left(a+4b+\sqrt{a^2+4ab}\right)}$;

e) $\dfrac{\left(xy^{-1}+1\right)^2}{xy^{-1}-x^{-1}y}\cdot\dfrac{x^3y^{-3}-1}{x^2y^{-2}+xy^{-1}+1}\div\dfrac{x^3y^{-3}+1}{xy^{-1}+x^{-1}y-1}$;

f) $\left(\dfrac{\sqrt[4]{x^3}-y}{\sqrt[4]{x}-\sqrt[3]{y}}-3\sqrt[12]{x^3y^4}\right)^{-1/2}\left(\dfrac{\sqrt[4]{x^3}+y}{\sqrt[4]{x}+\sqrt[3]{y}}-\sqrt[3]{y^2}\right)$;

g) $(4x-1)\left\{\dfrac{1}{8x}\left[\left(\sqrt{8x-1}+4x\right)^{-1}-\left(\sqrt{8x-1}-4x\right)^{-1}\right]\right\}^{1/2}$;

h) $\dfrac{\sqrt{4\left(x-\sqrt{y}\right)+yx^{-1}}\cdot\sqrt{9x^2+6\sqrt[3]{2yx^3}+\sqrt[3]{4y^2}}}{6x^2+2\sqrt[3]{2x^3}-3\sqrt{yx^2}-\sqrt[6]{4y^5}}$;

i) $\dfrac{\sqrt{x-2\sqrt{x+3}}+4}{x^{1/2}-(x-3)^{1/2}-\sqrt{3x+x^2}+\sqrt{x^2-9}}-\dfrac{1}{\sqrt{x}+\sqrt{x-3}}$;

j) $\sqrt{\dfrac{a-8\sqrt[6]{a^3b^2}+4\sqrt[3]{b^2}}{\sqrt{a}-2\sqrt[3]{b}+2\sqrt[12]{a^3b^2}}}+3\sqrt[3]{b}$.

IV. 27. Write in a simpler form the radicals:

a) $\sqrt{4\sqrt{2}+2\sqrt{6}}$; b) $\sqrt{17+\sqrt{288}}$;

c) $\sqrt{7-4\sqrt{3}}$; d) $\sqrt{17-4\sqrt{9+\sqrt{90}}}$.

IV. 28. Calculate the sum

$S=\sqrt{2}+\sqrt{2^2}+\sqrt{2^3}+...+\sqrt{2^{2010}}$.

IV. 29. Calculate the sum

$S=\left(1-\dfrac{1}{4}\right)+\left(2-\dfrac{1}{4}\right)+\left(3-\dfrac{1}{4}\right)+...+\left(n-\dfrac{1}{4}\right)$, were n is a positive integer and solve the equation:

$\sqrt{x_1-1}+\sqrt{x_2-2}+\sqrt{x_3-3}+...+\sqrt{x_n-n}+\dfrac{n(2n+1)}{4}=x_1+x_2+...+x_n$.

IV. 30. Consider the equalities:

a) $\left(\sqrt{5}-2\right)^n=\sqrt{m+1}-\sqrt{m}$;

b) $\left(\sqrt{17}-4\right)^n=\sqrt{m}-\sqrt{m-1}$;

c) $\left(5\sqrt{2}+7\right)^n=\sqrt{m+1}-\sqrt{m}$, where $m,n\in\mathbf{N}$.
Express m in terms of n .

IV. 31. Solve the equations:

a) $\sqrt{x-1+2\sqrt{x-2}} - \sqrt{x-1-2\sqrt{x-2}} = 1$;

b) $\sqrt{x+7+4\sqrt{x+3}} + \sqrt{x+7-4\sqrt{x+3}} = 6$;

c) $\sqrt{x+3+2\sqrt{x+2}} + \sqrt{x+3-2\sqrt{x+2}} = 2$;

d) $\sqrt{x+1-4\sqrt{x-3}} + \sqrt{x+1+4\sqrt{x-3}} = 4$;

e) $\sqrt{x+3-4\sqrt{x-1}} + \sqrt{x+8-6\sqrt{x-1}} = 1$.

IV. 32. Solve the irrational equations:

a) $\sqrt{x-2} + \sqrt{2x-3} = 5$;   b) $\sqrt{2x-1} + \sqrt{2-x} = 2$;

c) $\sqrt{3x+1} + \sqrt{4-3x} = 3$;   d) $\sqrt{x-1} + \sqrt{6-x} = 3$;

e) $\sqrt{x+8} - 2 = \sqrt{x}$;   f) $\sqrt{x+7} + \sqrt{x+2} = \sqrt{3x+19}$;

g) $\sqrt{5x} - \sqrt{2x-1} = \sqrt{3x+1}$;   h) $\sqrt{3x+7} + \sqrt{x-3} = 2\sqrt{x+1}$;

i) $x = 6\left(\sqrt{x-2} - 1\right)$;   j) $\sqrt{\dfrac{x+4}{x-4}} + 2\sqrt{\dfrac{x-4}{x+4}} = \dfrac{11}{3}$;

k) $\sqrt{24-10x} = 3 - 4x$;   l) $\sqrt{2x+1} + \sqrt{1-x} = 2$;

m) $\sqrt{2x+3} + \sqrt{10x-12} = \sqrt{5x-1} + \sqrt{7x-7}$.

IV. 33. Solve the equations:

a) $\sqrt{x+2} + \sqrt{x+7} = 5$;

b) $\sqrt{15-x} + \sqrt{3-x} = 6$;

c) $\sqrt{x+1} - \sqrt{9-x} = \sqrt{2x-12}$;

d) $\sqrt{x+9} + \sqrt{x+1} = \sqrt{4x+16}$;

e) $\sqrt{3x+4} + \sqrt{x-4} = 2\sqrt{x}$;

f) $\sqrt{x+3} + \sqrt{x-1} = \sqrt{4x-2+2\sqrt{5}}$;

g) $\sqrt{x+4} + \sqrt{x-1} = \sqrt{4x-5+4\sqrt{6}}$;

h) $\sqrt{x-5} + \sqrt{x-12} = \sqrt{11+\sqrt{72}}$;

i) $\left(\sqrt{x+2} + \sqrt{x+1}\right)^3 + \left(\sqrt{x+2} + \sqrt{x+1}\right)^2 = 2$.

IV. 34. Solve the equations:

a) $2 \cdot \sqrt{5 \cdot \sqrt[6]{x+1} + 4} - \sqrt{2 \cdot \sqrt[6]{x+1} - 1} = \sqrt{20 \cdot \sqrt[6]{x+1} + 5}$;

b) $\sqrt{10x + 2\sqrt{25x^2 - 1}} - \sqrt{10p - 2\sqrt{25x^2 - 1}} = 2\sqrt{5x - 1}$;

c) $\sqrt{6x + 2\sqrt{9x^2 - 4}} - \sqrt{6x - 2\sqrt{9x^2 - 4}} = 2\sqrt{3x - 2}$.

IV. 35. Solve the equation

$$\frac{1}{x - 5\sqrt{x} + 6} + \frac{1}{x - 3\sqrt{x} + 2} + \frac{1}{x - \sqrt{x}} = \frac{1}{x - 3\sqrt{x}} ,$$

$x \in R \setminus \{ 0, \ 1, \ 4, \ 9 \}$.

IV. 36. Solve the equations:

a) $\sqrt[3]{x+1} + \sqrt[3]{x+2} + \sqrt[3]{x+3} = 0$;

b) $\sqrt[3]{12 + x^2} + \sqrt[3]{12 - x^2} = 2\sqrt[3]{3}$;

c) $\sqrt[3]{\sqrt{x} + 3} + \sqrt[3]{13 - \sqrt{x}} = 4$;

d) $\sqrt[3]{(1-x)^2} + \sqrt[3]{(8+x)^2} = \sqrt[3]{(8+x)(1-x)} + 3$;

e) $\sqrt[3]{(x-1)^2} + \sqrt[3]{(3-x)^2} = \sqrt[3]{(x-1)(3-x)} + 1$;

f) $\sqrt[3]{9 - \sqrt{x+2}} + \sqrt[3]{7 + \sqrt{x+2}} = 4$;

g) $\sqrt[3]{x+1} + \sqrt[3]{x+2} = \sqrt[3]{2x+3}$;

h) $\sqrt[3]{x+2} + \sqrt[3]{x-14} = \sqrt[3]{x-6}$;

i) $2\sqrt[3]{x + 4 - 4\sqrt{x}} - \sqrt[3]{x + 4 + 4\sqrt{x}} = 3 \cdot \sqrt[3]{x - 4}$.

IV. 37. Solve the equation

$$\sqrt[3]{x^2 + 2} + \sqrt[3]{4x^2 + 3x - 2} = \sqrt[3]{3x^2 + x + 5} + \sqrt[3]{2x^2 + 2x - 5}$$

IV. 38. Solve the equations:

a) $\sqrt[3]{3x-1} + \sqrt{3x+7} = 6$; **b)** $\sqrt[3]{1+x} - \sqrt{2+x} = -1$;

c) $\sqrt[3]{8x+11} - \sqrt{3x-2} = 1$; **d)** $\sqrt[4]{x-7} - \sqrt[3]{x+4} = -1$;

e) $\sqrt[4]{x+8} + \sqrt[4]{x-8} = 1$; **f)** $\sqrt[4]{x-1} + \sqrt[4]{2-x} = 1$.

IV. 39. Solve the equation $\dfrac{x \cdot \sqrt[5]{x} - 1}{x \cdot \sqrt[5]{x^3} - 1} - \dfrac{\sqrt[5]{x^3} - 1}{\sqrt[5]{x} - 1} = 16$.

IV. 40. Solve the equation $\sqrt[4]{28 + 5x} + \sqrt[4]{54 - 5x} = 4$.

IV. 41. Solve the equations:

a) $\sqrt{x \sqrt[5]{x}} - 26 \cdot \sqrt[5]{x \sqrt{x}} = 27$; b) $\dfrac{\sqrt{x+4} + \sqrt{x-4}}{2} + 6 = x + \sqrt{x^2 - 16}$;

c) $\dfrac{4}{\sqrt[5]{x}+2} + \dfrac{\sqrt[5]{x}+3}{5} = 2$; d) $\dfrac{1}{\sqrt{x}+\sqrt[3]{x}} + \dfrac{1}{\sqrt{x}-\sqrt[3]{x}} = \dfrac{1}{3}$;

e) $\dfrac{\sqrt[3]{x^4}-1}{\sqrt[3]{x^2}-1} - \dfrac{\sqrt[3]{x^2}-1}{\sqrt[3]{x}+1} = 4$; f) $\dfrac{x \cdot \sqrt[5]{x}-1}{\sqrt[5]{x^3}-1} + \dfrac{\sqrt[5]{x^3}-1}{\sqrt[5]{x}-1} = 16$;

g) $\dfrac{\sqrt[7]{x-\sqrt{2}}}{2} - \dfrac{\sqrt[7]{x-\sqrt{2}}}{x^2} = \dfrac{x}{2} \cdot \sqrt[7]{\dfrac{x^3}{x+\sqrt{2}}}$;

h) $\dfrac{\sqrt{1+x} + \sqrt{1-x}}{\sqrt{1+x} - \sqrt{1-x}} = \dfrac{8}{\sqrt[3]{225x}}$;

i) $\dfrac{\sqrt[3]{(10-x)^2} + \sqrt[3]{(10-x)(x-1)} + \sqrt[3]{(x-1)^2}}{\sqrt{\sqrt[3]{10-x} + \sqrt[3]{x-1}}} = \dfrac{7}{\sqrt{3}}$;

j) $42 \cdot \sqrt[12]{x^{-7}} - \sqrt[4]{x^{-1} \cdot \sqrt[3]{x^2}} = 5 \cdot \sqrt[12]{x^{11}}$;

k) $5 \cdot \sqrt[3]{x \cdot \sqrt[5]{x}} + 3 \cdot \sqrt[5]{x \cdot \sqrt[3]{x}} = 8$.

IV. 42. Solve the equations:

a) $\sqrt[n]{(1+x)^2} + \sqrt[n]{(1-x)^2} = 4 \cdot \sqrt[n]{1-x^2}$, $n \in N^*$;

b) $\sqrt{1+\sqrt{x}} - 2\sqrt{1-\sqrt{x}} = \sqrt[4]{1-x}$;

c) $2\sqrt{2+\sqrt[3]{x}} - 2\sqrt{2-\sqrt[3]{x}} = 3\sqrt[4]{4-\sqrt[3]{x^2}}$;

d) $6 \cdot \sqrt[3]{2x+1} - \sqrt[3]{3x-2} = 5 \cdot \sqrt[6]{(2x+1)(3x-2)}$;

e) $5 \cdot \sqrt[n]{1+\sqrt[5]{x}} - 2 \cdot \sqrt[n]{1-\sqrt[5]{x}} = 3 \cdot \sqrt[2n]{1-\sqrt[5]{x^2}}$;

f) $\sqrt[7]{2x+1} + \sqrt[7]{x+2} = 2 \cdot \sqrt[7]{(2x+1)(x+2)}$.

IV. 43. Solve the equations:

a) $\dfrac{\sqrt{x^2+6x}}{\sqrt{x+1}}+\sqrt{x+5}=\dfrac{4+\sqrt{21}}{\sqrt{x+1}}$;

b) $\dfrac{\sqrt{32+x}+\sqrt{10-x}}{\sqrt{32+x}+\sqrt{10-x}}=\dfrac{21}{x+11}$;

c) $\dfrac{(32-x)\sqrt[3]{x+3}-(x+3)\sqrt[3]{32-x}}{\sqrt[3]{x+3}-\sqrt[3]{32-x}}=-30$.

IV. 44. Solve the equation

$$\sqrt{6+3\sqrt{2+\sqrt{2+x}}}+\sqrt{2-\sqrt{2+\sqrt{2+x}}}=2x .$$

IV. 45. Solve the inequations:

a) $2\sqrt{x^2-3x-10}<39-2x$; b) $\sqrt{x+2}<\sqrt{x-2}+\sqrt{x-3}$;

c) $\sqrt{x^2-3x+2}>2-x$; d) $\sqrt{x^2-4x}>x-3$;

e) $\sqrt{2-\sqrt{x+2}}<\sqrt{3+x}$; f) $\dfrac{\sqrt{x}}{3-x}<1$; g) $\dfrac{2\sqrt{x+2}}{1-2\sqrt{2-x}}<1$;

h) $\sqrt{x^2-53x+196}<x-13$; i) $\sqrt[3]{x^3-3x^2+5x-6}>x-2$;

j) $\sqrt{x-2}-\sqrt[3]{x-3}>1$.

IV. 46. Solve the systems of equations:

a) $\begin{cases}x\sqrt{y}+y\sqrt{x}=30\\x\sqrt{x}+y\sqrt{y}=35\end{cases}$; b) $\begin{cases}\dfrac{1}{\sqrt{x}}+\dfrac{1}{\sqrt{y}}=\dfrac{4}{3}\\xy=9;\end{cases}$ c) $\begin{cases}x\sqrt{y}+y\sqrt{x}=6\\x^2y+y^2x=20\end{cases}$;

d) $\begin{cases}\sqrt{3x-y+12}+\sqrt{3x+y+5}=4\\\sqrt[4]{3x-y+12}+\sqrt[4]{3x+y+5}=2\end{cases}$;

e) $\begin{cases}\sqrt{\dfrac{3x-2y}{2x}}+\sqrt{\dfrac{2x}{3x-2y}}=2\\x^2-8y^2=18(1-y)\end{cases}$;

f) $\begin{cases} \sqrt[3]{x} + \sqrt[3]{y} = 5 \\ \sqrt[3]{x^2} - \sqrt[3]{xy} + \sqrt[3]{y^2} = 7 \end{cases}$; g) $\begin{cases} \sqrt[4]{x} + \sqrt[4]{y} = 3 \\ x + y = 17 \end{cases}$;

h) $\begin{cases} \sqrt{x + \sqrt{y}} + \sqrt{x - \sqrt{y}} = 2 \\ \sqrt{y + \sqrt{x}} + \sqrt{y - \sqrt{x}} = 1 \end{cases}$.

IV. 47. Find the real numbers $x_1, x_2, ..., x_n$ such that

$$\sqrt{x_1 - 1} + 2\sqrt{x_2 - 2^2} + ... + n\sqrt{x_n - n^2} = \frac{1}{2}(x_1 + x_2 + ... + x_n).$$

Chapter V

QUADRATIC FUNCTIONS

AND APPLICATIONS

General form $f : \mathbf{R} \to \mathbf{R}$, $f(x) = ax^2 + bx + c$, $a, b, c \in \mathbf{R}$, $a \neq 0$.
Factored form $f(x) = a(x - x_1)(x - x_2)$, where x_1 and x_2 are the
roots of the equation $ax^2 + bx + c = 0$.
Vertex form (standard form) $f(x) = a(x - h)^2 + k$, $V(h, k)$ or

$$f(x) = a\left(x + \frac{b}{2a}\right)^2 - \frac{\Delta}{4a}, \quad \Delta = b^2 - 4ac,$$

$$V\left(-\frac{b}{2a}, -\frac{\Delta}{4a}\right) \quad \text{or} \quad V\left(\frac{x_1 + x_2}{2}, \ f\left(\frac{x_1 + x_2}{2}\right)\right).$$

The roots of the equation $ax^2 + bx + c = 0$, $a \neq 0$ are
$$x_{1,2} = \frac{-b \pm \sqrt{b^2 - 4ac}}{2a}.$$ The nature of the roots is given by the value
of the discriminant.

a) If $\Delta = b^2 - 4ac > 0$, there are two different real roots.
b) If $\Delta = b^2 - 4ac = 0$, there are two equal real roots.
c) If $\Delta = b^2 - 4ac < 0$, there are two conjugate complex roots.

Vietas formulas (formulae between coefficients and roots)
$$x_1 + x_2 = -\frac{b}{a} \quad \text{and} \quad x_1 + x_2 = -\frac{c}{a}.$$

V. 1. Factor:
a) $x^2 - 49$; b) $25x^2 - 1$; c) $81x^2 - 4$; d) $(x + 8)^2 - (x + 3)^2$;
e) $49(x - 2)^2 - 81(x - 3)^2$; f) $x^2 + 7x + 12$; g) $x^2 - 9x - 36$;
h) $x^2 + 6x - 16$;

V. 2. Write in the vertex form the quadratic functions:

a) $f(x) = x^2 - 2x + 8$; b) $f(x) = -x^2 + 5x + 2$;

c) $f(x) = \dfrac{2}{3}x^2 - 4x + 1$; d) $f(x) = 0.5x^2 - 0.2x + 0.3$;

e) $f(x) = -3x^2 + x + 0.1$; f) $f(x) = 5x^2 - \dfrac{2}{3}x + 1$.

V. 3. Determine the axe of symmetry and the vertex of each parabolas associated to the functions:

a) $f(x) = -2x^2 + 3x + 1$; b) $f(x) = 3x^2 - x + 1$;

c) $f(x) = -4x^2 + 3x - 2$; d) $f(x) = 5x^2 - 2x - 1$;

e) $f(x) = 0{,}4x^2 - 0{,}1x + 0{,}3$; f) $f(x) = \dfrac{1}{3}x^2 - \dfrac{1}{2}x + \dfrac{1}{4}$.

V. 4. Calculate the maximum or minimum for the functions:

a) $f(x) = -3x^2 + 2x - 1$; b) $f(x) = 2x^2 + x - 3$;

c) $f(x) = -x^2 + 4x + 2$; d) $f(x) = -2x^2 + 4x + 3$;

e) $f(x) = 5x^2 - 2x + 3$; f) $f(x) = x^2 - 4$.

V. 5. Draw the graph for the functions:

a) $f(x) = x^2 - 3x + 2$; b) $f(x) = -2x^2 - 7x - 5$;

c) $f(x) = -x^2 + 10x - 25$; d) $f(x) = 2x^2 - 3x + 1$;

e) $f(x) = -x^2 + 16$; f) $f(x) = -x^2 + 5x$.

V. 6. Prove that the functions

a) $f : (-\infty, 3] \to [-1, \infty)$, $f(x) = x^2 - 6x + 8$;

b) $f : [3, \infty) \to [-1, \infty)$, $f(x) = x^2 - 6x + 8$;

c) $f : \left(-\infty, \dfrac{3}{2}\right] \to \left(-\infty, \dfrac{9}{4}\right]$, $f(x) = -x^2 + 5x - 4$;

d) $f : \left[\dfrac{5}{2}, \infty\right) \to \left(-\infty, \dfrac{9}{4}\right)$, $f(x) = -x^2 + 5x - 4$ are bijections and

find their inverses. Draw the graph for the inverse and parent function.

Notice: The function $f : \mathbf{R} \to \mathbf{R}$, $f(x) = x^2 - 6x + 8$ is not an inversible function, it has an inverse, but this inverse is *not a function.*

V. 7. The graph of the quadratic function $f(x) = -5(x-1)(x+7)$ intersects x-axis at the points A and B. Find the length of the segment AB.

V. 8. Consider the linear function $f : \mathbf{R} \to \mathbf{R}$, $f(x) = (a+2)x - 4a + 3$, where $a \in \mathbf{N}$. Find a such that the point $A(a, -a+9)$ lies on the graph of f.

V. 9. Let $f : \mathbf{R} \to \mathbf{R}, f(x) = \left| ax^2 - 3x + 2c \right|$ a function. If $0 < a < 2$ and $c < 0$, determine a and c such that $f(0) = 4$ and $f(1) = 6$.

V. 10. Determine the quadratic function $f(x) = ax^2 + bx + c$ such that the graph passes through the points:
a) $A(1, 0)$, $B(-1, 6)$, and $C(2, 3)$.
b) $A(-2, 12)$, $B(-1, 6)$ and crosses x-axis at the point $C(2, 0)$.
c) $A(1, 0)$, $B(-1, 10)$, and $C(2, -2)$.
d) $A(-2, 0)$, $B(1, 6)$, and the maximum is 6.
e) $V(1, 2)$ is the vertex and crosses y-axis at the point $C(0, -3)$.
f) $A(-1, 13)$, $B(2, 10)$, and the minimum is 9.

V. 11. If $f(x)$ is a quadratic function such that $f(-1) = 6$, $f(0) = 1$, and $f(1) = 0$, find $f(2)$.

V. 12. Determine the quadratic function $f(x) = ax^2 + bx + c$ such that the point $V(3, 5)$ is the vertex and it crosses the y-axis at the point $A(0, -4)$.

V. 13. Two parabolae have the vertexes $V_1(2, \ a)$ and $V_2(2, \ b)$. If one of their common points of intersection is $A_1(3, \ 5)$, find the second common point of intersection.

V. 14. The inequality $4(x^2 + ax + 2) - 3(x^2 + 2x + b) > 0$ has the solution $x \in (-\infty, -2) \cup (4, \infty)$. Find $a, b \in R$.

V. 15. Find $a, b \in R$ such that the inequality $(x-3)^2 + 2(x-1)^2 + ax + b \le 0$ has the solution $x \in [-1, 4]$.

V. 16. Let A, B be the points of intersection of the graph of the function $f : R \to R$, $f(x) = -x^2 + (m+1)x - m$ to x-axis and the point V the vertex of the parabola. Find $m \in (1, \infty)$ such that the area of the triangle VAB is one.

V. 17. Find the value of x such that the function $f(x) = (x - a_1)^2 + (x - a_2)^2 + ... + (x - a_n)^2$ has the minimum value, where $a_1, a_2, ..., a_n \in R$.

V. 18. Determine the quadratic function $f(x) = ax^2 + bx + c$ knowing that it has a maximum five units and passes through the points $A(1, 1)$ and $B(-1, -11)$.

V. 19. Determine the parameter $m \in R$ such that the roots of the equations fulfil the indicated conditions.

a) $mx^2 + 2(m-1)x + 5 = 0$, $\qquad x_1 - x_2 = 2$;

b) $(m-1)x^2 + (3m-2)x + 3 = 0$, $\qquad x_1 = x_2$;

c) $(m+2)x^2 + (m-1)x + m = 0$, $\qquad x_1 = 3x_2$;

d) $3x^2 + (m-2)x + m + 6 = 0$, $\qquad x_1^2 + x_2^2 = \dfrac{8}{3}$;

e) $(m+4)x^2 - (m+6)x - m + 4 = 0$, $\qquad x_1 x_2 = x_1 + x_2$;

f) $mx^2 - (m-1)x - m + 1 = 0$, $\qquad x_1^3 + x_2^3 = 4$;

g) $(m+3)x^2 + (m-2)x + m + 1 = 0$, $\qquad x_1 + x_2 + 2x_1 x_2 = \dfrac{3}{2}$;

h) $x^2 - 2mx + m + 3 = 0$, $\qquad (x_1 - 1)^2 + (x_2 - 1)^2 = 6$;

i) $x^2 + (1-2m)x + m^2 - 2m - 3 = 0$, $\qquad \dfrac{x_1}{x_2} + \dfrac{x_2}{x_1} \leq 2$.

V. 20. Calculate in terms of the parameter $m \in \mathbb{R}$, the expressions:

a) $x_1 + x_2$; **b)** $x_1 x_2$; **c)** $x_1^2 + x_2^2$; **d)**; $x_1^3 + x_2^3$, **e)** $x_1 - x_2$; **f)** $x_1^4 + x_2^4$; **g)** $x_1^{-1} + x_2^{-1}$; **h)** $x_1^{-2} + x_2^{-2}$; **i)** $x_1^{-3} + x_2^{-3}$; **j)** $\sqrt{x_1} + \sqrt{x_2}$; **k)** $\sqrt[4]{x_1} + \sqrt[4]{x_2}$, where x_1 and x_2 are the roots of the equation $x^2 - 2mx + m - 1 = 0$.

V. 21. Let x_1 and x_2 be the roots of the equation $x^2 - x - 1 = 0$. Calculate the value of the expression

$$E = \frac{x_1^2 - 3x_1 + 1}{x_1^2 - 4x_1 + 2} + \frac{x_2^2 - 3x_2 + 1}{x_2^2 - 4x_2 + 2}.$$

V. 22. The equation $(1 - 2m)x^2 + 2x - 3 = 0$ has real roots. Find the largest numerical value that m may have.

V. 23. Find the quadratic equation with the roots:

a) $x_1 = 3$, $x_2 = -3$; **b)** $x_1 = -2$, $x_2 = 0$; **c)** $x_1 = -4$, $x_2 = 2$.

d) $x_1 = \dfrac{1 + \sqrt{3}}{2}$, $x_2 = \dfrac{1 - \sqrt{3}}{2}$; **e)** $x_1 = \dfrac{1 + i\sqrt{2}}{3}$, $x_2 = \dfrac{1 - i\sqrt{2}}{3}$.

V. 24. Consider the equation $x^2 + mx + 2m - 1 = 0$ with the roots x_1 and x_2. Find a quadratic equation at y with the roots y_1 and y_2 such that:

a) $y_1 = \dfrac{1}{x_1}$, $y_2 = \dfrac{1}{x_2}$; **b)** $y_1 = 1 + \dfrac{1}{x_1}$, $y_2 = 1 + \dfrac{1}{x_2}$;

c) $y_1 = \dfrac{x_1 + x_2}{x_1}$, $y_2 = \dfrac{x_1 + x_2}{x_2}$; **d)** $y_1 = \dfrac{x_1}{x_2}$, $y_2 = \dfrac{x_2}{x_1}$;

e) $y_1 = x_1^{-2}$, $y_2 = x_2^{-2}$; **f)** $y_1 = x_1^{-3}$, $y_2 = x_2^{-3}$.

V. 25. Solve the equation:
$$\frac{1}{x^2-6x+1}+\frac{2}{x^2-6x+2}+\frac{3}{x^2-6x+3}+...+\frac{2010}{x^2-6x+2010}=2010.$$

V. 26. Solve the equation
$$x^2+(x+2)^2+...+(x+2010)^2=(x+1)^2+(x+3)^2+...+(x+2011)^2.$$

V. 27. Solve the inequations:

a) $x^2+7x+3>-3$; **b)** $-x^2+x-2\geq-4x+4$;

c) $3x^2+x-2\leq-7x+9$; **d)** $5x^2-3x+1\geq-x^2+3$;

e) $\dfrac{x^2-3}{4}-\dfrac{x-1}{3}>\dfrac{2x-6}{3}-x$; **f)** $\dfrac{3x+1}{x-2}-\dfrac{2x+1}{x+2}\leq\dfrac{1}{2}$;

g) $\dfrac{2x^2+5(x+1)-10}{3x^2+x-2}\leq1$; **h)** $\dfrac{x-2}{x+2}\geq\dfrac{2(x+1)-5}{x+3}$;

i) $\dfrac{3x-3}{x^2-6x+5}>2$; **j)** $\dfrac{x^2+2x+13}{x^2-3x+2}\geq2$;

k) $\dfrac{x}{x-1}>\dfrac{x+1}{x}$; **l)** $\dfrac{x+2}{x+3}\geq\dfrac{2x+1}{x-1}$; **m)** $\left|\dfrac{x^2-5x+4}{x^2-4}\right|\leq1$;

n) $-1<\dfrac{x^2+3x+2}{x^2-4x+3}\leq2$; **o)** $\dfrac{(x-1)(x-2)(x-3)}{(x+1)(x+2)(x+3)}>1$;

p) $\dfrac{6}{(x+2)(x-3)}-\dfrac{1}{x+2}\geq2$.

V. 28. Solve the systems of inequations:

a) $\begin{cases}x^2-3x>4\\-5x^2+6x-1<0\end{cases}$; **b)** $\begin{cases}2x-7\leq0\\x^2-7x+12\leq0\\-x^2+2x+3<0\end{cases}$;

c) $\begin{cases}3(x-1)\geq2(x-2)\\2x^2+4x>0\end{cases}$; **d)** $\begin{cases}-4x+3\geq x-2\\(x+3)(x+2)\geq(x-3)(x-2)\\x^2-3x+2\geq0\end{cases}$;

e)
$$\begin{cases} \dfrac{|2x-1|+|1-2x|}{4-x^2} \le 1 \\ \dfrac{|2x-1|+|1-2x|}{4-x^2} \ge -1 \end{cases} ;$$

V. 29. For the function $f : \mathbf{R} \to \mathbf{R}$, $f(x) = x^2 - 4x + 3$ calculate:

a) $f([0, 3])$; b) $f([0, 4))$; c) $f([1, 5])$; d) $f([-\infty, 0])$; e) $f([3, +\infty))$;
f) $f((-\infty, 3])$; g) $f((-1, +\infty))$.

V. 30. Find the range for the functions:

a) $f : \mathbf{R} \to \mathbf{R}, f(x) = x^2 - x + 1$; b) $f : \mathbf{R} \to \mathbf{R}, f(x) = x^2 + 5x - 6$;

c) $f : \mathbf{R} \to \mathbf{R}, f(x) = |x^2 - 4| - 5$;

d) $f : \mathbf{R} \to \mathbf{R}, f(x) = \dfrac{x^2 - x - 3}{x^2 + x + 2}$;

e) $f : \mathbf{R} \to \mathbf{R}, f(x) = \dfrac{x^2 - 4x + 3}{x^2 - 2x + 3}$;

f) $f : \mathbf{R} \to \mathbf{R}, f(x) = \dfrac{-x^2 - 2x - 3}{x^2 + 2x + 1}$.

V. 31. Consider the function

$f : \mathbf{R} \to \mathbf{R}, \ f(x) = \dfrac{3x^2 + ax + b}{x^2 + 1}$. Find a and b such that the range of the function is the interval $[-3, 5]$.

V. 32. For the fraction $F = \dfrac{x^2 + (m+1)x + m + 2}{x^2 + x + m}$ find the value of m such that the fraction F is well defined and it is positive for any $x \in \mathbf{R}$.

V. 33. Let $f : \mathbf{R} \to \mathbf{R}, \ f(x) = x^2 - 6x + 9 + m$ be a quadratic function, where $m \in \mathbf{R}$. Find the value of m such that $f(x) \ge 0$ for any $x \in \mathbf{R}$ and for $m = 0$ find the value of x such that

$(f \circ f)(x) = 0$.

V. 34. The roots of the equation

$x^2 - 2(m-2)x - 2m + 3 = 0$ are x_1 and x_2. Find the value of $m \in \mathbf{R}$

such that $\dfrac{x_1}{x_2^2} + \dfrac{x_2}{x_1^2} \ge \dfrac{1}{x_1} + \dfrac{1}{x_2}$.

V. 35. The roots of the equation $mx^2 - 2(m+1)x + 8 = 0$ are

x_1 and x_2. Find the value of $m \in \mathbf{R}$ such that $\dfrac{x_1^2 + x_2^2}{x_1 x_2} > x_1 + x_2$.

V. 36. Consider the equations: $x^2 - (m+2)x + m^2 + 1 = 0$

with the roots x_1 and x_2 and $y^2 - (2m-1)y + m^2 + 2m = 0$ with the

roots y_1 and y_2. Find the value of m such that:

$(x_1 + x_2)(y_1 + y_2) \ge x_1 x_2 + y_1 y_2$.

V. 37. The roots of the equation

$4(m+1)x^2 - 2(m+1)x + m - m^2 = 0$ are x_1 and x_2. Find the

value of $m \in \mathbf{R} \setminus \{-1\}$ such that $-2 \le \dfrac{1}{x_1} + \dfrac{1}{x_2} + \dfrac{1}{2x_1 x_2} \le 0$.

V. 38. Consider the equation $x^2 - 2mx + m + 5 = 0$. Find the

value of m, such that

a) one of the roots is double to the other,

b) $\dfrac{x_1^2 + x_2^2}{x_1 + x_2} \ge x_1 x_2$, where x_1 and x_2 are the roots of the equation.

V. 39. Consider the family of quadratic functions

$f_m(x) = x^2 + 2(m-1)x + m - 1$. Find the value of m such that the

vertices of the parabolas lie above the x-axis.

V. 40. Consider the family of quadratic functions

$f_m(x) = x^2 - 2(m+2)x + m + 2$, $m \in \mathbf{R}$.

a) Show that the vertices of the parabolas lie on a parabola.

b) For what value of m, the vertices of the parabolas lie under the x-axis?

V. 41. For the family of quadratic functions:
$f_m(x) = x^2 - 2(m-1)x + m - 2$, $m \in \mathbf{R}$ show that the vertices of the parabolas lie on a parabola.

V. 42. Consider the family of quadratic functions
$f_m(x) = mx^2 + 2(m+1)x + m + 2$, $m \in \mathbf{R} \setminus \{0\}$.
a) Show that the vertices of the parabolas lie on the equation of the line $y = x + 1$.
b) If A and B are the x intercept and V and F are the vertices and the feet of the pependicular from V onto x-axis, show that for any $m \in \mathbf{R} \setminus \{0\}$ $AB = 2 \, FV$.
c) Show that all parabolas from the family pass through a fixed point and find the coordinates of the point.

V. 43. Consider the family of quadratic functions
$f_m(x) = (m+2)x^2 + (m+1)x - 3$, $m \in \mathbf{R} \setminus \{-2\}$.
a) For $m = -1$ find the intervals where the function decreases and increase.
b) Find $m \in \mathbf{R} \setminus \{-2\}$ such that the vertices of the parabolas lie on the first bisector.
c) Find $m \in \mathbf{R} \setminus \{-2\}$ such that $f_m(x) > 0$ for any $x \in \mathbf{R}$.

V. 44. Find the parameter $m \neq -1$ such that the graph of the parabola $f(x) = (m+1)x^2 - (2m+3)x + m - 1$ intersects the x-axis in two distinct points.

V. 45. Prove that for any $m \neq 0$, the graph of the function $f(x) = mx^2 - 2(m-3)x + m - 6$ intersects the x-axis in two distinct points.

V. 46. Find the numerical value of m such that the functions $f(x) = x^2 - 2x - 4$ and $g(x) = -x^2 - 2mx - 6$ have the same vertex.

V. 47. Show that the graphs of the functions
$f(x) = 2x^2 - 2x + 3$
and $g(x) = x^2 - 2x + 4$ have two common points.

V. 48. Consider the families of functions
$f_m(x) = x^2 - 2(m-2)x + m^2 + 3m$. For what value of m the graph of f_m lie above the horizontal line $y = -3$?

V. 49. Consider the quadratic function
$f_m(x) = (m-1)x^2 + 2(m+2)x + m + 1$, $(m \neq 1)$. Find the parameter m such that the graph of the function is below the x-axis.

V. 50. Consider the families of functions
$f_m(x) = x^2 - 2(m-1)x + m^2 - 3m$.
a) Find the value of m such that the vertices of the parabolas f_m lie under the x-axis.
b) Find the value of m such that the graph of the parabola f_m is tangent to Ox axes.
c) Show that the vertices of the parabolas f_m lie on a line for any $m \in \mathbf{R}$.

V. 51. Find the numeric value of $m \neq 1$, such that
$(m-1)x^2 + 2mx + m + 2 > 0$, for any $x \in \mathbf{R}$.

V. 52. Find the value of m, such that
$(m-2)x^2 + 2(2m-3)x + m - 2 < 0$, for any $x \in \mathbf{R}$.

V. 53. For what value of m, the inequation
$(m-1)x^2 + 2(m+2)x + m + 2 > 0$ does not have any solution?

V. 54. The roots of the equation
$$2x^2 - 2(m+1)x + m^2 - 2m + 3 = 0$$
are x_1 and x_2, $m \in \mathbf{R}$. Show that:

a) $(x_1 - 1)^2 + (x_2 - 1)^2 = 2$;

b) $2 - \sqrt{3} \le \dfrac{x_1}{x_2} \le 2 + \sqrt{3}$,

c) Find a relation between x_1 and x_2 which does not depend of m.

V. 55. The roots of the equation $f_m(x) = 0$ where
$$f_m(x) = 2x^2 - 2(m+3)x + m^2 - 4m + 24 \quad \text{are} \quad x_1 \text{ and } x_2, \ m \in \mathbf{R}.$$
Show that:

a) $(x_1 - 5)^2 + (x_2 - 5)^2 = 5$;

b) $x_1, x_2 \in \left[5 - \sqrt{5}, \ 5 + \sqrt{5}\right]$;

c) $-\sqrt{10} \le x_1 - x_2 \le \sqrt{10}$;

d) $10 - \sqrt{10} \le x_1 + x_2 \le 10 + \sqrt{10}$;

e) $x_1^2 + x_2^2 \le 55 + 10\sqrt{10}$;

f) $0.5 \le \dfrac{x_1}{x_2} \le 2$;

g) Determine the parameter $m \in \mathbf{R}$ such that $f_m(x) \ge 0$ for any $x \in \mathbf{R}$;

h) Find the smallest possible value of $f_m(x)$;

i) Find a relation between x_1 and x_2 which does not depend of m;

j) Find $m \in \mathbf{R}$ such that the equation has a root between 5 and 6.

V. 56. Solve the systems of equations:

a) $\begin{cases} x^2 - xy - y = 5 \\ x - 2y = 1 \end{cases}$;   b) $\begin{cases} x + 2y = 5 \\ x^2 + 2y^2 - xy = 7 \end{cases}$;

c) $\begin{cases} x + 2y = 7 \\ x^2 + 2y^2 - xy = 11 \end{cases}$;   d) $\begin{cases} \dfrac{x^2 - xy - 6}{x + y} = x - y \\ x + 3y = 10 \end{cases}$;

e) $\begin{cases} 2x - 3y = -8 \\ 2x^2 - xy + 5y^2 = 24 \end{cases}$; **f)** $\begin{cases} \dfrac{x^2 - xy - 1}{x - y} = 1 \\ 2x + y = 5 \end{cases}$,

g) $\begin{cases} \left(\dfrac{x}{y}\right)^2 + \left(\dfrac{x}{y}\right)^3 = 36 \\ (xy)^2 + xy = 12 \end{cases}$.

V. 57. Solve the homogeneous system of equations:

a) $\begin{cases} x^2 - 3y^2 = 1 \\ x^2 + xy = 6 \end{cases}$; **b)** $\begin{cases} 2y^2 - x^2 = -1 \\ xy = 6 \end{cases}$;

c) $\begin{cases} x^2 - 2xy + 3y^2 = 17 \\ 3x^2 - 2xy + y^2 = 11 \end{cases}$; **d)** $\begin{cases} 2x^2 - 3xy - 5y^2 = -6 \\ x^2 - 6y^2 = -5 \end{cases}$;

e) $\begin{cases} 2y^2 - 7x^2 = -5 \\ xy = -1 \end{cases}$; **f)** $\begin{cases} 2x^2 - 3xy + 7y^2 = 6 \\ 3x^2 - xy + y^2 = 3 \end{cases}$.

V. 58. Solve the symmetric systems of equations:

a) $\begin{cases} x^2 + y^2 = 5 \\ xy = 2 \end{cases}$; **b)** $\begin{cases} 2x^2 + 2y^2 - x^2 y^2 = 0 \\ xy = x + y \end{cases}$;

c) $\begin{cases} x^3 + y^3 = 9 \\ xy(x + y) = 6 \end{cases}$; **d)** $\begin{cases} xy + x + y = 11 \\ x^2 y + xy^2 = 30 \end{cases}$;

e) $\begin{cases} 5x^2 - 6xy + 5y^2 = 29 \\ 7x^2 - 8xy + 7y^2 = 43 \end{cases}$; **f)** $\begin{cases} x^{-2} + y^{-2} = 13 \\ x^{-1} + y^{-1} = 5 \end{cases}$;

g) $\begin{cases} x^3 + y^3 = 19 \\ x^2 y + xy^2 = -6 \end{cases}$; **h)** $\begin{cases} x^2 + y^4 = 20 \\ x^4 + y^2 = 20 \end{cases}$;

i) $\begin{cases} x^4 + x^2 y^2 + y^4 = 91 \\ x^2 + xy + y^2 = 13 \end{cases}$.

V. 59. Solve the systems of equations:

a) $\begin{cases} (x-0.8)^2 + (y-0.7)^2 = 1 \\ x+y = 2.9 \end{cases}$; b) $\begin{cases} x-y = 2 \\ x^3 - y^3 = 26 \end{cases}$;

c) $\begin{cases} x^3 + y^3 = -26 \\ x^3 y^3 = -27 \end{cases}$; d) $\begin{cases} \dfrac{x}{y} - \dfrac{y}{x} = -\dfrac{16}{25} \\ x^2 - y^2 = -16 \end{cases}$;

e) $\begin{cases} \left(\dfrac{x}{y}\right)^2 + \left(\dfrac{y}{x}\right)^2 = \dfrac{257}{16} \\ (xy)^2 + xy = 20 \end{cases}$; f) $\begin{cases} \dfrac{x+y}{x-y} - \dfrac{x-y}{x+y} = 4\dfrac{4}{5} \\ xy = 6 \end{cases}$;

g) $\begin{cases} (x-y)(x^2 - y^2) = 3a^3 \\ (x+y)(x^2 + y^2) = 15a^3 \end{cases}$, $a \neq 0$;

h) $\begin{cases} (x-y)(x-2y)(x-3y) = -105 \\ (y-x)(y-2x)(y-3x) = 60 \end{cases}$;

i) $\begin{cases} x+y = -3 \\ x+z = -2 \\ xy + xz + yz = 2 \end{cases}$; j) $\begin{cases} xy + xz = 7 \\ xy + yz = 15 \\ xz + yz = 16 \end{cases}$;

k) $\begin{cases} x-y+z = 6 \\ x^2 + y^2 + z^2 = 14 \\ x^3 - y^3 + z^3 = 36 \end{cases}$.

V. 60. Prove that $\sqrt{\sqrt[3]{x} + \sqrt[3]{y} + \sqrt[3]{z}} > \sqrt[3]{\sqrt{x} + \sqrt{y} + \sqrt{z}}$, for any x, y, z positive integers.

V. 61. A manufacturer produces CD-s at a cost of $2 apiece. The CD-s are sold for $5 apiece. At that price, consumers have bought 4,000 CD-s a month. The manufacturer intends to rise the unitary price and he estimates that for each $1 increase, he will sell 400 CD less every month.
a) Express the manufacturer's monthly profit as a function of price at which CD-s are sold.

b) What is the price to maximize the profit? What is the maximum profit?

V. 62. A ferry boat carries 300 cars daily from a place to another place charging $8 per car. The captain realized that if he dropped the charge by $0.25, he would have 10 cars more on board of his ferry boat and he would make $2,200 in fares every day. What fare should they charge to maximize the revenue?

V. 63. An apple orchard has an average yield of 30 bushels per tree only if there are at most 40 trees per acre. For more than 40 trees pe acre, the average yeld decreases by half a bushel per tree for every tree over 40. Find the number of trees which maximize the yield per acre.

V. 64. An object is thrown in the air. The height (in metres) of the object is given by the equation $h(t) = 25t - 5t^2$, where t is time in seconds. What is the maximum height reached by the object?

V. 65. A fence is 100 feet and it forms three sides of a rectangular garden. One side is occupied by a building. Find the necessary dimensions of the fence to enclose the largest area.

V. 66. On the segemt $AB = a$, consider the point M and the equilateral triangles with the sides AM and MB. Find the minumium of area of the two triangles.

V. 67. Let AD be the median of $\triangle ABC$. Find a point $M \in (AD)$ such that $S = MA^2 + MB^2 + MC^2$ is minum.

V. 68. Consider the circle with the radius r. On a diameter AB of the circle, consider the point M. Construct the circles with the diameter AM and MB. Find the position of the point M such that the area between the two circles and the circle with the diameter is maximum.

V. 69. On the segemt $AB = 2a$, consider the point M and the circles with the diameters AM and MB. Find the minumium of the area of the two circles and AM.

V. 70. The area of a rectangle is 100 m². Find the dimensions of the rectangle such that the perimeter is minimum.

V. 71. In $\triangle ABC$, $AB = 8$ cm and $AC + BC = 10$ cm. Find the altitude of the triangle such that its area is maximum.

Chapter VI

COMPLEX NUMBERS

$z = a + bi$, where $a \in \mathbf{R}$, $b \in \mathbf{R}$, $i^2 = -1$ is the algebraic representation of a complex number.

The set of the complex number is $\mathbf{C} = \{a + bi \mid a \in \mathbf{R}, \ b \in \mathbf{R}, \ i^2 = -1\}$, where a is the real part of the complex number z or $a = \text{Re}(z)$;

b is the imaginary part of the complex number z or $b = \text{Im}(z)$.

$\bar{z} = a - bi$ is the *conjugate* of the complex number z.

$|z| = \sqrt{a^2 + b^2}$ is the *modulus* (*absolute value*) of the complex number z. If $z = a + bi$ is a complex number, the point $M(x, y)$ is called the geometric imagine of z, therefore the distance from the origine to M is $|z|$.

VI. 1. Perform the following complex numbers:

a) $z_1 = (3 + 3i) + (2 - i)$; **b)** $z_2 = (3 + i) - (2 + 4i)$;

c) $z_3 = (3 + 2i)(1 + 4i)$; **d)** $z_4 = (3 + i) \div (2 + i)$;

e) $z_5 = (1 - i) \div (1 + 2i)$.

VI. 2. Compute:

a) $(1 + i)(1 - i)^2 (-1 - i)^3 (-1 + i)^4$;

b) $(1 + i) + (1 - i)^2 + (-1 - i)^3 + (-1 + i)^4$;

c) $i + i^3 + i^5 + i^7$; **d)** $\dfrac{1}{i} + \dfrac{1}{i^3} + \dfrac{1}{i^5} + \dfrac{1}{i^7}$;

e) $i^{2010} + i^{2011} + i^{2012} + i^{2013}$.

VI. 3. Calculate $(1 - i)(1 - i^2)(1 - i^3) \cdot \ldots \cdot (1 - i^{2010})$.

VI. 4. Let $z_1 = 4 - 2i$, $z_2 = 5 + 3i$ be two complex numbers, calculate: **a)** $z_1 + z_2$; **b)** $z_1 - z_2$; **c)** $z_1 \cdot z_2$; **d)** $\dfrac{1}{z_1}$; **e)** $2 \cdot z_1 + 4 \cdot z_2$.

VI. 5. Is the number $1+i$ a root of the equation $z^4 + 4 = 0$?

VI. 6. Consider the complex numbers:

a) $z_1 = -4 + 3i$; b) $z_2 = 3 - 2i$; c) $z_3 = -2 + i$; d) $z_4 = 1 - 2i$.

Compute $\operatorname{Re}(z)$, $\operatorname{Im}(z)$, and $|z|$ for each complex number.

VI. 7. If $z \in \mathbf{C}$ show that:

a) $z \cdot \bar{z} \in \mathbf{R}$, $z + \bar{z} \in \mathbf{R}$, $\dfrac{z}{\bar{z}} + \dfrac{\bar{z}}{z} \in \mathbf{R}$;

b) $z - \bar{z} \in \mathbf{C} - \mathbf{R}$, $\dfrac{z}{\bar{z}} - \dfrac{\bar{z}}{z} \in \mathbf{C} - \mathbf{R}$.

VI. 8. If $z_1 = 1 + 2i$, $z_2 = -3$, $z_3 = 5 + 5i$, and $z_4 = -8 - 6i$,

then calculate: a)$\dfrac{z_3 - \bar{z}_3}{z_1 \bar{z}_1}$; b) $\dfrac{z_1 z_3}{z_2 z_4}$; c) f) $\dfrac{z_1 \cdot z_2}{z_3 + z_4}$.

VI. 9. If $z = -2 - 4i$ then calculate: a) $\bar{z}$; b) $|z|$.

VI. 10. Write each complex expression in standard form:

a) $\dfrac{1+i}{2-5i}$; b) $\dfrac{-3+i}{2+3i}$; c) $\dfrac{-3+2i}{2+i}$;

d) $\dfrac{(2+i)^3 - (2-i)^3}{(2+i)^3 + (2-i)^3}$; e) $\dfrac{(1+i)^4 + (1-i)^4}{(1+2i)^4 + (1-2i)^4}$.

VI. 11. Find the real numbers x and y such that:

a) $(1 + 3i)x + (2 - i)\,y = 3 + 2i$;

b) $(x + i)x + (1 + 3yi)y = 2 + 4i$;

c) $3\sqrt{x^2 - 2y} + (1-i)x^2 = 2(1 + 2i)y - 12i$;

d) $4x + 5 + (y - 2)i = -19 + 5i$;

e) $\overline{3x - 5i} + 2x + y = \overline{3 + i} + 2yi$;

f) $\dfrac{x - 3 + (y - 3)i}{x + 2 + (y + 4)i} = i$;

g) $\dfrac{x-2+(y-1)i}{y-3} = -1+3i$.

VI. 12. Find the real numbers a and b such that:

a) $\dfrac{11+i}{1-11i} = a+bi$; **b)** $\dfrac{5+2i}{2-5i} = a+bi$.

VI. 13. Calculate $|z|$ for the complex numbers:

a) $z = \left(\sqrt{5+\sqrt5}+i\sqrt{5-\sqrt5}\right)^4$; **b)** $z = \left(\sqrt{3+\sqrt2}+i\sqrt{3-\sqrt2}\right)^4$.

VI. 14. Show that the number $z = \left(1+i\sqrt3\right)^3$ is an integer number.

VI. 15. Calculate $|5-12i|-|12+5i|$.

VI. 16. If $z_1 = 1+2i$ and $z_2 = 2-i$, show that $z_1\bar z_2 + z_2\bar z_1 \in \mathbf{R}$.

VI. 17. Find the complex number z such that $\dfrac{\bar z+5+7i}{z} = 6$.

VI. 18. If $z\cdot\bar z = 5$ and $(z-1)(\bar z-1) = 10$, find the number z.

VI. 19. Solve the equation $z^4 - 10z^2 + 41 = 0$ and compute the expressions

$$E_1 = \frac{z_1+z_2+z_3+z_4}{z_1z_2z_3z_4} \quad \text{and} \quad E_2 = \frac{z_1z_2z_3+z_1z_2z_4+z_2z_3z_4}{z_1z_2z_3z_4}.$$

VI. 20. Solve in **C** the equations:
a) $9x^2+6x+10 = 0$; **b)** $x^2+5x+7 = 0$;
c) $2x^2-3x+2 = 0$; **d)** $-x^2+3x-8 = 0$.

VI. 21. Prove that the complex numbers $1+3i$ and $1-3i$ are the solutions of the equation $x^3-4x^2+14x-20 = 0$.

VI. 22. Prove that the complex numbers $-\dfrac{1}{2}+i\dfrac{\sqrt{3}}{2}$ and

$-\dfrac{1}{2}-i\dfrac{\sqrt{3}}{2}$ are the solutions of the equation $x^4 + x^2 + 1 = 0$.

VI. 23. If x_1 and x_2 are the roots of the equation $x^2 + x + 1 = 0$, then calculate:

a) $\left(x_1^3 + x_1^2 - 1\right)^{2002} + \left(x_2^3 + x_2^2 - 1\right)^{2002}$;

b) $\left(x_1^4 + x_1^3\right)^{2002} + \left(x_2^4 + x_2^3\right)^{2002}$;

c) $\left(x_1^4 + x_1^3 + x_1^2 + 1\right)^{2010} + \left(x_2^4 + x_2^3 + x_2^2 + 1\right)^{2010}$.

VI. 24. If ε is one of the roots of the unity, $x^3 - 1 = 0$ and $\varepsilon \neq 1$, calculate the following complex numbers:

a) $z_1 = \varepsilon^4 + \varepsilon^2 + 1$; b) $z_2 = \dfrac{1+\varepsilon}{(1-\varepsilon)^2} + \dfrac{1-\varepsilon}{(1+\varepsilon)^2}$;

c) $z_3 = \dfrac{\varepsilon^3 - 1}{1 - 2\varepsilon^2}$; d) $z_4 = (a+b+c)(a+b\varepsilon+c\varepsilon^2)(a+b\varepsilon^2+c\varepsilon)$.

VI. 25. Prove the identities:

a) $\left(\dfrac{-1+i\sqrt{3}}{2}\right)^6 + \left(\dfrac{-1-i\sqrt{3}}{2}\right)^6 = 2$;

b) $\left(\dfrac{-1+i\sqrt{3}}{2}\right)^5 + \left(\dfrac{-1-i\sqrt{3}}{2}\right)^5 = -1$;

c) $\left(\dfrac{\sqrt{3}-i}{2}\right)^6 + \left(\dfrac{\sqrt{3}+i}{2}\right)^6 = -2$.

VI. 26. Calculate $z^{13} + \dfrac{1}{z^{13}}$ if z is a root of the equation

$z + \dfrac{1}{z} = 1$.

VI. 27. Let $z_1 = \dfrac{1-i\sqrt{3}}{a+(a+1)i}$ and $z_2 = \dfrac{b+i}{b-1-2i}$ be two complex numbers, where $a \in \mathbf{R}$, $b \in \mathbf{C}$.

a) Find a such that $z_1 \in \mathbf{R}$.

b) Find the points from the complex plane such that z_2 is a real number.

VI. 28. Calculate the complex number:

$$z = \left(\frac{1+i}{1-i}\right)^{2010} + \left(\frac{1+i}{1-i}\right)^{2011} + \left(\frac{1+i}{1-i}\right)^{2012}.$$

VI. 29. If x_1 and x_2 are the roots of the equation $x^2 + x + 1 = 0$ then calculate the value of the expression

$$E = \frac{(x_1+1)^{2011}}{x_1^2+1} + \frac{(x_2+1)^{2011}}{x_2^2+1}.$$

VI. 30. Let z_1, z_2 be two complex numbers such that $|z_1| = |z_2| = 1$. Show that $\dfrac{z_1+z_2}{1+z_1 z_2}$ is a real number.

VI. 31. Let $z \in C$ be a complex number such that $z \neq \bar{z}$ and $\dfrac{1-z+z^2}{1+z+z^2} \in \mathbf{R}$. Prove that $|z| = 1$.

VI. 32. Let $z_1 = x_1 + iy_1$ and $z_2 = x_2 + iy_2$ be two complex numbers such that $\bar{z}_1 = \dfrac{1-z_2}{1+z_2}$. Show that:

a) $y_1 y_2 > 0$;

b) If $z_1 + z_2 \in R$ find a relation between y_1 and y_2.

VI. 33. Reprezent the geometric image of the following complex numbers:

a) $z_1 = -2 + 3i$; b) $z_2 = 3 - 2i$; c) $z_3 = -2$; d) $z_4 = 2i$;

e) $z_5 = (2-2i)(3-i)$; **f)** $z_2 = \dfrac{3+2i}{1-i}$.

VI. 34. The geometric image of the complex numbers $z = 1 + 3i$ and $z = 3 - i$ in the complex plane is represented by the points A and B. Prove that ΔOAB is a right triangle.

VI. 35. Show that the geometric images of the complex numbers $z_1 = -1 - 2i$, $z_2 = 1 + 4i$, and $z_3 = 2 + 7i$ are three collinear points.

VI. 36. Let $z_1 = 1 + i$, $z_2 = 4 - 2i$, and $z_3 = 5 - i$ be three complex numbers. Show that the geometric imagine in the complex plane is a right triangle.

VI. 37. The points A, B, C, D are the geometric imagine in the complex plane of the numbers $z_1 = 1 - 2i$, $z_2 = 1 + 2i$, $z_3 = -2 + i$, and $z_4 = -2 - i$. Show that they lay on a circle and find the equation of the circle, moreover A, B, C, D is an isosceles trapezoid.

VI. 38. Solve for the complex variable z and express z in the form $a+bi$:

a) $|z| = |z + 1 - i| = |\bar{z} + 2 - 4i|$;

b) $\left|\dfrac{z+1}{z}\right| = 2$ and $\left|\dfrac{z-i}{1-i}\right| = 1$;

c) $\left|\dfrac{z-1}{z-i}\right| = \dfrac{1}{2}$ and $\left|\dfrac{z+1}{z}\right| = \sqrt{2}$.

VI. 39. Find the set of points in the complex plane such that:

a) $|z - i| = 2$; **b)** $|z - 1| + |z + 1| = 4$; **c)** $|z + i| < 2$;

d) $|z - 2| - |z + 2| < 2$; **e)** $1 < |2z + 3 - 2i| \le 2$; **f)** $|i - z| > 4$.

VI. 40. Find the set of complex numbers in the complex plane that satisfy the following inequality $|z+i-2| \le 2$.

VI. 41. Solve the equations:
a) $x^3 - 1 = 0$; **b)** $x^3 - 8 = 0$; **c)** $27x^3 + 1 = 0$;
d) $64x^3 - 27 = 0$; **e)** $x^4 - 16 = 0$; **f)** $x^4 - 81 = 0$;
g) $x^6 - x^3 - 2 = 0$; **h)** $x^6 + 3x^3 + 2 = 0$.

VI. 42. Let $z \in C$ be a complex number and $|z| < \dfrac{1}{2}$.
Show that $|(1+2i)z + 2i|z|| < 3$.

VI. 43. If $\varepsilon \ne 1$ is one of the roots of the equation $x^3 - 1 = 0$, prove that:
a) $(1+\varepsilon)^{2010} + (1+\varepsilon^2)^{2010} = 2$;

b) $(1+\varepsilon)^n + (1+\varepsilon^2)^n + (\varepsilon + \varepsilon^2)^n = \begin{cases} -3 \text{ if } n = 3k \\ 0 \text{ if } n \ne 3k \end{cases}, k \in \mathbf{N}$;

c) $|z-1|^2 + |z-\varepsilon|^2 + |z-\varepsilon^2|^2 = 3(1+|z|^2)$, $\forall z \in C$.

VI. 44. Prove that the number
$z = (\sqrt[4]{a} + i \cdot \sqrt[4]{b})(\sqrt[4]{a} + i^2 \cdot \sqrt[4]{b})...(\sqrt[4]{a} + i^{4k} \cdot \sqrt[4]{b})$ is real,
where a, b are positive real numbers.

VI. 45. Find $a \in \mathbf{R}$ such that:
a) $\left|\dfrac{2+4i}{1+ai}\right| = 2$; **b)** $\dfrac{a+2i}{3+ai} \in \mathbf{R}$.

VI. 46. Consider the function
$f : C \to C$, $f(z) = z^2 + 3z - \bar{z}$.
a) Compute: $f(-1)$, $f(i)$, $f(2-i)$, $f(\bar{z})$, $f(|z|)$;
b) Verify the egality $f(z) - f(\bar{z}) = 4(\bar{z} - z)$.

VI. 47. Find the sets of the points in the complex plane such that:

a) $A = \left\{ (x,y) \in \mathbf{R} \times \mathbf{R} \middle| \ \mathrm{Re}\left(\dfrac{z-2}{z-6}\right) = \mathrm{Im}\left(\dfrac{z-2}{z-6}\right), \ z = x+iy \right\};$

b) $B = \left\{ (x,y) \in \mathbf{R} \times \mathbf{R} \middle| \ \mathrm{Re}\left(\dfrac{z-2}{z-1}\right) = 0, \ z = x+iy \right\};$

c) $C = \left\{ (x,y) \in \mathbf{R} \times \mathbf{R} \middle| \ \dfrac{i+\bar{z}}{i-\bar{z}} \in \mathbf{R}, \ z = x+iy \right\}.$

VI. 48. Find the sets of the points in the complex plane such that $\ \log_{\frac{1}{2}} \dfrac{|z-3|+4}{3|z-3|-2} < 1.$

VI. 49. Determine $m \in \mathbf{R}$ such that the equation $(m+2)x^2 + (m-1)x + m = 0$ has only complex roots.

VI. 50. Factor each polinomoal in the set of complex numbers:
a) $x^4 - x;$ **b)** $x^3 + 8;$ **c)** $x^4 + x^2;$ **d)** $x^5 + 27x^2.$

VI. 51. Find $z \in C$ such that:
a) $z + 4\bar{z} = 20 + 9i;$ **b)** $|z+i| - z = 21 + 2i;$ **c)** $|z| + z = 2 - 4i.$

VI. 52. Find the complex numbers z such that $z^2 = \bar{z}.$

VI. 53. Consider the function
$f : C \setminus \{1\} \to C, \ f(z) = \dfrac{z^2 - z - 2}{z - 1}.$ Find the points from the plane such that $f(z) \in \mathbf{R}.$

VI. 54. Solve in C the equations $z^2 = x + iy$, where
a) $z^2 = -3 - 4i;$ **b)** $z^2 = 1 - 4\sqrt{3}i;$
c) $z^2 = 3 + 4i;$ **d)** $z^2 = 7 + 24i.$

VI. 55. Solve for the complex variable z:

a) $z^2 - 4iz - 3 = 0$; b) $z^2 - (1-i)z + 2 + i = 0$;

c) $(2-i)z^2 - (4-12i)z + 14 - 12i = 0$; d) $z^2 - 5iz - 7 + i = 0$;

e) $z^2 + 4iz + 5 = 0$; f) $z^2 - (5-2i)z + 6 - 4i = 0$;

g) $iz^2 + (1-2i)z - 1 + 3i = 0$; h) $z^2 - (5-2i)z + 5 - 5i = 0$.

VI. 56. Find $z = x + iy$, $x, y \in \mathbf{R}$ from the equations:

a) $(1+i)z + (5-3i)\bar{z} = 20 - 4i$;

b) $(2+2i)z - 3\,\mathrm{Re}(z) = -18 + 20i$.

VI. 57. Solve for the complex variable z and express z in the form $a + bi$:

a) $(1+3i)z - 11i = (3-2i)z + 13$;

b) $2z + i \cdot \bar{z} = 9 + 3i$.

VI. 58. Solve the equation: **a)** $z^3 = -11 + 2i$;

b) $z^3 = -9 - 46i$ where $z = a + bi$, $a, b \in \mathbf{Z}$, $a, b \neq 0$, $i^2 = -1$.

VI. 59. Simplify the fractions:

a) $\dfrac{z^2 - 8(1-i)z + 63 - 16i}{z^2 - 6z + 25}$; b) $\dfrac{z^2 - iz - 1 - i}{z^2 - 2z + 2}$;

c) $\dfrac{z^3 - (3+i)z^2 + (2+3i)z - 2i}{z^3 - (3-i)z^2 + (2-3i)z - 2i}$.

VI. 60. Solve the system of equations ($x, y, z_1, z_2 \in \mathbf{C}$):

a) $\begin{cases} (1+3i)x - (1-2i)y + 1 - i = 0 \\ 2(2+i)x - (2+i)y - (7-4i) = 0 \end{cases}$;

b) $\begin{cases} (2+3i)x + (2-3i)y = 4 - 2i \\ (1+i)x + (1-i)y = 2 \end{cases}$;

c) $\begin{cases} 4x + (3-2i)y = 5 - i \\ (3+2i)x + 4y = 5 + i \end{cases}$; d) $\begin{cases} iz_1 - 5z_2 = -6 - 15i \\ 2z_1 - 3iz_2 = 9 - i \end{cases}$.

VI. 61. Consider the complex function f having the property $f(x)+f(\varepsilon x)=4-\varepsilon x$, $\forall x \in C$, where ε is a complex root of the equation $x^3=1$. Prove that $f(x)+f(\varepsilon^2 x)=4-x$, $\forall x \in C$ and find $f(x)$.

VI. 62. Let be the complex number $z = \cos t + i\sin t$.

Prove that $\cos nt = \dfrac{z^{2n}+1}{2z^n}$ and $\sin nt = \dfrac{z^{2n}-1}{2iz^n}$, $n \in N^*$.

Calculate the product $A = \sin 10^0 \sin 50^0 \sin 70^0$.

.

VI. 63. Prove that $z_1, z_2, ..., z_n \in C \setminus \{0\}$ and $|z_1|=|z_2|=...=|z_n|$ then the expression

$$\left(1+\frac{z_2}{z_1}\right)\left(1+\frac{z_3}{z_2}\right)\cdot...\cdot\left(1+\frac{z_n}{z_{n-1}}\right)\left(1+\frac{z_1}{z_n}\right)$$ is a real number.

VI. 64. Solve the equation $(2-i)z^3 - 3 - i = 0$.

VI. 65. Solve the equation $\left(\dfrac{1+iz}{1-iz}\right)^n = \dfrac{1+ia}{1-ia}$, $n \in N$, $a \in R$ and show that all the roots of the equation are real.

VI. 66. Solve the equation

$$\left(\frac{x-i}{x+i}\right)^{n-1}+\left(\frac{x-i}{x+i}\right)^{n-2}+...+\left(\frac{x-i}{x+i}\right)^{1}+1=0, \; n \in N.$$

VI. 67. Solve the equation

$$\left(z+i\sqrt{1-z^2}\right)^n+\left(z+i\sqrt{1-z^2}\right)^n = 0, \text{ where } n \in N.$$

VI. 68. Consider the function $f:C \to C$, $f(z)=3z+|z|$. Prove that f is a bijection and find its inverse.

Chapter VII

EXPONENTIAL FUNCTION

$f : \mathbf{R} \to (0, \infty)$, $f(x) = a^x$, $a > 0$ and $a \neq 1$.

f decreases for $a \in (0, 1)$ and f increases for $a \in (1, \infty)$,

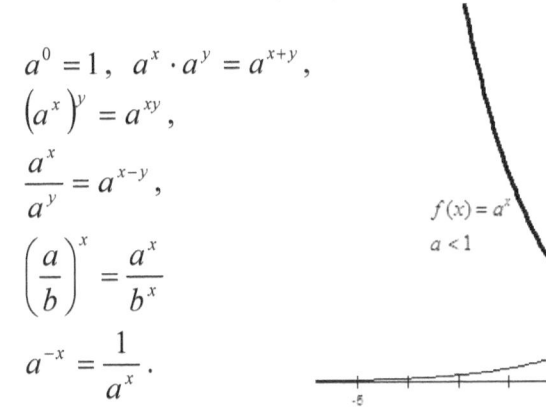

$a^0 = 1$, $a^x \cdot a^y = a^{x+y}$,

$\left(a^x\right)^y = a^{xy}$,

$\dfrac{a^x}{a^y} = a^{x-y}$,

$\left(\dfrac{a}{b}\right)^x = \dfrac{a^x}{b^x}$

$a^{-x} = \dfrac{1}{a^x}$.

LOGARITHMIC FUNCTION

If $a^x = y$, a solution of this equation is $x = \log_a y$, therefore

$a^{\log_a y} = y$.

$f : (0, \infty) \to \mathbf{R}$, $f(x) = \log_a x$, $a > 0$ and $a \neq 1$,

f decreases for $a \in (0, 1)$ and f increases for $a \in (1, \infty)$,

$f^{-1} : \mathbf{R} \to (0, \infty)$, $f^{-1}(x) = a^x$.

Laws of the Logarithmic function:

a) $\log_a 1 = 0$,

b) $\log_a xy = \log_a x + \log_a y$,

c) $\log_a \dfrac{x}{y} = \log_a x - \log_a y$,

d) $\log_a x^p = p \log_a x$, $p \in \mathbf{R}$

and $\log_a \sqrt[n]{x} = \dfrac{1}{n} \log_a x$, $n \in \mathbf{N}^*$,

$$\log_a a_1, a_2, ..., a_n = \log_a a_1 + \log_a a_2 + ... + \log_a a_n, \quad a_1, a_2, ..., a_n > 0,$$

$$\log_a \prod_{k=1}^{n} a_k = \sum_{k=1}^{n} \log_a a_k.$$

e) $\log_A B = \dfrac{\log_a B}{\log_a A}$, $\quad A, B > 0$, $\quad A \neq 1$.

Also $\log_{10} x = \log x$, $\log_e x = \ln x$, where
e = 2.718281828459045... which is the value of $(1 + 1/n)^n$ when n gets bigger and bigger.

VII. 1. Specify the domains of the functions:

a) $f(x) = \sqrt{(0.5)^x - 16}$; b) $f_1(x) = \sqrt{(0.2)^x - 25}$;

c) $f_2(x) = \sqrt[4]{5^x - 0{,}04}$; d) $f_3(x) = \sqrt[3]{3^{-3x} - \sqrt{3}}$;

e) $f_4(x) = \sqrt{a^x - a^{-x}}$; $a \in \mathbf{R}$; f) $f_5(x) = \sqrt[5]{4^x - 0{,}05}$.

VII. 2. Find the domains and ranges of the functions:

a) $f(x) = 2^{\sqrt{1-x^2}}$; b) $f(x) = 2^{-\sqrt{x^2-1}}$; c) $f_2(x) = \sqrt{3^x - 27}$;

d) $f_3(x) = \dfrac{1}{3^{|x|}}$; e) $f_4(x) = 4^x - 2^{x+3} + 18$.

VII. 3. Solve the equation $2^{|x+1|} - |2^x - 1| = 2^x + 1$.

VII. 4. Draw the graph of the following exponential functions:

a) $f : \mathbf{R} \to (0, \infty)$, $f(x) = 3^x$;

b) $f : \mathbf{R} \to [2, \infty)$, $f(x) = 2^x + 2^{-x}$;

c) $f : \mathbf{R} \to \mathbf{R}$, $f(x) = \dfrac{3^x - 3^{-x}}{2}$;

d) $f : \mathbf{R} \to (-1.5, \infty)$, $f(x) = 2 \cdot 3^{x-3} - 1.5$;

e) $f : \mathbf{R} \to (-\infty, 2)$, $f(x) = -3 \cdot 2^x + 2$.

VII. 5. Find the domain for the functions:

a) $f(x) = \log|x + 1|$; b) $f_1(x) = \log_{x+1}(9 - x^2)$;

c) $f_2(x) = \log_2 \sqrt[3]{x^2 - 5x - 6}$; d) $f_3(x) = \log_{x+1} \log_3 (10 - x^2)$;

e) $f_4(x) = \log_3 \left(\log_2 \left(\dfrac{6 + \sqrt[4]{x}}{4} \right) - \dfrac{\log_2 x}{4} \right)$;

f) $f_5(x) = \sqrt{\log_{\frac{1}{2}} \dfrac{x-1}{x+5}}$.

VII. 6. Graph $y = 2^x$ and $y = -2^{x-3} + 4$ on the same axis.

VII. 7. Graph $y = -3 \cdot \left(\dfrac{1}{2} \right)^{x-3} + 2$ and state the basic

(parent) function. Describe each transformation applied and what is the shape, domain, range, asymptotes, and y-intercept.

VII. 8. Given the function represented by the equation $y = 2^x$ state the equation of the graph that has been reflected in the x-axis, given a vertical stretch by a factor of 2, translated 3 units left and 6 units down.

VII. 9. Graph $y = \log_{\frac{1}{3}} x$ and $y = 2 \log_{\frac{1}{3}} (x-1) - 3$ on the same system of axis.

VII. 10. Graph $y = 8 \log_{\frac{1}{2}} \sqrt[4]{x+2} - 1$, state the basic

function and find a simplier form of the equation. Describe each transformation applied and what is the shape, domain, range, asymptotes, and y-intercept.

VII. 11. Given the function represented by the equation $y = \log_2 x$ state the equation of the graph that has been reflected in the x-axis, given a vertical stretch by a factor of 2, translated 3 units left and 5 units up.

VII. 12. Draw the graph of the following logarithmic functions and on the same graph draw the parent function specifying the transformations done:

a) $f:(1,\ \infty)\to \mathbf{R}$, $f(x)=\dfrac{1}{3}\log_2(x-1)-4$;

b) $f:\left(\dfrac{1}{3},\ \infty\right)\to(-2,\ \infty)$, $f(x)=\left|\log_2(3x-1)\right|-2$;

c) $f:\mathbf{R}\setminus\{0\}\to\mathbf{R}$, $f(x)=\log_2|x|+1$;

d) $f:(-\infty,\ 5)\to\mathbf{R}$, $f(x)=-3\log_3\left(1-\dfrac{x}{5}\right)-4$;

e) $f:(-2,\ \infty)\to\mathbf{R}$, $f(x)=\log_3\sqrt{x+2}-3$;

f) $f:(2,\ \infty)\to\mathbf{R}$, $f(x)=-5\log_2\dfrac{x-2}{3}+1$.

VII. 13. Compute: **a)** $\log_5 625$; **b)** $\log_3\dfrac{1}{2187}$;

c) $\log_{\frac{1}{2}}\left[\log_{\sqrt{6}}(\log_3 729)\right]$; **d)** $\log_2\left[\log_3\left(\log_{\frac{1}{2}}\dfrac{1}{512}\right)\right]$;

e) $\log_{\frac{49}{4}}\left(\log_4 6^{\log_6 128}\right)$; **f)** $\log_{\frac{2}{3}}\left(\log_5 3^{\log_3 \sqrt[3]{25}}\right)$; **g)** $\log_2\log_2\sqrt{\sqrt[4]{2}}$;

h) $\log_3\dfrac{81}{\sqrt{3}}-2\cdot\log_3 27$; **i)** $\dfrac{\log 20+\log 3-\log 6}{\log_{125}4+\log_{125}10-\log_{125}5}$;

j) $\dfrac{\log_3 6+\log_3 2-\log_3 4}{2\log_3 4-\log_3 2-\log_3 32}$.

VII. 14. Simplify the expressions:

a) $\dfrac{\log_6 x^3+\log_6 x^4}{\log_5 x^4+\log_5 x^3}$;

b) $\left(\log_a b+\log_b a+2\right)\left(\log_a b-\log_{ab} b\right)\log_b a-1$;

c) $\dfrac{\log_4\sqrt{a}+\log_4\sqrt{a^3}+\log_4\sqrt{a^5}}{\log_4\sqrt[3]{a^2}+\log_4\sqrt[3]{a^4}+\log_4\sqrt[3]{a^6}}$;

d) $\dfrac{1-\log_a^3 2}{\left(\log_a 2 + \log_2 a + 1\right)\log_a \dfrac{a}{2}}$;

e) $\dfrac{\log_a \sqrt[3]{a^2-1}\cdot\log_{a^{-2}} \sqrt{a^2-1}}{\log_{a^4}\left(a^2-1\right)\log_{\sqrt[6]{a}} \sqrt[4]{a^2-1}}$, $a>1$; $a\neq\sqrt{2}$;

f) $\left(b^{\frac{\log_{100} a}{\log_{10} a}} \times a^{\frac{\log_{100} b}{\log_{10} b}} \right)^{2\log_{ab}(a+b)}$;

g) $\left(9^{\frac{2}{\log_5 9}} + 3^{\frac{3}{\log_6 9}} \right)\left[\left(\sqrt{5}\right)^{\frac{2}{\log_{25} 5}} - 125^{\log_{25} 6} \right]$;

h) $\left(x^{1+\frac{1}{2\log_4 x}} + 8^{\frac{1}{3\log_{x^2} x}} + 1 \right)^{\frac{1}{2}}$;

i) $\dfrac{\left(3^{\frac{9}{\log_2 27}} + 5^{\log_{25} 49} \right)\left(2\cdot 81^{\frac{2}{\log_2 9}} - 8^{\log_4 9} \right)}{\left(3+5^{\log_{25} 16}\right)\cdot 5^{\log_5 3}}$;

j) $\dfrac{\log_a b\cdot\log_{ab} b^2}{b^{2\log_b(\log_a b)}-1}\cdot\dfrac{\log_a b + \log_a\left(b^{\frac{1}{3}\log_a b^3} \right)}{\log_a b^2 - \log_{ab} b^2}$;

k) $\log_2 2x^2 + \log_2 x\cdot x^{\log_x(\log_2 x+1)} + 8\log_4^2 x + 2^{3\log_2(\log_2 x)}$;

l) $\left(\sqrt{\log_a \sqrt[4]{ab}+\log_b \sqrt[4]{ab}} + \sqrt{\log_a \sqrt[4]{\dfrac{b}{a}}+\log_b \sqrt[4]{\dfrac{a}{b}}} \right)\cdot\sqrt{\log_a b}$.

VII. 15. Prove that

a) $\log_{ab} c = \dfrac{\log_a c\times\log_b c}{\log_a c+\log_b c}$; **b)** $\dfrac{\log_a x}{\log_{ab} x}=1+\log_a b$.

VII. 16. Calculate $\log_6 16$ in terms of $a = \log_{12} 27$.

VII. 17. Calculate $\log_3 18$ in terms of $a = \log_3 12$.

VII. 18. If $a = \log_{30} 3$ and $b = \log_{30} 5$, calculate $\log_{30} 16$ in terms of a and b.

VII. 19. If $a = \log_3 7$, $b = \log_7 5$, and $c = \log_5 4$, calculate $\log_3 12$ in terms of a and b.

VII. 20. If $\log_{10} 5 = a$ and $\log_{10} 3 = b$ calculate $\log_{30} 8$ in terms of a and b.

VII. 21. If $a = \log_{60} 3$ and $b = \log_{60} 5$, calculate the number $\log_3 4 + \log_{12} 2$ in terms of a and b.

VII. 22. If $\log_8 b = \dfrac{1}{1 - \log_8 a}$ and $\log_8 c = \dfrac{1}{1 - \log_8 b}$, calculate $\log_8 a$ in terms of $\log_8 c$.

VII. 23. If $a = \log_{14} 42$ and $b = \log_{14} 36$, calculate the number $\log_{49} 8$ in terms of a and b.

VII. 24. Let a, b, and c be positive non zero real numbers.
a) Prove that if $x = \log_a bc$, $y = \log_b ac$, and $x = \log_c ab$, then $x + y + z = xyz$;
b) If $2^a = 35$, $5^b = 14$, and $7^c = 10$, then $abc - 2 = a + b + c$.

VII. 25. Calculate:

a) $E = \left(N^{\frac{1}{\log_2 N}} \times N^{\frac{1}{\log_4 N}} \times N^{\frac{1}{\log_8 N}} \times \ldots \times N^{\frac{1}{\log_{512} N}} \right)^{\frac{1}{9}}$;

b) $F = 3^{\log_{3^{-1}} 2^{-1}} + \log_{\sqrt{2}} \dfrac{4}{\sqrt{3} + \sqrt{2}} + \log_{2^{-1}} \dfrac{1}{5 + 2\sqrt{6}}$.

VII. 26. Find a simple form for the expression

$$E = \left(\sqrt{3}\right)^{\frac{1}{\log_{\sqrt{2}} 3}} + \left(\sqrt{3}\right)^{\frac{1}{\log_{\sqrt{3}} 3}} + \left(\sqrt{3}\right)^{\frac{1}{\log_{\sqrt{4}} 3}} + \ldots + \left(\sqrt{3}\right)^{\frac{1}{\log_{\sqrt{n}} 3}}.$$

VII. 27. If $2\log_a(2x+3y) = \log_a 3x + 2\log_b 2y$,

then calculate the ratio $\dfrac{x}{y}$.

VII. 28. Prove that if $\log_3(2x+y) - 1 = \dfrac{\log_3 x + \log_3 y}{2}$,

then $4x^2 + y^2 = 5xy$, $x > 0$, $y > 0$.

VII. 29. For the non zero positive real numbers a and b,

$a \neq b$, prove the inequality $\log_2(a+b) > 1 + \dfrac{1}{2}(\log_2 a + \log_2 b)$.

VII. 30. Prove the inequality

$$\log_{a_3} \sqrt[n]{a_1 a_2} + \log_{a_4} \sqrt[n]{a_2 a_3} + \ldots + \log_{a_1} \sqrt[n]{a_{n-1} a_n} + \log_{a_2} \sqrt[n]{a_n a_1} \geq 2,$$

where $a_1, a_2, \ldots, a_n \in (0, \infty) \setminus \{1\}$.

VII. 31. Find x in the following exponential equations:

a) $\left(\dfrac{5}{9}\right)^{2x-3} = (1.8)^{3x-2}$; b) $2^{\frac{1}{\sqrt{x-1}}} \times \left(\dfrac{1}{2}\right)^{\frac{1}{\sqrt{x+1}}} = 2^{\frac{2\sqrt{x}}{x+\sqrt{x}}}$;

c) $9^{\sqrt{x-5}} - 27 = 6 \times 3^{\sqrt{x-5}}$; d) $2,5^{\frac{4+\sqrt{9-x}}{\sqrt{9-x}}} \times 0,4^{1-\sqrt{9-x}} = 5^{10} \times 0,1^5$;

e) $\sqrt[3]{5} \cdot 0,2^{\frac{1}{2x}} - 0,04^{1-x} = 0$.

VII. 32. Solve the equations:

a) $|x-2|^{x^2+4x} = (x-2)^5$;

b) $|x-3|^{x^2+3x} = (x-3)^4$;

c) $\left(x^2 - 5x + 6\right)^{x^2+3x} = 1$.

VII. 33. Solve the equations:

a) $\dfrac{7-2^x}{2^x-5} = \dfrac{2}{2^x-5} - 1$; b) $\dfrac{2\cdot5^x-3}{5^x-5} = \dfrac{7}{5^x-5} + 4$;

c) $\dfrac{3\cdot2^x-7}{2^x-8} = \dfrac{17}{2^x-8} + 2$; d) $\dfrac{41}{3^x-9} = \dfrac{5\cdot3^x-4}{3^x-9} - 5$;

e) $2^{2x+1} + 2^{2+x} = 160$; f) $2\cdot3^x + 3^{1-x} = 7$;

g) $2^{2x} - 3\cdot2^{x+1} - 8 = 0$; h) $2^x + 8\cdot2^{x-2} + 8^{\frac{x+2}{3}} = 7\cdot16^{\frac{2x-1}{4}}$;

i) $2^x + 4^{\frac{x+1}{2}} + 2\cdot2^{x+1} = 7\cdot8^{\frac{2x-1}{3}}$; j) $2^x + 4\cdot2^{x-1} + 8^{\frac{x+2}{3}} = 7\cdot2^{2x-1}$.

VII. 34. Solve the equations:

a) $2^{2-x^2}\cdot2^{7x} = 8^{-2}$; b) $\left(\dfrac{3}{4}\right)^{x-1}\sqrt[x]{\dfrac{4}{3}} = \dfrac{\sqrt{3}}{2}$; c) $\left(\dfrac{4}{3}\right)^{x-1}\sqrt{\dfrac{3}{4}} = \dfrac{2}{\left(\sqrt[4]{3}\right)^{3x-4}}$;

d) $5^{x^2-1}\cdot2^{2x^2} = 4\cdot3^{x^2-1}$; e) $\sqrt{5}\cdot0.04^{\frac{1}{4x}} - 0.04^{1-x} = 0$;

f) $\sqrt{2^x\cdot\sqrt[3]{4^x\cdot0.125^{\frac{1}{x}}}} = 4\cdot\sqrt[3]{2}$; g) $3\cdot4^x + \dfrac{1}{3}\cdot9^{2+x} = 6\cdot4^{x+1} - \dfrac{1}{2}\cdot9^{1+x}$.

VII. 35. Solve the equations:

a) $3^{x+1} + 3^x = 36$; b) $2^{x+1} - 5\cdot2^{x-2} - 12 = 0$;

c) $2^{x+2} - 2^{x-1} = 56$; d) $5^x + 5^{x-1} + 5^{x-2} = 155$;

e) $6^{2x+2} + 6^{2x-1} = 1302$; f) $4\cdot3^x - 5\cdot3^{x-1} + 3^{x-2} = 66$;

g) $3^{x-1} + 3^{x-2} - 4\cdot3^{x-4} = 32$; h) $2^{2x-1} + 2^{2x-2} + 2^{2x-3} = 224$;

i) $2^{x+3} + 2^{x+2} + 3\cdot2^{x+1} + 2^x = 38$;

j) $11\cdot2^{x+4} - 9\cdot2^{x+3} - 7\cdot2^{x+2} - 5\cdot2^{x+1} - 3\cdot2^x = 504$.

VII. 36. Solve the equations:

a) $2^{x+1} + 2^x + 8\cdot2^{x-1} = 6^{x+1} + 6^x$; b) $3^{x+1} + 5^{x+2} = 3^{x+5} - 3\cdot5^{x+3}$;

c) $2^{x+3} + 3\cdot2^{x+1} = 7^x + 7^{x-1}$;

d) $14 \cdot 7^{2x-1} - 7^{2x} + 7^{2x+1} = 5 \cdot 5^{2x-1} + 2 \cdot 5^{2x} + 5^{2x+1}$ '

VII. 37. Solve the equations:

a) $3 \cdot 5^{x-2008} + 5^{x-2009} + 2 \cdot 5^{x-2012} = 2012$;

b) $9 \cdot 6^{x-2000} + 6^{x-2001} + 3 \cdot 6^{x-2002} + 4 \cdot 6^{x-2003} = 2003$.

VII. 38. Solve the equations:

a) $2 \cdot 2^{2x} - 3 \cdot 2^x + 1 = 0$; **b)** $\dfrac{2^x + 10}{4} = \dfrac{9}{2^{x-2}}$;

c) $9^{x^2-1} - 36 \times 3^{x^2-3} + 3 = 0$; **d)** $3^{x+3} + 9 = \sqrt{9^{4x-2}} + 3^{4x}$;

e) $9^{\sqrt{x+5}} - 27 = 6 \cdot 3^{\sqrt{x+5}}$; **f)** $2^{x+\sqrt{x^2-4}} - 5 \cdot \sqrt{2}^{\,x-2+\sqrt{x^2-4}} = 6$.

VII. 39. Solve equations:

a) $6^{2x+1} - 6^{2x-1} = 210$; **b)** $9^{1+x^2} - 9^{1-x^2} = 80$;

c) $5^{2x-1} + 2 \cdot 5^{x-1} = 5^x + 2$; **d)** $27^x - 13 \times 9^x + 13 \times 3^{x+1} - 27 = 0$;

e) $\sqrt[5]{3}^{\,x} + \sqrt[10]{3}^{\,x-10} = 84$; **f)** $3^{3x} + 3^{-3x} + 3^{x+1} + 3^{1-x} = 8$;

g) $3 \cdot 5^{2x-1} - 2 \cdot 5^{x-1} = 5^{-1}$.

VII. 40. Solve equations:

a) $\left(2 + \sqrt{3}\right)^x + \left(2 - \sqrt{3}\right)^x = 14$;

b) $\left(5 + 2\sqrt{6}\right)^x + \left(5 - 2\sqrt{6}\right)^x = 10$;

c) $\left(4 + \sqrt{15}\right)^x + \left(4 - \sqrt{15}\right)^x = 62$;

d) $\left(\sqrt{7 + 4\sqrt{3}}\right)^x + \left(\sqrt{2 - 4\sqrt{3}}\right)^x = 14$;

e) $\left(6 + \sqrt{35}\right)^x + \left(6 - \sqrt{35}\right)^x = 2\sqrt{35}$;

f) $\left(2 + \sqrt{3}\right)^{x^2-2x+1} + \left(2 - \sqrt{3}\right)^{x^2-2x-1} = 4 \cdot \left(2 - \sqrt{3}\right)^{-1}$;

g) $\left(49 + 10\sqrt{24}\right)^x + \left(49 - 10\sqrt{24}\right)^x = 980$;

h) $\left(3 + \sqrt{5}\right)^x + \left(3 - \sqrt{5}\right)^x = 3 \cdot 2^x$.

VII. 41. Solve the exponential equations:

a) $6 \cdot 9^x - 13 \cdot 6^x + 6 \cdot 4^x = 0$;

b) $3^{2x+4} + 45 \cdot 6^x - 9 \cdot 2^{2x+2} = 0$;

c) $7 \cdot 4^{x^2} - 9 \cdot 14^{x^2} + 2 \cdot 49^{x^2} = 0$;

d) $5^{2x-1} + 2^{2x} + 2^{2x+2} = 5^{2x}$;

e) $6 \cdot 9^{\frac{1}{x+1}} + 6 \cdot 4^{\frac{1}{x+1}} - 13 \cdot 6^{\frac{1}{x+1}} = 0 \quad x \neq -1$;

f) $27^x + 12^x = 2 \cdot 8^x$;

g) $2^{2x} - 6^{x-1} - 2 \cdot 3^{2x} = 0$;

h) $2 \cdot 3^{4x} = 6^{2x} + 3 \cdot 4^{2x}$;

i) $3^{3x} - 3 \cdot 18^x - 12^x + 3 \cdot 2^{3x} = 0$.

VII. 42. Prove that the number $\log_5 10$ is not a rational number.

VII. 43. Solve the equations:

a) $2 \log_x 9 = 2$; b) $\log_7 (2x+1) = 1$;

c) $\log_{x+1} (x^2 + 3x) = 2$; d) $\log_3 (10 - x) = 2$;

e) $\log_6 (x+3) - \log_6 (2x-5) = 1$;

f) $\log(3x-2) = 2$; g) $\log_6 (x^2 + x + 1) = \log_6 (2x+3)$;

h) $\log_{2x+1} (x^2 + x + 2) = \log_{2x+1} (x+3)$;

i) $\log_{x+0.5} (5x+3) = \log_{x+0.5} (3x+9)$.

VII. 44. Solve the equations:

a) $\log_x (x^2 - 3x) = 1$; b) $x \log_{x+2} 7 \cdot \log_{\frac{1}{\sqrt[3]{7}}} (x+2) = \frac{x-4}{x}$;

c) $3 \log_8 (x-2) = \log_2 \sqrt{2x-1}$;

d) $\log_3 (x^3 + 1) - 0.5 \log_3 (x^2 + 2x + 1) = \log_3 7$;

e) $\log_{10} x + \log_{\sqrt{10}} x + \log_{\sqrt[3]{10}} x + ... + \log_{\sqrt[10]{10}} x = 5,5$;

f) $\log_4 x + \log_{16} x = \log_{\frac{1}{4}} \sqrt{3}$; g) $\log_3 \left[\log_{\frac{1}{2}} (x+1) \right] = 1$;

h) $\log_4 (\log_2 x) + \log_2 (\log_4 x) = 2$.

VII. 45. Solve the equations:

a) $\left(\log_2 x\right)^2 + \log_2(4x) = 4$;   **b)** $\log_2\left(\log_3\left(x^2+1\right)\right) = 1$;

c) $\log_2\left(\log_3\left(x^2+1\right)\right) = 1$;   **d)** $\log_5^2 x - \log_5 x - 2 = 0$;

e) $\log_5^2 x + \log_5 7 \cdot \log_7 x = 2$;

f) $\log\sqrt{1+x} + 3\log\sqrt{1-x} = \log\sqrt{1-x^2} - 2$;

g) $\left(3\log_a x - 2\right)\log_x^2 a = \log_{\sqrt{a}} x - 3$, $\left(a > 0, \ a \neq 1\right)$;

h) $\log_5^4(x-2)^2 + \log_5^2(x-2)^3 = 2$.

VII. 46. Solve the equations:

a) $\log_6(2+x) + 2\log_6\sqrt{3+x} = 1$;

b) $3\log_2(2+x) + \log_2(x-4)^3 = 0$;

c) $\log_4 x + \log_2 x = \log_{\frac{1}{3}} \dfrac{1}{27}$;

d) $3\log_8(x-2) = 2\log_4\sqrt{2x-1}$;

e) $\dfrac{\log_a(x-1)}{\log_a(x^2-4)} = \dfrac{1}{2}$;   **f)** $\dfrac{\log_5\left(\sqrt{2x-3}+1\right)}{\log_5\left(\sqrt{2x-3}+7\right)} = \dfrac{1}{2}$.

VII. 47. Solve the equations :

a) $\log_x\sqrt{5} + \log_x 5x = \left(\log_x\sqrt{5}\right)^2 + 2.25$;

b) $\left(\log_{\frac{1}{3}} 9x\right)^2 + \log_3 \dfrac{x^2}{27} = 8$;

c) $\sqrt{\log_a x} + \sqrt{\log_x a} = \dfrac{10}{3}$;

d) $\dfrac{1+2\log_9 2}{\log_9 x} = 1 + \log_{\sqrt{x}} 3 \cdot \log_9(12-x)$;

e) $\dfrac{2 - 4\log_{12}^2}{\log_{12}(x+2)} = 1 + \dfrac{\log_6(8-x)}{\log_6(x+2)}$;

f) $\dfrac{\log_{4\sqrt{x}} 2}{\log_{2x} 2} + \log_{2\sqrt[3]{x}} 2 \cdot \log_{1/2} 2 \cdot \sqrt[3]{x} = 0$;

g) $\log_{2x+1}\left(x^3 - 9x + 8\right) \cdot \log_{x-1}(2x+1) = 3$.

VII. 48. Solve the equations:

a) $\log_8(4^{x^2-1} - 1) + \dfrac{2}{3} = \log_8(2^{x^2+2} - 7)$;

b) $\log_2\left(9^{x-2} + 7\right) = 2 + \log_2\left(3^{x-2} + 1\right)$;

c) $\log_3\left(\sqrt[3]{9}\right) + \log_8\left(9^{x+1} - 1\right) = 1 + \log_8\left(3^{x+1} + 1\right)$;

d) $\log(3^x + x - 5) = x\log 30 - x$;

e) $\log 2 + \log(4^{x-2} + 9) = 1 + \log(2^{x-2} + 1)$;

f) $\log_2\left(4^x + 1\right) = x + \log_2\left(2^{x+3} - 6\right)$.

VII. 49. Solve the equations:

a) $\sqrt{\log_x \sqrt{4x}} + \log_x 4 = 0$; b) $\sqrt{\log_x \sqrt{3x}} \cdot \log_3 x = -1$;

c) $\sqrt{\log_2\left(2x^2\right) \cdot \log_4(16x)} = 3 \cdot \log_4 x$;

d) $2 \cdot \sqrt[3]{2\log_{16}^2 x} - \sqrt[3]{\log_2 x} - 6 = 0$; e) $\sqrt{(2x)^{\log\sqrt{2x}}} = 10$;

f) $\sqrt{\log_5 x} + \sqrt[3]{\log_5 x} = 2$.

VII. 50. Solve the equations:

a) $\log_x x\sqrt[3]{4} + 3\log_x\left(x\sqrt[3]{2}\right) + \log_x^2 \sqrt[3]{4} = 13$;

b) $\log_x^2 6 + 3\log_{\frac{1}{6}}^2 \dfrac{1}{x} + \log_{\frac{1}{\sqrt{x}}} \dfrac{1}{6} + \log_{\sqrt{6}} x + \dfrac{3}{4} = 0$;

c) $\log_x \sqrt{5} + \log_x 5x = \left(\log_x \sqrt{5}\right)^2 + 2.25$;

d) $2 - \log_{a^2}(1+x) = 3\log_a \sqrt{x-1} - 2\log_{a^2}\left(x^2 - 1\right)$, $a \in (0,\infty)\setminus\{1\}$;

e) $\log_x^3 + \log_3 x = \log_{\sqrt{x}} 3 + \log_3 \sqrt{x} + 0.5$;

f) $\log_a x + \log_{a^2} x + \log_{a^3} x = 11$, $a \in (0,\infty)\setminus\{1\}$;

g) $2\log_x \sqrt[4]{2} - \left(\log_x \sqrt{2}\right)^2 = \log_2 8 - \log_x(2x)$.

VII. 51. Solve the equations:

a) $\log_{3x} \dfrac{3}{x} + \log_3^2 x = 1$; **b)** $3\log_x 4 + 2\log_{4x} 4 + 3\log_{16x} 4 = 0$;

c) $5\log_{\frac{x}{9}} x + 2\log_9 x^3 + 8\log_{9x^2} x^2 = 2$;

d) $30\log_{9x} \sqrt[3]{x} + 7\log_{81x} x^3 - 3\log_{\frac{x}{3}} x^2 = 0$;

e) $3\log_x a^2 + 3\log_{a^2x} a - \dfrac{14\log_{ax} a}{\log_a x} = 0$, $a \in (0,\infty)\setminus\{1\}$;

f) $\log_{\frac{2}{x^2}} x + \log_{\frac{2}{x}} x^2 - 4\log_{2x} x^2 = 0$.

VII. 52. Solve the equations:

a) $x^{2+\log_3 x} = 3^8$; **b)** $\log_2 3 + 2\log_4 x = x^{\frac{\log_9 16}{\log_3 x}}$;

c) $\sqrt{x^{\log_{10} \sqrt{x}}} = 10$; **d)** $\left(\sqrt[4]{x}\right)^{\log x+7} = 10^{\log x+1}$;

e) $x^{2\log^3 x - 1.5\log x} = \sqrt{10}$.

VII. 53. Solve in **R** the ecuations:

a) $343^x + 512^x + 729^x = 3 \cdot 504^x$;

b) $9^x + 36^x + 2 \cdot 32^x = 4 \cdot 24^x$.

VII. 54. Solve the systems of exponential equations:

a) $\begin{cases} x^{x+y} = y^{x-y} \\ x^2 y = 1 \end{cases}$; **b)** $\begin{cases} y = 2^x \\ y^x = 16 \end{cases}$; **c)** $\begin{cases} 2^x \cdot 3^y = 18 \\ 2^y \cdot 3^x = 12 \end{cases}$;

d) $\begin{cases} \dfrac{2^{2x+1}+1}{2^{2x}+2} - 4^x = \dfrac{y}{2^{x+1}+4} \\ 2^{3x+2} + y^2 = 4 \end{cases}$; **e)** $\begin{cases} x^{y+1} = 27 \\ x^{2y-5} = \dfrac{1}{3} \end{cases}$;

f) $\begin{cases} 3^x + 3^y = 12 \\ 3^{x+y} = 27 \end{cases}$; **g)** $\begin{cases} 4^{x+y} = 175 + 9^{x-y} \\ 5 \cdot 2^{x+y} + 2 \cdot 3^{x-y+1} = 134 \end{cases}$;

h) $\begin{cases} x^{\sqrt[4]{x}+\sqrt{y}} = y^{\frac{8}{3}} \\ y^{\sqrt[4]{x}+\sqrt{y}} = x^{\frac{2}{3}} \end{cases}$; i) $\begin{cases} 4^{\sqrt[3]{x}}3^{\sqrt{y}} = 144 \\ 4^{2\sqrt[3]{x}} + 3^{2\sqrt{y}} = 337 \end{cases}$;

j) $\begin{cases} 3^{\sqrt[3]{x}} \cdot 5^{\sqrt[4]{y-2}} = 45, \\ 9^{\sqrt[3]{x}} + 25^{\sqrt[4]{y-2}} = 106 \end{cases}$; k) $\begin{cases} z^x = x \\ z^y = y \\ y^y = x \end{cases}$.

VII. 55. Solve the systems of logarithmic equations:

a) $\begin{cases} \log_x (x+y) = 2 \\ \log_x (4x-2y) = 2 \end{cases}$; b) $\begin{cases} \log_x y + \log_y x = 2 \\ x^2 - 2y = 8 \end{cases}$;

c) $\begin{cases} \log_3 x + 2\log_3 y = 5 \\ 3\log_3 x - \log_9 y^2 = 1 \end{cases}$; d) $\begin{cases} 2^x + 2^y = 20 \\ \log_2 x + \log_2 y = 3 \end{cases}$;

e) $\begin{cases} \log_3 x + \log_2 y = 2 \\ 3^x - 2^y = 23 \end{cases}$; f) $\begin{cases} x^{\log_3 y} + 2y^{\log_3 x} = 27 \\ \log_3 x - \log_3 y = -1 \end{cases}$;

g) $\begin{cases} \log_3 \dfrac{\sqrt[4]{x^3}}{\sqrt[3]{y}} + \log_3 \dfrac{\sqrt[4]{y^3}}{\sqrt[3]{x}} = \dfrac{5}{3} \\ x^2 + y^2 = 738 \end{cases}$;

h) $\begin{cases} 9^{\sqrt[4]{xy^2}} - 27 \cdot 3^{\sqrt{y}} = 0 \\ \dfrac{1}{4}\log_4 x + \dfrac{1}{2}\log_4 y = \log_4 \left(4 - \sqrt[4]{x}\right) \end{cases}$;

i) $\begin{cases} 2^x \cdot 3^y + 2^z = 124, \\ 3^y \cdot 2^z + 2^x = 436, (x, y, z \in \mathbf{Z}); \\ 2^z \cdot 2^x + 3^y = 91; \end{cases}$ j) $\begin{cases} x + y = 7 \\ \log_3 x + \log_3 y = 1 + 2\log_3 2 \end{cases}$;

k) $\begin{cases} \log_3 x + 2\log_3 y = 5 \\ 3\log_3 x - \log_3 y = 1 \end{cases}$; l) $\begin{cases} x^2 + y^2 = 34 \\ \log_5 x + \log_5 y = \log_5 15 \end{cases}$;

m) $\begin{cases} xy = 27 \\ x^{\log_3 y} = 9 \end{cases}$; n) $\begin{cases} \log_{10}\left(x^2 + y^2\right) = 2 \\ \log_2 x + 2 = \log_2 3 + \log_2 y \end{cases}$;

o) $\begin{cases} \log_{12} x \cdot \left(\dfrac{1}{\log_x 2} + \log_2 y \right) = \log_2 x \,; \\ \log_2 x \cdot \left[\log_3 (x+y) \right] = 3\log_3 x \end{cases}$

p) $\begin{cases} \left(\log_a x \right)^2 + \left(\log_a y \right)^2 = 2m, \\ x^{\log_a x} + y^{\log_a y} = 2 \cdot \dfrac{p^2 + q^2}{p^2 - q^2} a^m, \end{cases} \quad (p \neq q).$

VII. 56. Solve the system of equations

$\log_{10} \left(x^3 - x^2 \right) = \log_5 y^2$

$\log_{10} \left(y^3 - y^2 \right) = \log_5 z^2$

$\log_{10} \left(z^3 - z^2 \right) = \log_5 x^2 \quad$ where $x, y, z > 1$.

VII. 57. Solve the exponential inequations:

a) $4^x - 6 \cdot 2^x + 8 \leq 0$; **b)** $0.25^{\sqrt{x}} + 2 > 3 \cdot 0.5^{\sqrt{x}}$;

c) $4^x \leq 3 \cdot 2^{\sqrt{x+x}} + 4^{1+\sqrt{x}}$; **d)** $0.008^x + 5^{1-3x} + 0.2^{3+3x} < 30.04$;

e) $8 \cdot 3^{\sqrt{x}+\sqrt[4]{x}} + 9^{1+\sqrt[4]{x}} \geq 9^{\sqrt{x}}$; **f)** $3^{2|x-1|} + 3 < 4 \cdot 3^{|x-1|}$;

g) $\sqrt{2^3 + 2^{1+\sqrt{3-x}}} - 2^{\sqrt{12-4x}} + 2^{1+\sqrt{3-x}} > 5$;

h) $x^2 + (2 \cdot 3^x - 5)x + 9^x - 5 \cdot 3^x + 4 > 0$;

i) $\dfrac{1}{3} \leq \dfrac{3^x + 1}{3^{2x} + 3} \leq \dfrac{1}{2}$; **j)** $\sqrt[4]{x^{3x}} < \sqrt{x}^{x^2 - x + 1}$.

VII. 58. Solve the logarithmic inequations:

a) $\log_3 \left(x^2 + 5x + 4 \right) < 0$; **b)** $\log_3 \left[-\log_{1/2} \left(2 - \log_4 x \right) - 1 \right] \leq 1$;

c) $\log_2 \left(2^x - 2 \right) \cdot \log_{1/2} \left(2^{x+1} - 4 \right) < -2$;

d) $\dfrac{1}{\log_{\frac{1}{2}} \sqrt{x+3}} \leq \dfrac{1}{\log_{\frac{1}{2}} \sqrt{x+1}}$;

e) $\dfrac{\ln \left(2x^2 - 3x + 1 \right)}{x^2 - 3x} \geq 0$; **f)** $\log_{\frac{x}{2}} 8 + 3 \log_{\frac{x}{4}} 2 < \dfrac{2 \log_2 x^2}{\log_2 x^2 - 4}$;

g) $\log_3 \log_4 \dfrac{2x-1}{x-1} - \log_{\frac{1}{3}} \log_{\frac{1}{4}} \dfrac{x-1}{2x-1} < 0$;

h) $\log_{\frac{1}{2}} \log_8 \dfrac{x^2 - 2x}{x-3} \le 0$; **i)** $\log_2 \log_{1/2}\left(2^x - 4^x\right) \le 1$;

j) $\left(\log_2 x\right)^4 - \left(\log_{1/2} \dfrac{x^3}{8}\right)^2 + 45 < 4\left(\log_{1/2} x\right)^2 + 9\log_2 x^2$.

VII. 59. Prove the inequality:

$$(ab)^{\sqrt{\log_a c \cdot \log_b c}} + (bc)^{\sqrt{\log_b a \cdot \log_c a}} + (ac)^{\sqrt{\log_a b \cdot \log_c b}} \ge a^2 + b^2 + c^2, \text{ for any}$$
$a, \ b, \ c > 1$.

VII. 60. Solve the equations:

a) $3^x + 4^x = 5^x$; **b)** $\sqrt{6^x} + \sqrt{8^x} = \sqrt{10^x}$;

c) $3 \cdot 5^{2x-1} - 7 \cdot 2^{4x-3} = 19$.

VII. 61. Solve the equations:

a) $\log_3\left(2^x + 1\right) = \log_2\left(3^x - 1\right)$;

b) $\log_5\left(4^x + 1\right) = \log_4\left(5^x - 1\right)$;

c) $e^x - e^{-x} = 2\ln\!\left(x + \sqrt{x^2 + 1}\right)$.

VII. 62. Solve the equation $\left(2^{\log_5 x} + 3\right)^{\log_5 2} = x - 3$.

VII. 63. Consider $a, b, c \in (0,1)$, $a \le b \le c$. Compare the

numbers: $A = \log_b a + \log_c b + \log_a c$ and
$$B = \log_a b + \log_b c + \log_c a.$$

VII. 64. Find the real root(s) of the equation
$25^{[x]} + 5^x = 6 \cdot 5^{[x]}$, where $[x]$ is the integer (floor) of x.

Chapter VIII

ARITHMETIC AND GEOMETRIC PROGRESSIONS

ARITHMETIC PROGRESSIONS

$a_1, a_2, a_3, ..., a_n$; a_n is the nth term, d is the ratio or common difference.

$$a_n = a_{n-1} + d,$$

$$a_n = a_1 + (n-1)d,$$

$$a_k = \frac{a_{k-1} + a_{k+1}}{2}, \quad k = 2,3,4,.., n-1,$$

$$a_1 + a_n = a_i + a_{n-i+1}, \quad i = 1,2,3,.., n'$$

$$S_n = a_1 + a_2 + ... + a_n = \sum_{k=1}^{n} a_k$$

$$S_n = \frac{(a_1 + a_n)n}{2} = \frac{[2a_1 + (n-1)d]n}{2}$$

VIII. 1. The general term of a sequence is given. Write the first five terms.

a) $t_n = 2n$; **b)** $t_n = 10 - n$; **c)** $t_n = 3n$; **d)** $t_n = 2^n$;

e) $t_n = 1 + 2n$; **f)** $t_n = 11n$; **g)** $a_n = 3n + 1$;

h) $a_n = n^2 - 2$; **i)** $a_n = \frac{(-1)^n}{n!}$; **j)** $y_n = (-1)^{n-1}\frac{1}{n} + (-1)^n n$;

k) $z_n = \frac{n}{n^2 + 1}$; **l)** $c_n = \left(1 + \frac{1}{n}\right)^n$; **m)** $a_n = \frac{1 + (-1)^n + (-1)^{n+1}}{n}$.

VIII. 2. Find the formula for the general term for each of the sequences:

a) $1, -\frac{1}{2}, \frac{1}{3}, -\frac{1}{4}, ...$; **b)** $\frac{1}{2}, \frac{1}{5}, \frac{1}{10}, \frac{1}{17}, \frac{1}{26}, ...$;

c) $\frac{2 \cdot 4}{1 \cdot 3}, \frac{4 \cdot 6}{3 \cdot 5}, \frac{6 \cdot 8}{5 \cdot 7}, ...$; **d)** $\frac{2}{5}, \frac{6}{25}, \frac{18}{125}, ...$;

e) $b_1 = 1$, $b_{n+1} = b_n - \dfrac{1}{n(n+1)}$; **f)** $a_1 = 1$, $a_{n+1} = (n+1)a_n$;

g) $x_1 = \dfrac{2}{3}$, $x_{n+1} = x_n + \dfrac{1}{(n+2)(n+3)}$.

VIII. 3. Describe each sequence and write an expression for the general term, t_n:

a) 1, 3, 5, 7, 9,...; **b)** 6, 12, 18, 24, 30,...;
c) 4, 9, 14, 19, 24,...; **d)** 10, 100, 1000, 10000,...;
e) 5, 50, 500, 5000,...

VIII. 4. Find the indicated terms in the sequences:

a) $t_n = 12 + 2n$, t_{10} and t_{12}; **b)** $t_n = 6n + 7$, t_2 and t_5;
c) $t_n = n^2 - 4$, t_4 and t_9; **d)** $t_n = (-2)^n$, t_2 and t_5;
e) $t_n = \dfrac{n+1}{2n+1}$, t_3 and t_6; **f)** $t_n = 1 + \dfrac{1}{3^n}$, t_3 and t_4.

VIII. 5. In a sequence $a_1 = 1$, $a_2 = 6$, $a_3 = 15$, $a_4 = 28$. Find the formula for general term.

VIII. 6. In a sequence the kth is given by the formula $t_k = 2^k - 2k + 1$. Determine the sum of the first n terms.

VIII. 7. Find the first three terms of a series for which:

a) $S_n = 3n$; **b)** $S_n = 3n^2 - n$; **c)** $S_n = 2n^2 - 3n$;
d) $S_n = 5n - 3$; **e)** $S_n = n^2 + 2n$.

VIII. 8. For a given series, $S_5 = 44$. Find t_6 if S_6 is:
a) 40; **b)** 60; **c)** 76; **d)** 84.

VIII. 9. The sum of the first n terms of a series is given.

a) $S_n = n^2 - n$; **b)** $S_n = 3n^2 - 5n$; **c)** $S_n = 2^n - 1$;
d) $S_n = 2n^2 - 3n$; **e)** $S_n = 2(3^n - 1)$; **f)** $S_n = n^2 + 2n$.

Find: **i)** S_{n-1}; **ii)** t_n.

VIII. 10. The sum of the first n terms of a series is given.
Find: **i)** S_{n-1}; **ii)** t_n.
a) $S_n = n^2 - n$; **b)** $S_n = 3n^2 - 5n$; **c)** $S_n = 2^n - 1$;
d) $S_n = 2n^2 - 3n$; **e)** $S_n = 2(3^n - 1)$; **f)** $S_n = n^2 + 2n$.

VIII. 11. Given: $3 + 7 + 11 + 15 + \ldots$
a) Find: **i)** t_{20} and t_n; **ii)** S_{20} and S_n;
b) How many terms are less than 500?

VIII. 12. For the arithmetic sequence $-8, -3, 2, 7, \ldots$, find:
a) t_{17} and t_{33};
b) what term is 242.

VIII. 13. Find the missing terms in the arithmetic sequence:
a) _, _, 8, 2, _; **b)** _, 9, 16, _, _; **c)** 12, _, 22, _, _;
d) 3, _, _, 24, _; **e)** _, −4, _, _, − 19; **f)** 15, _, _, _, − 21.

VIII. 14. Find the missing terms for the arithmetic
progression
$a_1, a_2, a_3, 2, 6, 10, \ldots$

VIII. 15. Consider the sequence (a_n), $n > 1$, with the nth
term given by the formula $a_n = 3n^2 + 2$.
Which of the number is a term of this sequence?
a) 200; **b)** 256; **c)** 434; **d)** 4109.

VIII. 16. Which of the expressions S_n is the sum of the first
n terms of:

a) $1 + 4 + 7 + \ldots$? **i)** $S_n = \dfrac{n^2 - 7n}{2n}$; **iv)** $S_n = 3n^2 - 6n$;

b) $2 + 5 + 8 + \ldots$? **ii)** $S_n = \dfrac{3n^2 - n}{2}$; **v)** $S_n = 2n^2 + 5n$;

c) $7 + 11 + 15 + \ldots?$ iii) $S_n = 3n^2 + 2n$;   vi) $S_n = \dfrac{7n^2 + 3n}{2}$.

d) $5 + 12 + 19 + \ldots?$

VIII. 17. Consider the sequence (a_n), where $a_0 = 2$, $a_1 = 3$, for any positive integer k, such that $a_{k+1} = 3a_k - 2a_{k-1}$. Find the formula for the general term.

VIII. 18. Consider the sequence of real numbers (a_n), where $a_1 = 1$, $a_2 = 4$ and $a_{k+1} = 2a_k - a_{k-1} + 2$ for any k, a positive integer, find the formula for general term.

VIII. 19. In an arithmetic sequence, the third term is 11 and the eighth term is 46. Find the sequence.

VIII. 20. The third term of an arithmetic sequence is 14 and the ninth term is -1. Find the twentieth term.

VIII. 21. A sequences is given by $a_1 = 1$ and $a_{n+1} = a_n + n^2 - n + 1$; find a_{100}.

VIII. 22. Find the first five terms of the arithmetic progression (a_n) and write the formula for the nth term:
a) $a_1 = 2$, $d = -3$;   **b)** $a_{10} = 20$, $d = 2$;
c) $a_5 = 27$, $a_{27} = 60$;   **d)** $a_3 = 14$, $a_6 = 29$.

VIII. 23. Find the sum of the 15 terms of an arithmetic sequence if the middle term is 92.

VIII. 24. How many terms are in each sequence?
a) $2, 6, 10, \ldots, 122$;   **b)** $-9, -4, 1, \ldots, 171$;
c) $4, 15, 26, \ldots, 444$;   **d)** $17, 12, 7, \ldots, -83$;
e) $5, 12, 19, \ldots, 222$;   **f)** $4, 12, 20, \ldots, 796$.

VIII. 25. For the arithmetic series $5 + 11 + 17 + 23 +...$, find:
a) the 30th term;
b) the sum of the first 30 terms.

VIII. 26. For an arithmetic progression $(a_n)_{n>1}$ find a_{10} and S_{12} if $a_5 = 55$ and $a_{15} = 5$.

VIII. 27. The 9th term of an arithmetic sequence is 45. If the common difference is $- 5$, find the first three terms.

VIII. 28. Calculate the sum $S_n = 1 + 5 + 9 + ... + 101$.

VIII. 29. Find the sum of all positive multiples of 7 that are less than 100.

VIII. 30. If $5 + 3x$, 8, and $1 + 2x$ are consecutive terms in an arithmetic sequence, find x.

VIII. 31. Find x such that the numbers $x - 2, x^2, x + 6$ are in an arithmetic progression.

VIII. 32. If $x + 2y + 1$, $2x - 4y - 2$, and $2x - 5y - 7$ are consecutive terms in an arithmetic sequence, find the relation between x and y.

VIII. 33. Find the 50th term of the arithmetic progression (a_n) if $a_{48} = \sqrt{18 - 8\sqrt{2}}$ and $a_{49} = \sqrt{27 - 10\sqrt{2}}$.

VIII. 34. For the arithmetic series $44 + 41 + 38 + 35 + ...$, find the sum of the first:
a) 15 terms; **b)** 30 terms; **c)** 60 terms.

VIII. 35. If the sum of the first and the fifth term of an arithmetic progression is equal to 8 and the product of the third term and the fifth term is equal 80, find the sum of the first twenty terms of the progression.

VIII. 36. The sum of the first 5 terms of an arithmetic series is 85 and the sum of the first 6 terms is 123. Find the series.

VIII. 37. For an arithmetic progression $(a_n)_{n>1}$
$a_1 + a_2 + a_3 = 14$ and $a_2 + a_3 + a_4 = 28$, find a_1, d, and S_{10}.

VIII. 38. For an arithmetic series the fifth term is 16 and the sum of 20 terms is 650. Find a_1 and d.

VIII. 39. Find a_1 and d for an arithmetic progression if:

a) $\begin{cases} a_2 - a_6 + a_4 = -2 \\ a_7 - a_5 = 10 \end{cases}$; b) $\begin{cases} a_1 a_2 a_3 = 312 \\ a_1 + a_2 + a_3 = 24 \end{cases}$.

VIII. 40. In an arithmetic progression $a_1, a_2, a_3, ..., a_{10}$, $S_{10} = 140$ and $a_2 a_9 = 147$. Find the progression.

VIII. 41. Find a_1 and d for an arithmetic progression if $a_1 + a_2 + a_3 + a_4 = 36$ and $a_1^2 + a_2^2 + a_3^2 + a_4^2 = 324$.

VIII. 42. The sum of the terms of an arithmetic progression is 1,050, the first term is 5 and the last term is 100. Find the common difference.

VIII. 43. In an arithmetic series, $a_1 = 6$ and $S_9 = 108$. Find the common difference and the sum of the first 20 terms.

VIII. 44. Find the general term of an arithmetic progression if the sum $S_n = 3n^2 - 5n$.

VIII. 45. Find a_1 and d for an arithmetic progression knowing: a) $S_n = 5n^2 + 6n$; b) $S_n = 3n^2 + 4n$.

VIII. 46. In an arithmetic progression the sum of three consecutive terms is 6, and the sum of the squares of the terms is 62. Find the numbers.

VIII. 47. Let a_n, a_{n+1}, a_{n+2} be three consecutive terms of an arthmetic sequence. Solve the equation
$$\left(a_{n+1} - a_{n+2}\right)x^2 + \left(a_{n+2} - a_n\right)x + \left(a_n - a_{n+1}\right) = 0.$$

VIII. 48. Let $a, b, c > 1$ real numbers such that $2\log_a b = 1 + \log_a c$. Show that the numbers
$1 - \log_{abc}(bc)$; $1 - \log_{bca}(ac)$; $1 - \log_{cba}(ab)$ are in an arithmetic progression.

VIII. 49. In an arithmetic progression $\dfrac{S_m}{S_n} = \dfrac{m}{n}$.

Prove that $\dfrac{a_m}{a_n} = \dfrac{2m-1}{2n-1}$.

VIII. 50. Find the formula for the nth term for an arithmetic progression (a_n), if $a_1 = \alpha$ and $\dfrac{S_m}{S_n} = \dfrac{m^2}{n^2}$, where $\alpha \in \mathbf{R}$.

VIII. 51. Prove that for any $a, b \in \mathbf{R}$, the numbers $(a+b)^2$, $a^2 + b^2$, $(a-b)^2$ are in an arithmetic progression.

VIII. 52. Prove that if the numbers
$\dfrac{1}{a+2b}, \dfrac{1}{3c+a}, \dfrac{1}{3c+2b} (a \neq -2b \neq -3c)$ are in arithmetic progression then the numbers $a^2, 4b^2, 9c^2$ are also in arithmetic progression.

VIII. 53. If a, b, c and m_a, m_b, m_c are the sides and the medians of the $\triangle ABC$, respectively and a^2, b^2, c^2 are in arithmetic progression, then m_a^2, m_b^2, m_c^2 are also in arithmetic progression.

VIII. 54. If the numbers a, b, c are positive numbers in an arithmetic progression, show that the numbers $\dfrac{1}{\sqrt{b}+\sqrt{c}}$, $\dfrac{1}{\sqrt{c}+\sqrt{a}}$, and $\dfrac{1}{\sqrt{a}+\sqrt{b}}$ are in an arithmetic progression.

VIII. 55. If the numbers $\dfrac{1}{1-x}$, $\dfrac{m}{x(1-x)}$, and $\dfrac{2x-m}{x(1-x)}$, $x \in \mathbf{R} \setminus \{0,\ 1\}$ are positive numbers in an arithmetic progression, find $m \in \mathbf{R}$ such that $x = 3$.

VIII. 56. Let a, b, c, d be non zero real numbers such that the numbers $a\sqrt[3]{\dfrac{a^2}{bcd}}$, $b\sqrt[3]{\dfrac{b^2}{acd}}$, $c\sqrt[3]{\dfrac{c^2}{abd}}$, $d\sqrt[3]{\dfrac{d^2}{abc}}$ are in an arithmetic progression. Prove that the numbers a^2, b^2, c^2, d^2 are in an arithmetic progression.

VIII. 57. Solve the equation $\dfrac{1}{x}+\dfrac{2}{x}+...+\dfrac{x-1}{x}=10$, where x is a non zero positive integer number.

VIII. 58. Solve the equation $3+7+11+...+x=210$.

VIII. 59. In an arithmetic progression $a_1, a_2, a_3,..., a_{2n+1}$ with positive terms and a non zero ratio r, prove that
$$\prod_{k=1}^{n} \frac{a_{2k-1}}{a_{2k}} \le \sqrt{\frac{a_1}{a_{2n+1}}}.$$

VIII. 60. In an arithmetic progression $a_1, a_2, a_3,..., a_n$ where d is the ratio, prove the identities:

a) $\dfrac{1}{\sqrt{a_1}+\sqrt{a_2}}+\dfrac{1}{\sqrt{a_2}+\sqrt{a_3}}+\ldots+\dfrac{1}{\sqrt{a_{n-1}}+\sqrt{a_n}}=\dfrac{n-1}{\sqrt{a_1}+\sqrt{a_n}}$;

b) $a_1^k-\binom{n}{1}\cdot a_2^k+\binom{n}{2}\cdot a_3^k+\ldots+(-1)^n\binom{n}{n}\cdot a_{n+1}^k=0$;

c) $\left(1-\dfrac{d^2}{a_2^2}\right)\cdot\left(1-\dfrac{d^2}{a_3^2}\right)\cdot\ldots\cdot\left(1-\dfrac{d^2}{a_{n-1}^2}\right)=\dfrac{a_1 a_n}{a_2 a_{n-1}}$

VIII. 61. In an arithmetic progression $a_1,a_2,a_3,\ldots,a_n$ where $d\neq 0$ is the ratio, calculate:

$$S_1=\dfrac{1}{a_1 a_2}+\dfrac{1}{a_2 a_3}+\ldots+\dfrac{1}{a_{n-1}a_n};$$

$$S_2=\dfrac{a_1+a_2}{a_1^2 a_2^2}+\dfrac{a_2+a_3}{a_2^2 a_3^2}+\ldots+\dfrac{a_{n-1}+a_n}{a_{n-1}^2 a_n^2}$$

$$S_3=\dfrac{a_2^3-r^3}{a_2^3+r^3}+\dfrac{a_3^3-r^3}{a_3^3+r^3}+\ldots+\dfrac{a_{n-1}^3-r^3}{a_{n-1}^3+r^3};$$

$$S_4=\dfrac{\sqrt{a_1}}{\sqrt{a_2}-\sqrt{a_1}}+\dfrac{\sqrt{a_2}}{\sqrt{a_3}-\sqrt{a_2}}+\ldots+\dfrac{\sqrt{a_n}}{\sqrt{a_n}-\sqrt{a_{n-1}}}.$$

VIII. 62. Calculate the sums

$$A_n=a_1 a_2 a_3\ldots a_k+a_2 a_3 a_4\ldots a_{k+1}+\ldots+a_{n+1}a_{n+2}a_{n+3}\ldots a_{n+k}$$

and $B_n=\dfrac{1}{a_1 a_2\ldots a_k}+\dfrac{1}{a_2 a_3\ldots a_{k+1}}+\ldots+\dfrac{1}{a_n a_{n+1}\ldots a_{n+k-1}}$, where (a_n) is an arithmetic progression.

VIII. 63. Let $a_1,a_2,\ldots,a_n$ be a sequence in an arithmetic progression and $S_{n+1}=\sum\limits_{i=1}^{n+1}a_i$. Prove that $\sum\limits_{k=0}^{n}C_n^k a_{k+1}=\dfrac{2^n S_{n+1}}{n+1}$.

GEOMETRIC PROGRESSIONS

$b_1, b_2, b_3, ..., b_n$; r b_n is the nth term, r is the ratio.

$$b_n = b_1 r^{n-1}$$

$$b_k^2 = b_{k-1} b_{k+1}, \ k = 2,3,4,.., n-1,$$

$$b_i b_n = b_i b_{n-i+1}, \ i = 1,2,3,.., n,$$

$$S_n = b_1 + b_2 + ... + b_n = \sum_{k=1}^{n} b_k$$

$$S_n = \frac{b_1(1-r^n)}{1-r}, \ r \neq 1,$$

If $|r| < 1$, $S = \dfrac{b_1}{1-r}$, where $S = b_1 + b_2 + ... = \displaystyle\sum_{k=1}^{\infty} b_k$.

VIII. 64. Find the first five terms for the geometric progression (b_n), knowing that:

a) $b_1 = 2$, $r = 3$; **b)** $b_2 = -6$, $r = -2$;

c) $b_2 = \sqrt{6}$, $b_5 = 9\sqrt{2}$; **d)** $b_3 = 20$, $b_9 = 1280$.

VIII. 65. Consider the sequences $(a_n)_{n \geq 1}$ and $(b_n)_{n \geq 1}$ where $a_n = \left(\dfrac{1}{2}\right)^n$ and $b_n = 9^n$. Show that the sequences are geometric progressions and find their ratios.

VIII. 66. In the system of coordinates xOy consider the points $A_n(\log_2 a_n, \log_3 b_n)$, $n \geq 1$. Write the equation of the line which passes through the points A_1 and A_2 and prove that the points $A_n(\log_2 a_n, \log_3 b_n)$ lie on the line $A_1 A_2$ for any $n \geq 1$.

VIII. 67. Consider the sequences $(x_n)_{n \geq 0}$ and $(y_n)_{n \geq 0}$ such that $x_0 = 4$, $x_{n+1} = \dfrac{3x_n + 1}{x_n + 3}$, $\forall n \in N$ and $y_n = \dfrac{x_n + 1}{x_n - 1}$, $\forall n \in N$.

Prove that $\left(y_n\right)_{n\geq0}$ is a geometric sequence and find the general formula for $\left(x_n\right)_{n\geq0}$ and $\left(y_n\right)_{n\geq0}$.

VIII. 68. Find the terms b_2, b_3, b_5, b_6 of the following geometric progression $\sqrt{2}$, b_2, b_3, $3\sqrt{6}$, b_5, b_6, $27\sqrt{2}$, ...

VIII. 69. The numbers $\sqrt{3}$, x, y, $18\sqrt{2}$ are the consecutive terms in a geometric progression. Find x and y.

VIII. 70. For the geometric progression $\left(b_n\right)_{n\geq1}$, $b_1 = -1$, $r = 3$, calculate b_5 and S_{10}.

VIII. 71. In a geometric progression $\left(b_n\right)_{n\geq1}$, $b_2 + b_3 = 14$ and $b_3 + b_4 = 28$. Find b_1, r, S_{10}.

VIII. 72. Insert two numbers between 5 and 135 so that the four numbers form a geometric sequence.

VIII. 73. What number must be added to each term of the sequence 1, 9, 33 to make the new sequence geometric?

VIII. 74. Find b_1 and r for a geometric progression if:

a) $\begin{cases} b_5 - b_1 = 80 \\ b_4 - b_2 = 24 \end{cases}$; b) $\begin{cases} b_5 + b_2 - b_4 = 114 \\ b_4 + b_1 - b_3 = 38 \end{cases}$.

VIII. 75. In a geometric sequence of positive numbers each term after the second is the sum of the two preceding terms. Find the common ratio.

VIII. 76. If in a geometric sequence of positive terms any term is equal to the sum of the next two terms, find the ratio.

VIII. 77. The positive numbers x, y, z form a geometric sequence. Prove that the numbers $\log x$, $\log y$, $\log z$ form an arithmetic sequence.

VIII. 78. Prove the identity

$$\left(1+x+x^2+...+x^{n-1}\right)\left(1+x+x^2+...+x^{n+1}\right)=\left(1+x+x^2+...+x^n\right)^2-x^n.$$

VIII. 79. The sum of the first three terms of a geometric progression is 21 and their product is 64. Find the numbers.

VIII. 80. The sum of the first and fifth terms of a geometric series is equal to twice the sum of the second and sixth terms. What is the ratio of the geometric series?

VIII. 81. The product of the first three terms of a geometric progression is $3\sqrt{3}$ and the sum of their cubes is $28+3\sqrt{3}$. Find the progression.

VIII. 82. The numbers a, b, c, and d are in a geometric progression, if the numbers $a + 1$, $b + 6$, $c + 6$, and $d - 4$ are in an arithmetic progression, find the numbers.

VIII. 83. The sum of three numbers in a geometric progression is 13. If 1 is added to the first and the second, and it is subtracted from the third, the numbers will be in a geometric progression. Find the numbers.

VIII. 84. The sum of the first three terms of a geometric series is 21 times the first term. The sum of the first and the third term is 17. What is the second term?

VIII. 85. The sum of an infinite geometric series is equal to three times its initial term. What is the ratio of the series?

VIII. 86. Determine the sum of the following infinite geometric series with first term b_1 and ratio r:

a) $b_1 = 7, r = -\dfrac{3}{4}$; **b)** $b_1 = -2, r = \dfrac{1}{3}$;

c) $b_1 = 3.865, r = 0.227$; d) $b_1 = -4, r = -\dfrac{1}{2}$;

e) $b_1 = 35, r = \dfrac{12}{17}$; f) $b_1 = 5, r = 0.33$.

VIII. 87. The lottery jackpot is $100,000. Each succeeding winning number pays 40% more than the preceding winning number. How much will be paid out in prizes if six numbers were drawn?

VIII. 88. Three numbers are in a geometric progression. Increasing the second number by 8, the numbers will be in an arithmetic progression and if, in the arithmetic progression, the last term is increased by 64, the numbers will be in a geometric progression. Find the numbers.

VIII. 89. In a geometric progression find four successive terms such as the second term is smaller than the first by 54, and the third term is larger than the forth by 6.

VIII. 90. The first term of an arithmetic progression is 1 and the sum of the first nine terms is equal to 369. The first and ninth terms of the arithmetic progression coincide with the

first and ninth terms of a geometric progression, respectively. Find the sum of the first twenty terms of the geometric progression.

VIII. 91. The sum of four consecutive terms of a geometric sequence is 60, and the fourth term is 4 times the second term. Find the fourth terms.

VIII. 92. The sum of the first five terms in a geometric progression is 62. At the same time the fifth, eighth, and eleventh term of this progression are the first, the second, and the tenth in an arithmetic progression, respectively. Find the first term of the geometric progression.

VIII. 93. The sum of the first three terms in a geometric progression is 65 and the sum of logarithms in the base 15 is 3. Find the firs term and the ratio.

VIII. 94. In a geometric progression of nine positive terms, the third and fifth terms are the smallest and larger solution of the equation $\frac{1}{2}\left(1+\frac{1}{2}\log_2(3x-2)\right)=\log_4\left(1+\sqrt{10x-11}\right)$.
Find the sum of the geometric progression.

VIII. 95. Let (a_n) and (b_n), $n \geq 1$ be an arithmetic and a geometric progression, such that $a_1 + a_2 + ... + a_{10} = 155$ and $b_1 + b_2 = 9$. Find the progressions, if a_1 is the ratio of the geometric progression and b_1 is the ratio of the arithmetic progression.

VIII. 96. If the sum of the first and the fifth term of an geometric progression with positive terms is 164 and the product of the third term and the fifth term is 2916, find the 10th term of the progression.

VIII. 97. The real positive numbers $b_1, b_2, b_3, ..., b_{2n+2}$, are in a decreasing geometric progression.
Prove that $\quad \dfrac{b_1}{b_2}\cdot\dfrac{b_3}{b_4}...\dfrac{b_{2n-1}}{b_{2n}} < \sqrt{\dfrac{b_1}{b_{2n+2}}}.$

VIII. 98. Evaluate the sum
$$S = 1 + 2\cdot3 + 3\cdot3^2 + 4\cdot3^3 + 5\cdot3^4 + ... + 100\cdot3^{99}.$$

VIII. 99. Evaluate the sum $\quad S_n = 3 + 33 + 333 + ... + \underbrace{333...3}_{n \text{ times}}$.

VIII. 100. Calculate the sums:
a) $S_1 = q + 2q^2 + 3q^3 + ... + nq^n$;
b) $S_1 = q + 4q^2 + 9q^3 + ... + n^2 q^n$, , $q \in \mathbf{R}$.

VIII. 101. If $a \in \mathbf{R}$, $a \neq 0$, calculate the numbers:

$$S_1 = \left(a + \frac{1}{a}\right)^2 + \left(a^2 + \frac{1}{a^2}\right)^2 + \ldots + \left(a^n + \frac{1}{a^n}\right)^2 \text{ and}$$

$$S_2 = \left(a - \frac{1}{a}\right)^2 + \left(a^2 - \frac{1}{a^2}\right)^2 + \ldots + \left(a^n - \frac{1}{a^n}\right)^2.$$

VIII. 102. Let $b_1, b_2, b_3, \ldots, b_n$ be a geometric progression and $r \neq 1$ the ratio. Calculate:

a) $A_n = \sum\limits_{k=2}^{n} \dfrac{\sqrt{b_{k-1}}}{\sqrt{b_k} - \sqrt{b_{k-1}}}$;

b) $B_n = \sum\limits_{k=1}^{n} \dfrac{b_k b_{k+1}}{b_k + b_{k+1}}$.

VIII. 103. Let $b_1, b_2, b_3, \ldots, b_n$ be a geometric progression and r the ratio. If $A = b_1 + b_2 + \ldots + b_n$

and $B = \dfrac{1}{b_1} + \dfrac{1}{b_2} + \ldots + \dfrac{1}{b_n}$, calculate $C = b_1 b_2 \ldots b_n$.

VIII. 104. In a geometric progression $b_1, b_2, b_3, \ldots, b_n$,

$b_7 = \left(x^4 + 1\right)^4$ and $b_{11} = \left(x^4 + 1\right)^6$, $x \in \mathbf{R}$.

Calculate the sum $S = \sum\limits_{k=1}^{n} \log_{b_1} b_k$.

VIII. 105. Consider the geometric progression $(b_n)_{n \geq 1}$.

Show that $\sum\limits_{k=1}^{2n} (-1)^{k-1} b_k^2 \geq \sum\limits_{k=2n+1}^{4n} (-1)^{k-1} b_k^2$, $n \geq 1$.

VIII. 106. Let $b_1, b_2, b_3, \ldots, b_n$ be n positive numbers in a geometric progression.

Prove that $b_1 b_2 \ldots b_n = \sqrt{(b_1 b_n)^n}$.

VIII. 107. Let $b_1, b_2, b_3, \ldots, b_n$ be in a geometric progression, $b_i < b_{i+1}$ for any $i = 1, 2, \ldots, n-1$, and p a non zero positive integer. Calculate the ratio of the sums

$$S_1 = \frac{1}{b_2^p + b_1^p} + \frac{1}{b_3^p + b_2^p} + \ldots + \frac{1}{b_{n+1}^p + b_n^p} \quad \text{and}$$

$$S_2 = \frac{1}{b_2^p - b_1^p} + \frac{1}{b_3^p - b_2^p} + \ldots + \frac{1}{b_{n+1}^p - b_n^p} \quad \text{does not depend on } n.$$

VIII. 108. Prove that in a geometric progression

$$\frac{S_{2n} - S_n}{S_{3n} - S_{2n}} = \frac{S_n}{S_{2n} - S_n}$$

VIII. 109. Let $b_1, b_2, , \ldots, b_n$ be a sequence of distinct positive numbers. Prove that this sequence is a geometric progression if and only if $\dfrac{b_1}{b_2} \displaystyle\sum_{k=1}^{n-1} \dfrac{b_n^2}{b_k b_{k+1}} = \dfrac{b_n^2 - b_1^2}{b_2^2 - b_1^2}$, $\forall n \geq 2$.

VIII. 110. Let $x_1, \; x_2, \; , \ldots, x_n$ be a sequence of distinct positive numbers, where $x_1 = a$ and

$$2\sum_{k=1}^{n} \sqrt{x_k} = (n+1)\sqrt{x_n}, \quad \forall n \geq 2. \text{ Calculate } \sum_{k=1}^{n} x_k .$$

VIII. 111. Consider the numbers $x_0, x_1, \ldots, x_n, \ldots$ such that

$$x_n = 3x_{n-1} - \frac{1}{p^2} x_{n-1}^3, \text{ where } |x_0| \leq 2p \text{ and } p \text{ is a positive}$$

number. Calculate x_n in terms of n and x_0.

VIII. 112. Consider the sequence $(x_n)_{n \in N}$, such that $x_0 = a$ and $x_1 = b$ where a, b are real numbers and $x_n = \dfrac{5}{7} \cdot x_{n-1} + \dfrac{2}{7} \cdot x_{n-2}$,

$\forall n \geq 2$. Find x_n in terms of a, b and n.

Some considerations of **Fibonacci numbers**

Let $\qquad F_{n+1} = F_n + F_{n-1}$, $n \in N \qquad$ (1)

be a recurrent sequence. Suppose that
$$F_0 = 0,\ F_1 = 1,\ F_2 = 1,\ F_3 = 2,\ F_4 = 3,\ F_5 = 5$$
Writing the terms of this sequence, we obtain
$$0,\ 1,\ 1,\ 2,\ 3,\ 5,\ 8,\ \dots \qquad (2)$$
called the Fibonacci sequence and F is called the nth Fibonacci number.

Fibonacci was one of the greatest European mathematicians of the Middle Ages. His full name was Leonardo of Pisa or Leonardo Pisano in Italian, he was born in Pisa (Italy), the city of the famous Leaning Tower, about 1175 AD.

The Lucas numbers are defined by the equations $L_1 = 1$,
$$L_n = F_{n+1} + F_{n-1} \qquad (3)$$
and satisfy the same recurrence $L_{n+1} = L_n + L_{n-1}$ where the first few numbers are
$$1,\ 3,\ 4,\ 7,\ 11,\ 18,\ 29,\ 47,\ 76,123,\ \dots$$

The French mathematician Edouard Lucas (1842-1891), was the first who called the series of numbers 0, 1, 1, 2, 3, 5, 8, 13 the Fibonacci Numbers.

Assuming that the sequence F_n has the form $F_n = \lambda^n$, where λ is a real parameter and substituting F_n in (1)
$$\lambda^{n+1} = \lambda^n + \lambda^{n-1} \qquad (4)$$
or equivalent $\lambda^{n-1}\left(\lambda^2 - \lambda - 1\right) = 0$

Since $F_n \neq 0$ $(n \in N^*)$, the last equality becomes $\lambda^2 - \lambda - 1 = 0$ and it has the roots
$$\lambda_1 = \frac{1 + \sqrt{5}}{2} \quad \text{and} \quad \lambda_2 = \frac{1 - \sqrt{5}}{2} \qquad (5)$$

Thus the sequences $F_n = \left(\dfrac{1 + \sqrt{5}}{2}\right)^n$, $F_n = \left(\dfrac{1 - \sqrt{5}}{2}\right)^n$

verify the equality (1). So we conclude that the equation (1) can have more solutions. In general there are an infinite number of sequences verifying (1). Easy to observe that (1) has the form

$$F_n = c_1 \left(\frac{1+\sqrt{5}}{2} \right)^n + c_2 \left(\frac{1-\sqrt{5}}{2} \right)^n \qquad (6)$$

where c_1, c_2 –are fixed real numbers, verifying (1), too. Also you can prove that any sequence verifying (1) has the form (6).

For $n = 0$ and $n = 1$ in (6), we obtain the linear system

$$\begin{cases} c_1 + c_2 = 0 \\ c_1 \left(\frac{1+\sqrt{5}}{2} \right) + c_2 \left(\frac{1-\sqrt{5}}{2} \right) = 1 \end{cases}$$

having the solutions $c_1 = \dfrac{1}{\sqrt{5}}$, $c_2 = -\dfrac{1}{\sqrt{5}}$

Finally, the general term of the Fibonacci sequence has the form

$$F_n = \frac{1}{\sqrt{5}} \left(\frac{1+\sqrt{5}}{2} \right)^n - \frac{1}{\sqrt{5}} \left(\frac{1-\sqrt{5}}{2} \right)^n, \quad n \in \mathbb{N}. \qquad (7)$$

Some proprieties of the Fibonacci sequence.

1. $F_1 + F_2 + ... + F_n = F_{n+2} - 1$ (8)

Proof: $F_1 = F_3 - F_2$
$F_2 = F_4 - F_3$
$\cdots\cdots\cdots\cdots$
$F_{n-1} = F_{n+1} - F_n$
$F_n = F_{n+2} - F_{n+1}$

Summing all equalities we get $F_1 + F_2 + ... + F_n = F_{n+2} - F_2$, since $F_2 = 1$ so (8) is shown.

2. $F_1 + F_3 + ... + F_{2n-1} = F_{2n}$.

3. $F_2 + F_4 + ... + F_{2n} = F_{2n+1} - 1$.
       (2. and 3. can be proven in a similar manner).

4. $F_1^2 + F_2^2 + ... + F_n^2 = F_n F_{n+1}$ (9)

Proof: It is easy to observe that

$$F_n F_{n+1} - F_{n-1} F_n = F_n \left(F_{n+1} - F_{n-1} \right) = F_n^{\;2}, \quad n \in \mathbb{N}.$$

From this we have successively the equalities:

$$F_1^{\;2} = F_1 F_2$$
$$F_2^{\;2} = F_2 F_3 - F_1 F_2$$
$$F_3^{\;2} = F_3 F_4 - F_2 F_3$$
$$\cdots\cdots\cdots\cdots$$
$$F_n^{\;2} = F_n F_{n+1} - F_{n-1} F_n$$

Summing all these equalities we get (9).

5.   Prove that $F_{n+m} = F_{n-1} F_m + F_n F_{m+1}$ (10)

where F_n is the nth term of the Fibonacci sequences.
Proof: Using (1) and (9) then (10) is easy to be proven.
To prove (10) using the mathematical induction is an option. I'll
proceed by mathematical induction with respect to $m \in \mathbb{N}$.
For $m = 1$, the equality (10) becomes $F_{n+1} = F_{n-1} F_1 - F_n F_2$, which is
obvious, then   (10) is true for $m = 1$. (For example when $m = 2$ the
formula (10) is true $F_{n+2} = F_{n-1} F_2 + F_n F_3 = F_{n-1} + 2F_n = F_{n+1} + F_n$).
Assuming that (10) is true for $m = k$ and $m = k + 1$.
I'll prove that (10) is true for $m = k + 2$, too.
Therefore being true the equalities $F_{n+k} = F_{n-1} F_k - F_n F_{k+1}$

$$F_{n+k+1} = F_{n-1} F_{k+1} - F_n F_{k+2},$$
$$F_{n+k+2} = F_{n-1} F_{k+2} - F_n F_{k+3}$$

By summing both equalities, we are actually obtaining
(10) for $m = k + 2$.

6. Prove that $F_{2n} = F_n L_n$ (sometimes called *double angle formula*),
Proof: If taking $m = n$ (10)   we obtain
$$F_{2n} = F_{n-1} F_n + F_n F_{n+1} = F_n \left(F_{n-1} + F_{n+1} \right) = F_n L_n,$$

7. Prove that $F_n^{\;2} = F_{n-1} F_{n+1} + (-1)^{n+1}, \quad n > 1$ (11)
Proof: By mathematical induction.

For $n = 2$ then (11) becomes $F_2^2 = F_1 F_3 - 1$, which is true. Thus (11) is true for $n = 2$.

Assuming that (11) is true for n, I'll prove that it is true for $n + 1$, too.

So $F_n^2 = F_{n-1} F_{n+1} + (-1)^{n+1}$ is true. Adding on both sides of the last equality the number $F_n F_{n+1}$ we will obtain

$$F_n^2 + F_n F_{n+1} = F_{n-1} F_{n+1} + F_n F_{n+1} + (-1)^{n+1} \text{ or}$$

$$F_n(F_n + F_{n+1}) = F_{n+1}(F_{n-1} + F_n) + (-1)^{n+1}.$$

Next using (1) we have $F_n F_{n+2} = +F_{n+1}^2 + (-1)^{n+1}$ and finally

$$F_{n+1}^2 = F_n F_{n+2} + (-1)^{n+2}.$$ Hence condition (11) is true for $n + 1$.

8. Prove that $2^{n+1} - 7F_{n-1}^2 = (F_{n-1} + 2F_n)^2$. (by mathematical induction).

Chapter IX

POLYNOMIALS AND ALGEBRAIC EQUATIONS

$$P(x) = a_n x^n + a_{n-1} x^{n-1} + a_{n-2} x^{n-2} + \ldots + a_1 x + a_0,$$

$a_n, \; a_{n-1}, \; a_{n-2}, \; + \ldots + a_1, \; a_0$ are the coefficients. The degree of the polynomial is n, then $a_n \neq 0$.

$$\textit{Vieta's formulas} \quad \begin{cases} x_1 + x_2 + \ldots + x_n = -\dfrac{a_{n-1}}{a_n} \\[2mm] x_1 x_2 + x_1 x_3 + \ldots + x_{n-1} x_n = (-1)^2 \dfrac{a_{n-2}}{a_n} \\[2mm] \cdots\cdots\cdots\cdots\cdots\cdots\cdots\cdots\cdots\cdots\cdots\cdots \\[2mm] x_1 x_2 \cdot \ldots \cdot x_n = (-1)^n \dfrac{a_0}{a_n} \end{cases}$$

Bézout's theorem (*reminder theorem*): The reminder of a polynomial when devided by $x - a$ is $r = P(a)$.

Factor theorem: If $P(x)$ is a polynomial, then $P(a) = 0$ if and only if $x - a$ is a factor of $P(x)$.

Fundamental theorem of algebra (d'Alambert-Gauss). Every polynomial with complex coefficients of the degree at last one has at least a complex root. An outcome of this theorem is: Every polynmial of degree n, $n > 0$ can be factored into n linear factors, that is, $P(x) = (x - x_1)(x - x_2)\ldots(x - x_n)$.

IX. 1. Divide $P(x) = x^4 + x^3 + 2x^2 + 1$ by $x + 1$ using synthetic division (*Horner's algorithm*).

IX. 2. Using synthetic division (*Horner's Algorithm*), find the quotient q and the reminder r when dividing the polynomial f by the polynomial g:

a) $f(x) = x^3 + 3x^2 - 7x - 6, \;\; g(x) = x - 2$;
b) $f(x) = 2x^4 + 3x^3 - 5x^2 - 3x + 1, \;\; g(x) = x - 3$;
c) $f(x) = 3x^4 + 6x^3 - 8x^2 + 3x - 3, \;\; g(x) = x - 1$;

d) $f(x) = x^4 - x^3 + 5x^2 - 6x + 3$, $g(x) = x + 1$;

e) $f(x) = 4x^3 + 6x^2 - 7x + 8$, $g(x) = 2x + 1$;

f) $f(x) = x^5 - x^4 + 3x^2 + 3x - 2$, $g(x) = x + 2$;

g) $f(x) = x^5 - 2x^4 + 3x^2 - 5$, $g(x) = x - 1$;

h) $f(x) = x^8 - 4x^6 + 6x^4 - 2x^2 + 2$, $g(x) = x + 1$;

IX. 3. Is $x - 1$ a factor of the polynomial
$f(x) = x^3 + 2x^2 - 6x + 3$?

IX. 4. Determine the remainder when $x^3 + 2x^2 - 6x + 1$ is divided by $x + 2$.

IX. 5. Consider the polynomial $P(x) = x^3 - 2x^2 - 5x + 6$.
a) Calculate $P(1)$;
b) Find the quotient and the reminder when devide P by $x - 1$;
c) Solve the equation $f(x) = 0$.

IX. 6. Determine a polynomial such that the graph passes through $(0, -12)$ with zeros $-3, -1, 1,$ and 2.

IX. 7. Determine the equation of the polynomial with zeros $\pm 1, -2$, and y-intercept of -6.

IX. 8. Use a polynomial to find three consecutive integers with a product of -1320.

IX. 9. The dimensions of a rectangular solid are $x - 1$, $x - 2$ and $x + 3$. The volume of the solid is 42 cm^3. Find the dimensions of the solid.

IX. 10. Two spheres have different radii. The radius of the smaller sphere is r. The volume, $V(r)$, of the larger sphere is related to the radius of the smaller sphere by the equation

$V(r) = \dfrac{4}{3}\pi\left(r^3 + 12r^2 + 48r + 64\right)$. What is the radius of the larger sphere?

IX. 11. Factor the polynomial $x^3 - 5x^2 - x + 5$.

IX. 12. Let $P(x) = x^3 - 5x^2 + 6x$. Find the zeros of the polynomial and sketch the graph.

IX. 13. Sketch the graph of each polynomial function:
a) $f(x) = -0.5(x+1)(x-2)(x-4)$;
b) $f(x) = 0.4(x+1)(x-2)^2(x-3)$;
c) $f(x) = x(x+1)^2(x-2)^3$.

IX. 14. Find a possible polynomial equation for the graph:

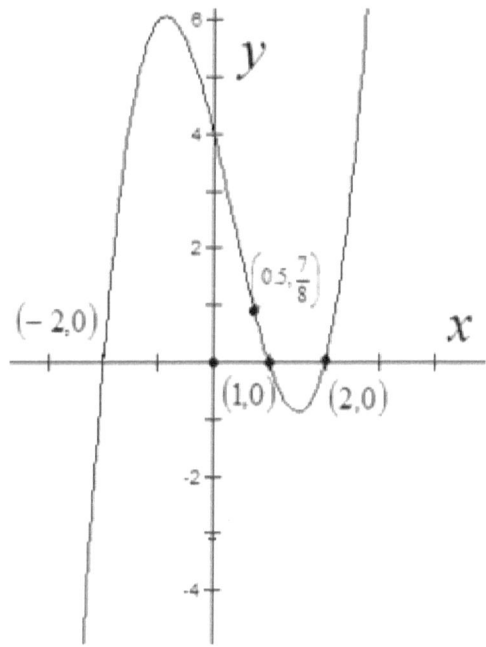

IX. 15. Given the graph of the function $y = f(x)$, sketch each transformation:

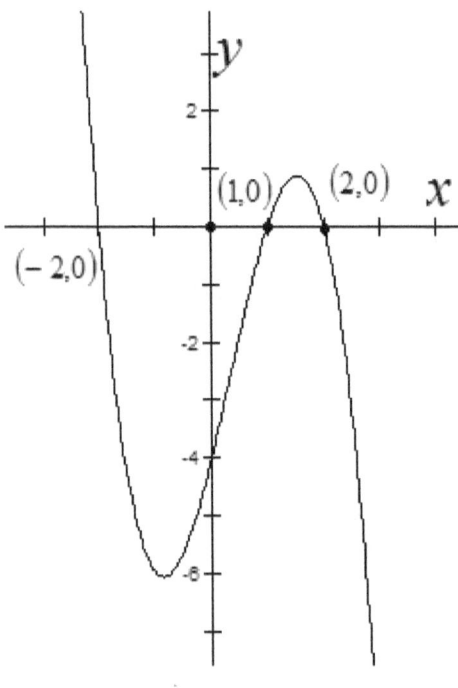

a) $y = 2f(x-2)$;
b) $y = f(-x)+3$;
c) $y = -f(3x)$.

IX. 16. Find the equation of a polynomial function with x-intercepts at -3, 1, 4 and degree 3 passing through the point $(2 , -20)$.

IX. 17. a) Find the family of polynomial functions that is degree 4 and have x-intercepts -3, 0, 2, 4;
b) Find the polynomial that passes through the point $(1 , -24)$.

IX. 18. Show that both $x-2$ and $x-3$ are factors of the polynomial $2x^3 - 11x^2 + 17x - 6$ and find another factor of this polynomial.

IX. 19. Determine the remainder when $x^3 + 2x^2 - x - 2$ is divided by $x^2 + 4x + 5$.

IX. 20. Consider the polynomial
$f(x) = x^4 - 5x^3 + 4x^2 + 5x - 6$.
a) Calculate $f(1)$ and $f(-1)$;
b) Calculate quotint and reminder when f is divided
   by $(x-1)(x+1)$;
c) Solve the equation $f(x) = 0$.

IX. 21. Find a linear polynomial P such that if P divided by $x+1$ to have a remainder of 2 and if divided by $x-1$ to have a remainder of -3.

IX. 22. If $f(x) = x^3 + 5x^2 - x - 5$, show that $f(-5) = 0$ and what is the linear and quadratic factor?

IX. 23. Using the Factor theorem, prove that
$x^3 - 6x^2 + 3x + 10$ is divisible by $x^2 - x - 2$.

IX. 24. Find the numeric value of m such that the polynomial
$f(x) = x^3 - 3x^2 + mx - 6$ is divisible by $x - 2$.

IX. 25. $2x + 3$ is a factor of $2x^3 + 5x^2 + 5x + p$. Calculate p.

IX. 26. Find the numeric value(s) of m such that the polynomial
$f(x) = x^3 - 4x^2 + 9x - 6$ is divisible by $x - m$.

IX. 27. Find the numeric value of m knowing that the polynomial $f(x) = x^6 - mx^4 + (m^2 + 4)x^2 - 2$ divided by $x - 1$ has the remainder of 5.

IX. 28. For the polynomial $(x^3 + kx^2 + x + 2) \div (x + 1)$, determine the value of k if the remainder is 3.

IX. 29. Find k if:

a) $x-2$ is a factor of $x^3 + kx^2 + 3x - 10$;

b) $x+4$ is a factor of $x^3 + 3x^2 - kx + 12$;

c) $x-3$ is a factor of $x^3 + 3x^2 + 2x + k$.

IX. 30. Find a quadratic polynomial $f(x) = ax^2 + bx + c$ such that

$$f(-1) = 0, \ f(1) = 2, \ f(2) = 9.$$

IX. 31. Find a quadratic polynomial $f(x) = ax^2 + bx + c$ such that $f(1) = f(2) = 0, \ f(-1) = 18$.

IX. 32. Consider the polynomial $f(x) = x^3 + ax + b$, a and b real patameters.

a) Find a and b such that the polynomial f is divisible by $x + 1$ and $f(1) = 4$.

b) For $a = 1$ and $b = 2$, solve the equation $f(x) = 0, x \in \mathbf{C}$.

c) For $a = 1$ and $b = 2$, solve the inequation $f(x) < 0$, $x \in \mathbf{R}$.

IX. 33. Using long division, find the quotient q and the reminder r when dividing the polynomial f to g:

a) $f(x) = 2x^3 + 2x^2 - 6x + 1, \ g(x) = x^2 - x + 2$;

b) $f(x) = x^4 + 2x^3 + 3x^2 + 7x + 1, \ g(x) = x + 3$;

c) $f(x) = x^4 + 6x^3 + 8x^2 + 1, \ g(x) = x^3 - x + 1$;

d) $f(x) = x^4 + x^3 + 7x^2 - 6x + 8, \ g(x) = x^2 - x + 1$;

e) $f(x) = x^5 - x^4 + 2x^3 - 3x^2 + x - 4, \ g(x) = x^2 - x + 3$;

f) $f(x) = x^5 - x^4 + 3x^2 - 4, \ g(x) = x^2 - x$;

g) $f(x) = x^8 - 3x^6 + 5x^4 - 2x^2 + 1, \ g(x) = x^3 - 4x + 1$.

IX. 34. The divisions $(2x^3 + 4x^2 - kx + 5) \div (x + 3)$ and $(6y^3 - 3y^2 + 2y + 7) \div (2y - 1)$ have the same remainder. Determine the value of k.

IX. 35. If the polynomial $4x^3 + ax^2 + bx + 11$ is divided by $x+2$, the remainder is -7 and if the polynomial is divided by $x-1$, the remainder is 14. Find the values of a and b.

IX. 36. If the polynomial $x^4 + ax^3 + x^2 + bx + 3$ is divided by $x-1$, the remainder is three and if the polynomial is divided by $x-2$, the remainder is five. Find the values of a and b.

IX. 37. Find the values of k when $f(x) = x^3 + 6x^2 + kx - 4$ gives the same remainder when divided by either $x-1$ or $x+2$.

IX. 38. When $f(x) = x^5 - 2x^4 - mx^3 - x^2 + nx - 2$ is divided by $x+1$ the reminder is -7 and divided by $x-2$ the reminder is 32. Find the values of m and n.

IX. 39. Let $P(x) = ax^3 + bx^2 + cx + d$ be a polynomial. Find a, b, c, and d such that $P(-1) = -1$, $P(0) = 0$, $P(1) = 1$, and $P(2) = 0$.

IX. 40. Consider the polynomial
$f(x) = 1 + (x+a)^1 + (x+a)^2 + ... + (x+a)^n$.
Calculate the sum of coefficients of f.

IX. 41. What is the degree of the polynomial
$f(x) = (m^2 + 3m + 2)x^4 + (m^3 + 2m^2 - m - 2)x^3 +$
$+ (m^2 + 4m + 3)x^2 + (m^2 - 1)x + 3, \ m \in \mathbf{R}$?

IX. 42. Consider the polynomial $P(x) = x^3 - 6x + 5$. Find the value of P for indicated value.

a) $\sqrt[3]{7 + \sqrt{41}} + \sqrt[3]{7 - \sqrt{41}}$;

b) $\sqrt[3]{4 - 2\sqrt{2}} + \sqrt[3]{4 + 2\sqrt{2}}$;

c) $\sqrt[3]{6 - 2\sqrt{7}} + \sqrt[3]{6 + 2\sqrt{7}}$.

IX. 43. Find the real number m, such that the polynomial $f(x) = 2x^4 - mx^3 - 5x^2 - 8x + 7$ divided by $x - 1$, the reminder is -3.

IX. 44. When the polynomial $x^3 - mx^2 - 7mx + 1$ si divided by $x - 2$, the reminder is -9. Find the value of m.

IX. 45. Find $a \in \mathbf{R}$ such that the polynomial $P(x) = x^4 + (a+2)x^3 + 3a^2x^2 - 5x - 14$ is divisible by $x + 2$.

IX. 46. Show that if $a, b \in \mathbf{R}$, $a \neq b$ the reminder of dividing a polynomial $P(x)$ of degree at least two by $(x-a)(x-b)$ is
$$r(x) = \frac{P(b) - P(a)}{b - a} x + \frac{bP(a) - aP(b)}{b - a}.$$

IX. 47. If a polynomial by degree at least three is divided by $x - 1$ the remainder is zero, divided by $x + 1$ the remainder is one, and if the polynomial is divided by $x - 2$, the remainder is -3. Find the reminder if the polynomial is divided by $(x-1)(x+1)(x-2)$.

IX. 48. The polynomial $f(x) = x^5 - mx^2 - mx + 1$ is divisible by the polynomial $g(x) = x^2 + 2x + 1$. Find the value of $m \in \mathbf{R}$.

IX. 49. The polynomial $f(x) = x^5 - 2x^4 + 3x^3 - 7x^2 + 8x + a$ is divisible by the polynomial $g(x) = x^3 - 3x^2 + 3x - 1$. Find the numeric value of $a \in \mathbf{R}$.

IX. 50. Find a, b, and $c \in \mathbf{R}$ such that the polynomial $f = (x+1)^6 + a(x+1)^3 + bx + c$ is divisible by $(x-1)^3$.

IX. 51. Let $P(x)$ be a polynomial of degree at least two. If $P(x)$ divided by $x + 1$ the remainder is 2 and $(x+1) P(x) + x P(x+3) = 1$, find the remainder when $P(x)$ is divided by $x^2 - x - 2$.

IX. 52. Prove that each polynomial:

a) $f(x) = x^{n+1} - (n+1)x + n$, $n \in \mathbb{N}^*$ is divisible by $(x-1)^2$;

b) $f(x) = nx^{n+1} - (n+1)x^n + 1$. $n \in \mathbb{N}^*$ is divisible by $(x-1)^2$;

c) $f(x) = (2n-1)x^{2n} - 2nx^{2n-1} + 1$. $n \in \mathbb{N}^*$ is divisible by $(x-1)^2$;

e) $f(x) = x^{n+2} - x^{n-1} - 3x + 3$. $n \in \mathbb{N}^*$ is divisible by $(x-1)^2$.

IX. 53. Find $m \in \mathbb{R}$ such that the polynomial
$f(x) = x(x+1)^{2n+1} + (m-1)x^n$ is divisible by $x^2 + x + 1$.

IX. 54. Prove that each of the polynomials:

a) $f(x) = (x^2 + x - 1)^{4n+1} - x$, $n \in \mathbb{N}^*$ is divisible by $x^2 - 1$;

b) $f(x) = (x^2 + x + 1)^{8n+1} - x$, $n \in \mathbb{N}^*$ is divisible by $x^2 + 1$;

c) $f(x) = (x^2 - x - 1)^{2n+2} + (x^4 + x - 1)^{2n+5}$, $n \in \mathbb{N}^*$
is divisible by $x^2 + 1$;

d) $f(x) = (2x^2 + x + 2)^{4n+1} - x$, $n \in \mathbb{N}^*$ is divisible by $x^2 + 1$;

e) $f(x) = x^{6n-1} + x + 1$, $n \in \mathbb{N}^*$ is divisible by $x^2 + x + 1$;

f) $f(x) = (x+1)^{4n+1} + (x+1)^{6n+5} + (x^2 + 1)^{6n+3}$, $n \in \mathbb{N}^*$
is divisible by $x^2 + x + 1$;

g) $f(x) = (x-1)^{2n+1} + (-1)^{n+1} + x^{n+2}$, $n \in \mathbb{N}^*$ is divisible by $x^2 - x + 1$.

IX. 55. Consider the polynomial
$f(x) = x^3 - 4x^2 + (6-i)x - 3 + i$.
Find roots of the polynomial if it has a real root.

IX. 56. Consider the polynomial $f(x) = (3x+1)^{6n+1} + 3x^2$.
Prove that the polynomial is divisible by $g(x) = 3x^2 + 3x + 1$ for any
$n \in \mathbb{N}^*$. The same question for $f(x) = (2x+1)^{3n+2} + 2x + 2$
and $g(x) = 4x^2 + 6x + 3$.

IX. 57. Find a polynomial of degree five knowing that $f(x)+1$ is divisible by $(x-1)^3$ and $f(x)-1$ is divisible by $(x+1)^3$.

IX. 58. If $f(x)$ is a polynomial by degree 5 such that $f(x)-2$ is divisible by $(x+1)^3$ and $f(x)$ is divisible by x^3, find $f(x)$.

IX. 59. Find the roots of the polynomial
$f = aX^3 + bX^2 + cX + d$, $a,b,c,d \in \mathbf{R}$, if $b+d=a+c$.

IX. 60. Consider the polyinomial $f = aX^3 + bX^2 + cX + d$, $a,b,c,d \in \mathbf{R}$. Prove that if $bc=ad$ then $x_1 + x_2 = 0$ and viceversa.

IX. 61. Solve each equation with the indicated condition and in some cases find the parameter m as well.

a) $x^3 + 5x^2 + 2x - 8 = 0$, $\qquad\qquad x_1 = 2x_2$;

b) $x^4 - 10x^3 + 35x^2 - 50x + 24 = 0$, $\qquad x_1 = 3x_2$;

c) $x^3 - 8x^2 + (3m+2)x - 2m = 0$, $\qquad 3x_1 = x_2 + x_3$;

d) $x^4 - 10x^3 + 5mx^2 - 50x + 24 = 0$, $\qquad x_1 - x_4 = x_3 - x_2$;

e) $x^4 - 4x^3 + (m+3)x^2 + 18x + 14 = 0$, $\qquad x_1 + x_2 = x_3 + x_4$;

f) $x^3 - 9x^2 + 6x + 56 = 0$, $\qquad\qquad x_1 - x_2 = 3$;

g) $x^4 + 3x^3 - mx^2 - 3x + 10 = 0$, $\qquad x_1 + x_2 = -3$;

h) $x^3 + (1-\sqrt{3})x^2 - (2+\sqrt{3})x + 2\sqrt{3} = 0$; $\quad x_1^2 = x_2^2 + x_3^2$;

i) $x^4 - mx^3 + 12x - 5 = 0$, $\qquad\qquad x_1 + x_2 = 2$.

IX. 62. If one of the roots of the polynomial $x^3 - x^2 + 5x + 3$ is $-1+\sqrt{2}$ find the sum of the other two roots.

IX. 63. Find a polinomial such that $x = -1+\sqrt{2}+\sqrt{3}$ is one of the roots.

IX. 64. For the polynomial $f(x) = x^3 - x^2 + mx - 1$, $m \in \mathbf{R}$, consider $S_n = x_1^n + x_2^n + x_3^n$, $n \in \mathbf{N}^*$ where $x_1, x_2, x_3 \in \mathbf{C}$ are the roots of the polynomial f. Show that $S_n - S_{n-1} + mS_{n-2} - 3S_{n-3} = 0$ and find $m \in \mathbf{R}$ such that $S_3 = 1$.

IX. 65. If $x_1, x_2, x_3 \in \mathbf{C}$ are the roots of the equation $x^3 + mx + n = 0$, $m, n \in \mathbf{R}$, show that:

a) $12(x_1^7 + x_2^7 + x_3^7) = 7(x_1^3 + x_2^3 + x_3^3)(x_1^2 + x_2^2 + x_3^2)^2$;

b) $3(x_1^2 + x_2^2 + x_3^2)(x_1^5 + x_2^5 + x_3^5) = 5(x_1^3 + x_2^3 + x_3^3)(x_1^4 + x_2^4 + x_3^4)$.

IX. 66. Solve the equation $3x^3 + 8x^2 + 13x + 6 = 0$ if the roots fulfil the condition $x_1 = \dfrac{1}{x_2} + \dfrac{1}{x_3}$.

IX. 67. Let $x^3 + x^2 + mx - 1 = 0$, $m \in \mathbf{R}$ be a equation.
a) Show that the equation has a root on the interval $[-1, \ 1]$;
b) Find m if the equation has a double root.
c) Find $m \in \mathbf{R}$ such that the roots of the equation fulfill the condition $\dfrac{1}{x_1^2} + \dfrac{1}{x_2^2} + \dfrac{1}{x_3^2} = \dfrac{4}{x_1 x_2 x_3}$.

IX. 68. Show that the equation
$x^4 + 2(a+1)x^3 + (2a^2 + 4a + 3)x^2 + 5x + 9 = 0$ can not have all real roots.

IX. 69. Find the numeric value of $m \in \mathbf{R}$ and solve the equation $2x^3 - x^2 - 7x + m = 0$ knowing that the sum of two roots is one.

IX. 70. Consider the polynomial $f(x) = x^3 + x^2 + ax + b$. Find the numeric value for a and b such that $f(x-1)$ divided by $x+1$ the reminder is -4 and the roots of the equation $f(x) = 0$ fulfill the condition $x_1^3 + x_2^3 + x_3^3 = 8$.

IX. 71. Consider the polynomial $f(x) = x^3 - 6x^2 + ax + b$. Find the numeric value for a and b such that $f(x-1)$ divided by $x-5$ the reminder is 21 and the roots of the equation $f(x) = 0$ fulfill the condition $x_1^3 + x_2^3 + x_3^3 = 36$.

IX. 72. Let $f(x) = x^3 + ax^2 + bx + c$, $a, b, c \in \mathbf{R}$. be a polynomial. Find he condition between the coefficients a, b, and c such that the roots of $f(x)$ are in a geometric progression.

IX. 73. Solve the equation $x^4 - 15x^3 + 70x^2 - 121x + m = 0$, if the roots are in a geometric progression.

IX. 74. Consider the polynomial $f(x) = 2x^4 - 15x^3 + 35x^2 - 30x + m$. Find $m \in \mathbf{R}$ such that the roots of f are in a geometric progression and find the roots of f.

IX. 75. For the equation $x^3 + x + m = 0$, find $m \in \mathbf{R}$ such that $x_1^5 + x_2^5 + x_3^5 = 1$, where x_1, x_2, x_3 are the roots of the equation. Find the value of the expression

$$E = \frac{x_1^2 + x_2^2}{x_3^2} + \frac{x_2^2 + x_3^2}{x_1^2} + \frac{x_1^2 + x_3^2}{x_2^2} \quad \text{when } m = 2.$$

IX. 76. Consider the polynomial
$f(x) = x^3 - mx^2 - 3x + m - 2$, $m \in \mathbf{R}$.
a) Calculate $S(m) = x_1^3 + x_2^3 + x_3^3$, where x_1, x_2, x_3 are the roots of the polynomial;
b) Find $m \in \mathbf{R}$ such that $S(m) \le (x_1 x_2 x_3)^2 + 2$;
c) Find $m \in \mathbf{R}$ such that f is divisible by $x + 2$ and solve the equation.

IX. 77. Consider the polynomial $f(x) = ax^3 + bx^2 + cx + d$, $a, b, c, d \in \mathbf{R}$ Find a, b, c, d such that $f(1) + f(2) + ... + f(n) = n^4$, for any n a natural number.

IX. 78. Prove that the polynomial:

a) $f(x) = x^3 - 3x - 2\sqrt{5}$ has the root $\sqrt[3]{\sqrt{5}+2} + \sqrt[3]{\sqrt{5}-2}$;

b) $f(x) = x^3 + 3x - 2$ has the root $\sqrt[3]{\sqrt{2}+1} - \sqrt[3]{\sqrt{2}-1}$;

c) $f(x) = x^3 - 3x^2 - 2\sqrt{3}$ has the root $\sqrt[3]{\sqrt{3}-\sqrt{2}} + \sqrt[3]{\sqrt{3}+\sqrt{2}}$.

IX. 79. Show that the number $x = \sqrt[5]{27} - \sqrt[5]{9}$ is a root of the equation $x^5 + 15x^3 - 225x - 18 = 0$.

IX. 80. Solve the equation $z^3 - (4+i)z^2 + (7+3i)z - 2i - 6 = 0$ in the set **C** (*complex numbers*), knowing that it has a real root.

IX. 81. Consider the polynomial $f(x) = x^3 + ax^2 + 4x + b$, where $a, b \in \mathbf{R}$. Find the numeric value for a and b such that $x_1 = \dfrac{1 - i\sqrt{3}}{2}$ is a root of the equation $f(x) = 0$.

IX. 82. The equation $x^3 - 3x^2 + ax + b = 0$ has rational coefficients and one of the roots is $-1 + i\sqrt{3}$, determine the value of a and b.

IX. 83. Consider the polynomial $f(x) = 2x^3 + mx^2 + 4x + 4$. Find m such that f has a double root.

IX. 84. Consider the polynomial $f(x) = x^4 + ax^3 + bx + c$. If $f(0) = f(1)$ and $x_1 = \dfrac{1 + i\sqrt{3}}{2}$ is a root, find the other two roots.

IX. 85. One of the roots of the polynomial $f(x) = x^4 - mx^3 + 13x^2 - 14x + n$, $m, n \in \mathbf{R}$ is $x_1 = 1 - i$. Find m and n and solve the equation $f(x) = 0$.

IX. 86. Consider the polynomial

$f(x) = x^4 - 5x^3 + ax^2 + bx + c$, $a, b, c \in \mathbf{R}$. If $f(0) = f(1)$ and $x_1 = 2 - i$ is a root, find the other two roots and a, b, c.

IX. 87. Solve each equation with the indicated root.

a) $x^3 - 3x^2 - 3x + 1 = 0$, $x_1 = 2 - \sqrt{3}$;

b) $x^3 - 3x^2 - 5x + 7 = 0$, $x_1 = 1 - 2\sqrt{2}$;

c) $x^4 - 2x^3 - 2x - 1 = 0$, $x_1 = 1 - \sqrt{2}$;

d) $x^3 - 7x^2 + 16x - 10 = 0$, $x_1 = 3 + i$;

e) $x^4 - 4x^3 + 11x^2 + 8x - 26 = 0$, $x_1 = 2 + 3i$;

f) $x^4 - 7x^3 + 19x^2 + mx + n = 0$, $m, n \in \mathbf{R}$, $x_1 = 2 + i$;

g) $x^4 + 2x^3 + mx^2 - 2x + n = 0$, $m, n \in \mathbf{R}$, $x_1 = -1 + i$;

h) $x^5 + x^4 - 16x^3 - 16x^2 + 4x + 4 = 0$, $x_1 = -\sqrt{3} - \sqrt{5}$;

i) $x^6 - 23x^4 + 76x^2 - 48 = 0$, $x_1 = -\sqrt{3} - \sqrt{7}$.

IX. 88. Solve the reciprocal quadratic equations:

a) $2x^3 + 3x^2 + 3x + 2 = 0$;

b) $2x^3 + 7x^2 + 7x + 2 = 0$;

c) $2x^4 + x^3 + x^2 + x + 2 = 0$;

d) $2x^4 - 5x^3 + 4x^2 - 5x + 2 = 0$;

e) $x^5 + 2x^4 + 3x^3 + 3x^2 + 2x + 1 = 0$;

d) $2x^5 + 4x^4 + 2x^3 + 2x^2 + 4x + 2 = 0$.

IX. 89. Consider the polynomial $P(x) = x^{2n} - x^n + x^3 + 1$. Find the remainder of $P(x)$ divided by $(x - 1)^2$.

IX. 90. Find the remainder when the polynomial

$f(x) = x^4 + x^{n-1} + ... + x + 1$, $n \geq 3$ is divided by the polynomial $g(x) = x^2(x - 1)$.

IX. 91. The roots of the equation $x^n = 1$ are x_1, x_2, ..., x_{n-1}, 1. Prove that:

a) $\dfrac{1}{1-x_1} + \dfrac{1}{1-x_2} + \dots + \dfrac{1}{1-x_{n-1}} = \dfrac{n-1}{2}$;

b) $(a+b)(a+bx_1)(a+bx_2)\dots(a+bx_{n-1}) = a^n + (-1)^{n+1} b^n$;

c) $(1-x_1)(1-x_2)\dots(1-x_{n-1}) = n$;

d) $(1+x_1)(1+x_2)\dots(1+x_{n-1}) = \dfrac{1+(-1)^{n+1}}{2}$.

IX. 92. Find the remainder of the polynomial
$f(x) = x^4 + x^{n-1} + \dots + x + 1$, $n \geq 3$, divided by the polynomial
$g(x) = x^2(x-1)$.

IX. 93. Consider the polynomial $P(x) = x^{2m} - x^n + x^3 + 1$.
Find the remainder of P (x) when divided by $(x-1)^2$.

IX. 94. Consider the polynomial
$f(x) = (m-1)x^4 - (m+3)x^3 - 7x^2 - (m+6)x + m - 4$, $m \in \mathbf{R}$.

a) Show that for any $m \in \mathbf{R}, f(x)$ is divisible by $x^2 + x + 1$.

b) Find $m \in \mathbf{R}$ such that the polynomial $f(x)$ has a double root.

c) Find $m \in \mathbf{R}$ such that $\dfrac{1}{x_1} + \dfrac{1}{x_2} + \dfrac{1}{x_3} + \dfrac{1}{x_3} \geq \dfrac{5}{4}$, where

x_1, x_2, x_3, and x_4 are the roots of the polynomial $f(x)$.

d) Find $m, \beta \in \mathbf{R}$ such that $x_0 = -1 + i\tan\beta$ is a root of $f(x)$ and
in this case solve the equation $f(x) = 0$.

IX. 95. If $P_n(x+1) - P_n(x) = x^n$ where $P_n(x) \in \mathbf{Q}[x]$ and
$P_n(0) = 0$, $(\forall)n \in \mathbf{N}$. Prove that:

a) $(\forall)k \in \mathbf{N}, \quad P_n(k) = 1^2 + 2^n + 3^n + \dots + (k-1)^n$;

b) $(\forall)n \geq 1, \quad P_n(1) = 0$.

IX. 96. Prove that all of the real roots of the equation
$x^{2n+1} - x^{2n} + x^{2n-1} + 2nx^n - n^2 = 0$ are positive numbers for
any $n \in \mathbf{N}^*$.

IX. 96. Consider the polynomial
$$f = x^5 + 2ax^3 + (2b+1)x + 2c + 1$$
where $a, b, c \in \mathbf{Z}$. Prove that f does not have any integer root.

IX. 97. Let f be a polinomial with real coefficients and the degree of f is n such that $\sum_{k=1}^{n} f(x^k) = \left(\sum_{k=1}^{n} x^k \right) \cdot f(x)$. Find f.

IX. 98. If the roots of the polynomial
$$P(x) = x^4 + a_1 x^{n-1} + \ldots + a_{n-1}x + (-1)^n$$ have the same module as $P(-1)$ is a real number.

IX. 99. Let ε one of the roots of the equation $x^n = 1$, where n is a positive integer. Prove that for any polynomial
$$f(x) = a_0 + a_1 x + \ldots + a_n x^n,$$ where $a_0, \ a_1, \ a_2, \ldots, a_n$ are real numbers, the sum $\sum_{k=1}^{n} f\left(\dfrac{1}{\varepsilon^k} \right)$ is a real number.

Chapter X

PERMUTATIONS AND COMBINATIONS.
NEWTON'S BINOMIAL.

Permutations of n objects taken k at a time is denoted variously by $P(n,k)$, A_n^k, etc.

$$P(n,k) = \frac{n!}{(n-k)!} = n \cdot (n-1) \cdot ... \cdot (n-k+1)$$

The number of ways of a *combinations* of n objects taken k at a time is denoted variously by $\binom{n}{k}$, $_n C_k$, C_n^k, etc. We shall use $\binom{n}{k}$,

$$\binom{n}{k} = \frac{n!}{k! \cdot (n-k)!} = \frac{n \cdot (n-1) \cdot ... \cdot (n-k+1)}{1 \cdot 2 \cdot 3 \cdot ... \cdot k}.$$

Binomial formula

$$(a+b)^n = \binom{n}{0} a^n + \binom{n}{1} a^{n-1} b + \binom{n}{k} a^{n-2} b^2 + ... + \binom{n}{k} b^n$$

$$(a+b)^n = \sum_{k=1}^{n} \binom{n}{k} a^k b^{n-k},$$

$t_{k+1} = \binom{n}{k} a^k b^{n-k}$ ($k = 0,1,2,...,n$) is the *general term* of the binomial formula.

X. 1. How many numbers can be formed using the set $\{1, \ 3, \ 5, \ 7, \ 9\}$?

X. 2. In how many ways can 6 students standing in a line can be arranged?

X. 3. Using the digits 0, 1, 2, 3, 4, 5 find:
a) the total number which can be formed;
b) how many numbers begin with the digit one?
c) how many numbers end with the digit one?
d) how many numbers begin with 50?
e) how many numbers do not begin with 50?

X. 4. Find the number of arrangements possible using the letters of the words SUCCESS, PARALELIPIPED, and MISSISSIPPI.

X. 5. A resort manager must assign 15 guards to three buildings: 6 in the first building, 5 in the second building, and 4 in the third building. In how many ways can this be done?

X. 6. In a maths contest, in how many ways can 14 students get the first three positions?

X. 7. In how many ways can 5 of 7 books be arranged on a bookshelf? In how many ways can these 7 books be arranged on the shelf?

X. 8. How many different numbers can be formed with the digits 0, 1, 2, 3, 4 if in each number any digit can be written at most once?

X. 9. A college promotion committee consists of 7 members. A favorable majority decision is reached if at least 4 members vote favorably. In how many ways can the committee reach a majority decision in favor of a problem?

X. 10. A club has 20 members. How many different 4-member committees are possible?

X. 11. A team consists of 8 boys and 4 girls. In how many ways can 6 people be chosen under each of the following restrictions?
a) only one girl is included,
b) at least one girl is included,
c) one girl, Alexandra, is included.

X. 12. In how many ways can be selected a team consisting of 2 teachers and 5 students if there are 4 teachers and 10 students?

X. 13. In a class there are 25 students. Determine how many ways a 5-person committees can be selected to organize a trip
a) with no restrictions
b) with Alex on the committee.

X. 14. How many lines can be drawn from 6 points if any 3 of them are not collinear?

X. 15. What are the maximum points of intersection between 3 lines?

X. 16. How many 5-digit numbers can be formed such that the in each number the digits are in an increasing order from left to right? But if they are in a decreasing order from left to right?

X. 17. From a group of 15 kids, 9 boys and 6 girls, a team of 7 kids has to be formed with at least 3 girls. Determine how many ways this group can be formed.

X. 18. At a party there are 9 boys and 7 girls. In how many ways can 4 pairs be formed?

X. 19. In a class there are 25 students, 15 boys, and 10 girls. Determine how many ways a team can be selected with
a) 4 boys and 6 girls
b) 8 students with at least 3 girls in each team.

X. 20. Prove the identity
$$\binom{15}{5}\cdot\binom{10}{3}+\binom{15}{4}\cdot\binom{10}{4}+\binom{15}{3}\cdot\binom{10}{5}+\binom{15}{2}\cdot\binom{10}{6}+\binom{15}{1}\cdot\binom{10}{7}+\binom{15}{0}\cdot\binom{10}{8}=$$
$$=\binom{25}{8}-\binom{15}{6}\cdot\binom{10}{2}-\binom{15}{7}\cdot\binom{10}{1}-\binom{15}{8}\cdot\binom{10}{0}.$$

X. 21. Compute $\binom{9}{n}$ if $\binom{18}{n}=\binom{18}{n+4}$.

X. 22. Solve the equations:
a) $\binom{x+9}{x+4}=5\dfrac{(x+7)!}{(x+4)!}$; **b)** $P(x,2)=42$;

c) $\dfrac{(x-1)!}{(x-3)} = 72$;   **d)** $3\binom{2x}{x+1} = 5\binom{x}{2x-1}$;   **e)** $13\binom{2x}{x+1} = 7\binom{2x+1}{x-1}$;

f) $11\binom{2x}{x} = 6\binom{2x+1}{x+1}$;   **g)** $\binom{x}{2} + \binom{x+1}{2} = x+2$;

h) $\binom{x}{1} + 2\binom{x}{2} + \binom{x}{3} = 460$;   **i)** $\binom{x+3}{x+1} = \binom{x+1}{x-1} + \binom{x+1}{x} + \binom{x}{x-2}$;

j) $\binom{x+3}{x+1} - 5\cdot\binom{3x}{2} + 19x^2 = 6$,   **k)** $6\cdot\binom{x+1}{1} + 6\cdot\binom{x+3}{3} = 13\cdot\binom{x+2}{2}$;

l) $\binom{n}{n-1} + \binom{n}{n-2} + \binom{n}{n-3} + \ldots + \binom{n}{n-10} = 1024$;

m) $\binom{n-10}{n-11} + \binom{n-9}{n-11} + \binom{n-8}{n-11} + \ldots + \binom{n-1}{n-11} = 65$.

X. 23. Prove the identities:

a) $P(n,k) = P(n-1,k) + kP(n-1,k-1)$;

b) $k\binom{n}{k} = n\binom{n-1}{k-1}$;   **c)** $\binom{n}{k} = \binom{n-1}{k} + \binom{n-1}{k-1}$;

d) $\binom{n}{k} = \binom{n-2}{k} + 2\binom{n-2}{k-1} + \binom{n-2}{k-2}$;

e) $\binom{n}{k} = \binom{n-3}{k} + 3\binom{n-3}{k-1} + 3\binom{n-3}{k-2} + \binom{n-3}{k-3}$;

f) $\binom{n}{k} - \binom{n-2}{k} - \binom{n-2}{k-2} = \dfrac{2(n-k)}{n-1}\binom{n-1}{k-1}$;

g) $(k-1)k\binom{n}{k} = (n-1)n\binom{n-2}{k-2}$.

X. 24. Calculate and simplify the expressions:

a) $P(5,3) + P(5,4)$;   **b)** $\binom{9}{6}$;   **c)** $\dfrac{\binom{6}{3}\cdot 3!}{P(6,2)}$;

d) $\dfrac{P(10,6) + 10!}{7!\binom{9}{6}}$;   **e)** $\dfrac{P(n+k,n-k+2) + P(n+k,n-k+1)}{P(n+k,n-k)}$;

f) $\dfrac{(n-k)!\,P(n-1,k-1)}{(k-1)!}$;   **g)** $\dfrac{\binom{n}{k}}{\binom{n}{k} + \binom{n}{k+1}}$;

h) $\dfrac{\binom{n}{k} - \binom{n-2}{k} - \binom{n-2}{k-2}}{\binom{n-1}{k-1}}$;   **i)** $\dfrac{\binom{n+1}{k}}{\binom{n+2}{k}} - \dfrac{\binom{n}{k}}{\binom{n+1}{k}}$.

X. 25. Prove that the number $\binom{1}{2n+1} + \binom{2}{2n+1} + \ldots + \binom{2n}{2n+1}$ is a perfect square.

X. 26. Solve the inequations:

a) $\binom{2n}{7} \geq \binom{2n}{5}$; **b)** $\binom{17}{n} \leq \binom{17}{n-2}$; **c)** $7\binom{n}{4} > 3\binom{1+n}{5}$;

d) $\binom{19}{n-1} < \binom{19}{n}$; **e)** $5\binom{n}{3} > 3\binom{2+n}{4}$; **f)** $\binom{15}{n} < \binom{15}{n-2}$;

g) $\binom{n+8}{n+3} \leq 5P(n+8,3)$; **h)** $11\binom{2n}{n} \geq 6\binom{2n+1}{n+1}$;

i) $(n+2) \cdot \binom{n+1}{n-1} \leq 6 \cdot \binom{n}{n-1} + 9n$.

X. 27. Solve the system of equations:

a) $\begin{cases} P(x, y+1) = 2P(x, y-1) \\ 21\binom{x}{y+1} = \binom{x}{y-1} \end{cases}$; **b)** $\begin{cases} P(2y,3x) = 8P(2y, 3x-1) \\ 9\binom{2y}{3x} = 8C_{2y}^{3x-1}\binom{2y}{3x-1} \end{cases}$;

c) $\begin{cases} P\binom{2x}{y-2} = 8P\binom{2x}{y-3} \\ 3\binom{2x}{y-2} = 8\binom{2x}{y-3} \end{cases}$; **d)** $\begin{cases} P(n,k) = 7P(n, k-1) \\ 6\binom{n}{k} = 5\binom{n}{k+1} \end{cases}$;

e) $\begin{cases} P(x,2) + P(y,2) = 20 \\ P(x,3) + P(y,3) = 61 \end{cases}$; **f)** $\begin{cases} 4\binom{x}{y+1} = 3\binom{x}{y} \\ 3\binom{x}{y} = 4\binom{x}{y-1} \end{cases}$;

g) $\begin{cases} 5P(x,y) = 2P(x+1, y) \\ 3\binom{x}{y} = 2\binom{x}{y-1} \end{cases}$.

X. 28. Prove that:

a) $\dfrac{\binom{2}{1} + \binom{4}{2} + \binom{6}{3} + \ldots + \binom{2n}{n}}{\binom{1}{1} + \binom{3}{2} + \binom{5}{3} + \ldots + \binom{2n-1}{n}} = 2$;

b) $\dfrac{\binom{2}{1} \cdot \binom{4}{2} \cdot \binom{6}{3} \cdot \ldots \cdot \binom{2n}{n}}{\binom{1}{1} \cdot \binom{3}{2} \cdot \binom{5}{3} \cdot \ldots \cdot \binom{2n-1}{n}} = 2^n$.

X. 29. Calculate the sums:

a) $S_1 = 1! \cdot 1 + 2! \cdot 2 + \ldots + n! \cdot n$;

b) $S_2 = \dfrac{1}{2!} + \dfrac{2}{3!} + \ldots + \dfrac{n}{(n+1)!}$;

c) $S_3 = \sum\limits_{k=2}^{n} \dfrac{k^2 - 2}{k!}$; **d)** $S_4 = \sum\limits_{k=1}^{n} \dfrac{(k-1)!}{(k+1)!}$;

e) $S_5 = \sum\limits_{k=2}^{n} \dfrac{k^2 + k + 1}{(k+1)! \, k(k+1)}$;

f) $S_6 = \dfrac{3}{1!+2!+3!} + \dfrac{4}{2!+3!+4!} + \dfrac{5}{3!+4!+5!} + ... + \dfrac{n+1}{(n-1)!+n!+(n+1)!}.$

X. 30. Prove the identity $\binom{k}{k} + \binom{k+1}{k} + ... + \binom{n+k-1}{k} = \binom{n+k}{k+1}.$

X. 31. Calculate the sums:

a) $S_1 = 1\cdot2\cdot3 + 2\cdot3\cdot4 + ... + n\cdot(n+1)\cdot(n+2);$

b) $S_2 = 1\cdot2\cdot3\cdot...\cdot k + 2\cdot3\cdot4\cdot...\cdot k + ... + n\cdot(n+1)\cdot...\cdot(n+k-1).$

X. 32. Calculate the sums: **a)** $\displaystyle\sum_{k=2}^{n}\binom{k}{2};$ **b)** $\displaystyle\sum_{k=2}^{n}\binom{k}{2}^2.$

X. 33. Find the positive integers a, b, c such that

$$\sum_{k=1}^{n} k(bk+a)(ak+c) = n(n+1)(n+2)(n+3)$$

is true for any positive integer n.

X. 34. Prove hat

$$\binom{n}{k}^p = \binom{n-1}{k-1}^p + \dfrac{n^p - k^p}{(n-k)^p}\binom{n-1}{k}^p, (\forall)k,n,p \in N^*, k < n.$$

X. 35. Show that $\sqrt{\binom{n+1}{k}\cdot\binom{n-1}{k}} \le \binom{n}{k} \le \dfrac{\binom{n+1}{k}+\binom{n-1}{k}}{2},$

n, $k \in N$, $n \ge 1$.

X. 36. Prove that for any integer positive n, the number $\binom{2n}{n}$ is divizible by $n + 1$.

X. 37. Prove the identity $\displaystyle\sum_{k=0}^{n} \dfrac{k\binom{n}{k+1}}{\binom{n}{k}} = \dfrac{n(n+1)}{2}.$

X. 38. Prove that the fraction $\dfrac{\binom{n-1}{n+1}}{\prod\limits_{k=1}^{n} k!}$ is a natural number for

any non zero positive integer n.

X. 39. Prove the identities:

a) $\dfrac{1}{\binom{2n}{k}} = \dfrac{2n+1}{2(n+1)}\left(\dfrac{1}{\binom{2n+1}{k}} + \dfrac{1}{\binom{2n+1}{k+1}} \right)$;

b) $\displaystyle\sum_{k=0}^{2n}(-1)^k \cdot k!(2n-k)! = \dfrac{(2n+1)!}{n+1}$.

X. 40. Expand:

a) $(1+2x)^4$; b) $\left(x+\sqrt{x^2-1}\right)^6 + \left(x-\sqrt{x^2-1}\right)^6$.

X. 41. Find the fifth term of the expansion $\left(\sqrt[3]{x} + \dfrac{1}{\sqrt[3]{x}}\right)^9$.

X. 42. Write down the expansion by the binomial theorem $\left(\dfrac{x}{2} + 2y\right)^5$, and collect the terms.

X. 43. a) Find, in ascending powers of x, the first three terms in the expansion of $(3x-2)^8$.

b) Find the independent term of x from the expansion of $\left(x^2 - \dfrac{2}{x^2}\right)^4$.

X. 44. Find the independent term of x from the expansion $\left(x + \dfrac{2}{\sqrt[5]{x}}\right)^{2010}$.

X. 45. a) Find the value of the expression $\left(\sqrt{2}+1\right)^6 + \left(\sqrt{2}-1\right)^6$.

b) If the coefficients of x^2 and x^3 in the expansion of $(3+ax)^9$ are the same, find the value of a.

X. 46. Find the exact value of $\left(\sqrt{3}+1\right)^5 - \left(\sqrt{3}-1\right)^5$. Hence show, without calculators, that the value of $\left(\sqrt{3}+1\right)^5$ lies between 152 and 153.

X. 47. Using the formula $b_1 \dfrac{r^n - 1}{r - 1}$ for the sum of the first n terms of a geometrical progression (r ratio), calculate the sum of the series $1 + (1+x) + (1+x)^2 + ... + (1+x)^{20}$. Using the binomial theorem verify that the coefficients of x, in the given expression and in your answer, are equal.

X. 48. In the expansion of $(1+x)^n$ the first terms are 1+3+4+.... Calculate the values of n and x, and the value of the fourth term of the expansion.

X. 49. Find the coefficient of x^{-16} in the expansion of $\left(x^2 - \dfrac{1}{x}\right)^{25}$.

X. 50. The expansion of $(1 + ax)^n$, where $n > 0$, by the binomial theorem is $1 + 20x + 45a^2 x^2 + bx^3 + ...$. Calculate n, a, and b.

X. 51. The first three terms of the binomial expansion are $1 - \dfrac{16}{\sqrt{3}} x + \dfrac{112}{3} x^2$. Find the forth term.

X. 52. In the binomial expansion of $\left(1 + \dfrac{1}{5}\right)^n$, the second and third terms are equal. Calculate the value of n.

X. 53. In the expansion $\left(1 + \sqrt[3]{x}\right)^n$ the binomial coefficient of the sixth and eleventh terms are equal. Find the value of n.

X. 54. In the expansion $\left(\dfrac{\sqrt{a}}{\sqrt[5]{x}} + \dfrac{\sqrt[3]{x}}{\sqrt{a}} \right)^n$ the binomial coefficient of the third term is 190. Find the value of n.

X. 55. In the expansion of $(k + x)^8$, where k is a positive number, the coefficients of x^2 and x^3 are equal. Find the value of k.

X. 57. Find the positive integer n, $n \square 1$ such that the fourth term of the expansion $\left(\dfrac{\sqrt[5]{x^4}}{\sqrt[n]{x^{n-1}}} + x \cdot {}^{n+1}\!\sqrt{x^{n-1}} \right)^8$ is $56 \cdot x^{\frac{11}{2}}$.

X. 58. Find the term in the expansion $\left(x + \dfrac{1}{x^2} \right)^{12}$ which does not contain x.

X. 59. Find the term in the expansion $\left(x - \dfrac{1}{x^2} \right)^{3n}$ which does not contain x.

X. 60. Find the sixth term in the expansion $\left(x^{\frac{1}{2}} + y^{\frac{1}{3}} \right)^n$ if the binomial coefficient of the third term from the end is 45.

X. 61. Find out which term of the binomial expansion $\left(\sqrt{x} + \dfrac{1}{\sqrt[4]{x^3}} \right)^n$ contains $x^{\frac{13}{2}}$ if the ninth term has the largest coefficient.

X. 62. In the binomial expansion $\left(x \cdot \sqrt[5]{\dfrac{x}{3}} - \dfrac{y}{\sqrt[7]{x^3}} \right)^n$ the sum of the binomial coefficients of the odd terms is 2,048. Find the term which contains x^3.

X. 63. Let a_k, a_{k+1}, a_{k+2}, a_{k+3}, $1 \le k \le n-3$ ($n \in \mathbf{N}$) be four consecutive binomial coefficients in the expansion $(x+y)^n$. Prove that

$$\frac{a_k}{a_k + a_{k+1}} + \frac{a_{k+2}}{a_{k+2} + a_{k+3}} = \frac{2a_{k+1}}{a_{k+1} + a_{k+2}}.$$

X. 64. In the expansion $\left(x^2 + \dfrac{1}{\sqrt[3]{x}} \right)^n$

a) the sum of the second, the third, and the fourth binomial terms is 41.

b) the sum of the first three binomial terms is 56. Find the value of n in each case.

X. 65. Find the coefficient of the expansion

$$\left(\sqrt[3]{\frac{x}{\sqrt{y}}} + \sqrt{\frac{y}{\sqrt[3]{x}}} \right)^{21}, \text{ such that } x \text{ and } y \text{ have equal exponents.}$$

X. 66. Find the positive number x, if the third coefficient of the expansion $\left(x + x^{\log x} \right)^5$ is 10^6, log is in the base 10.

X. 67. Find the integer positive n such that the tenth term of the expansion $(5 + m)^m$ is the largest.

X. 68. Find n and x in the expansion $\left(\sqrt{2^x} + \sqrt{2^{1-x}} \right)^n$ knowing that the sum of the first three coefficients is 22 and the sum between the third and the fifth term is 135.

X. 69. For what value of x, the third term of the expansion

$$\left(\frac{1}{\sqrt[7]{x^2}} + x^{\log \sqrt{x}} \right)^9 \text{ is } 36\,000?$$

X. 70. Find n and x in the expansion $\left(\sqrt{3^{x+1}} + \dfrac{1}{\sqrt{3^x}} \right)^n$ if the

sum of the first three binomial coefficients is 22 and the sum of the third an fifth terms is 420.

X. 71. In the expansion $\left(2 \cdot \sqrt[x]{2^{-1}} + \dfrac{4}{\sqrt[4-x]{4}} \right)^6$, the forth term is

240. Find the number real x.

X. 72. In the expansion $\left(\sqrt{x^{\frac{1}{1+\log x}}} + \sqrt[12]{x} \right)^6$, the forth term is

200. Find the number real x.

X. 73. In the binomial expansion $\left(\sqrt{x} + \dfrac{1}{2 \cdot \sqrt[4]{x}} \right)^n$ the first

three binomial coefficients form an arithmetic progresson. Find the rational terms in the expansion.

X. 74. The sixth term of the expansion

$\left[\sqrt{2^{\log(10-3^x)}} + \sqrt[5]{2^{(x-2)\log 3}} \right]^n$ is 21 and the coefficients of the second,

third, and fourth terms are in an arithmetic progression. Find the numerical value of x and n.

X. 75. In the expansion $\left(x \cdot \sqrt[3]{x} - \dfrac{2}{\sqrt[5]{x^2}} \right)^n$ the sum of the

binomial coefficients is 256. Find the term containing x^2.

X. 76. In the expansion $\left(\sqrt[5]{x} - \dfrac{1}{\sqrt[5]{x^3}} \right)^n$ the sum of the

coefficients of odd terms is 256. Find the term containing $\dfrac{1}{x^3}$.

X. 77. Find the rational terms of the expansions:

a) $\left(\sqrt[3]{3}+\sqrt{2}\right)^{5}$; b) $\left(\sqrt[3]{4}+\sqrt[5]{2}\right)^{20}$; c) $\left(\sqrt{3}+\sqrt[14]{5}\right)^{200}$; d) $\left(\sqrt[5]{2}-\sqrt[7]{3}\right)^{24}$.

X. 78. Find the irrational terms of the expansion $\left(\sqrt[5]{3}-\sqrt[7]{2}\right)^{24}$.

X. 79. Consider the binomial expansion $\left(\sqrt[3]{x}+\dfrac{1}{2\cdot\sqrt[5]{x^{2}}}\right)^{100}$.

Determine in each case:
a) the 4th binimial coefficient of the expression;
b) the term containing x^{26} ;
c) the sum of the coefficients of the expansion;
d) the middle term of the expansion;
e) the number of rational terms of the expansion when $x = 3$.

X. 80. Find the sum of the coefficients of the expansions:

a) $\left(9x^{3}-8y\right)^{20}$;   b) $\left(7x^{4}-5y\right)^{22}$.

X. 81. Find the largest term in the binomial expansion:

a) $\left(\dfrac{1}{2}+\dfrac{1}{2}\right)^{100}$;   b) $\left(\dfrac{9}{10}+\dfrac{1}{10}\right)^{100}$;   c) $\left(\sqrt{2}+\sqrt{5}\right)^{20}$.

X. 82. The coefficient of x^{2} in the expansion of $(2+x)(1-ax)^{5}$ is zero. Find the positive value of a.

X. 83. Solve for x the following equations:

a) $\displaystyle\sum_{k=0}^{9}\binom{9}{k}\cdot(-1)^{k}x^{9-k}=1$; b) $\displaystyle\sum_{k=0}^{13}\binom{13}{k}\cdot(-1)^{k}x^{13-k}2^{2k}=0$;

c) $\displaystyle\sum_{k=0}^{13}\binom{10}{k}\cdot2^{-k}\sin^{10-k}x=0$.

X. 84. Simplify:

a) $\displaystyle\sum_{k=0}^{n}\binom{n}{k}\sin^{2k}x\cos^{n-2k}x$;

b) $\displaystyle\sum_{k=0}^{n}\binom{n}{k}(-1)^{k}\cosh^{2k}x\cdot\sinh^{n-2k}x$.

X. 85. Prove that for any positive integer n,
$\left(2+\sqrt{2}\right)^{n}+\left(2-\sqrt{2}\right)^{n}$ is an integer positive number.

X. 86. Prove the identities:

a) $1-\binom{n}{2}+\binom{n}{4}-\binom{n}{6}+\ldots=2^{\frac{n}{2}}\cos\dfrac{n\pi}{4}$;

b) $\binom{n}{1}-\binom{n}{3}+\binom{n}{5}-\binom{n}{7}+\ldots=2^{\frac{n}{2}}\sin\dfrac{n\pi}{4}$;

c) $\left[1-\binom{n}{2}+\binom{n}{4}-\binom{n}{6}+\ldots\right]^{2}+\left[\binom{n}{1}-\binom{n}{3}+\binom{n}{5}-\binom{n}{7}+\ldots\right]^{2}=2^{n}$;

d)
$\left[1-3\binom{n}{2}+9\binom{n}{4}-27\binom{n}{6}+\ldots\right]^{2}+3\left[\binom{n}{1}-3\binom{n}{3}+9\binom{n}{5}-27\binom{n}{7}+\ldots\right]^{2}=2^{2n}$.

X. 87. Prove the identities:

a) $\binom{n}{0}+\binom{n}{3}+\binom{n}{6}+\ldots=\dfrac{1}{3}\left(2^{n}+2\cos\dfrac{n\pi}{3}\right)$;

b) $1-3\binom{n}{2}+9\binom{n}{4}-27\binom{n}{6}+\ldots=(-1)^{n}2^{n}\cos\dfrac{2n\pi}{3}$;

c) $\binom{n}{1}-3\binom{n}{3}+9\binom{n}{5}-\ldots=\dfrac{(-1)^{n+1}2^{n+1}}{\sqrt{3}}\sin\dfrac{2n\pi}{3}$.

X. 88. Prove the identity

$\binom{n}{1}-\dfrac{1}{3}\binom{n}{3}+\dfrac{1}{9}\binom{n}{5}-\dfrac{1}{27}\binom{n}{7}+\ldots=\dfrac{2^{n}}{3^{\frac{n-1}{2}}}\sin\dfrac{n\pi}{6}$.

X. 89. Prove the identity $\displaystyle\sum_{k=0}^{3n} \binom{6n}{2k}(-3)^k = 2^{6n}$.

X. 90. Prove the identities:

a) $\binom{n}{0}+\binom{n}{4}+\binom{n}{8}+\ldots = \dfrac{1}{2}\left(2^{n-1}+2^{\frac{n}{2}}\cos\dfrac{n\pi}{4}\right)$;

b) $\binom{n}{1}+\binom{n}{5}+\binom{n}{9}+\ldots = \dfrac{1}{2}\left(2^{n-1}+2^{\frac{n}{2}}\sin\dfrac{n\pi}{4}\right)$;

c) $\binom{n}{2}+\binom{n}{6}+\binom{n}{10}+\ldots = \dfrac{1}{2}\left(2^{n-1}-2^{\frac{n}{2}}\cos\dfrac{n\pi}{4}\right)$;

d) $\binom{n}{3}+\binom{n}{7}+\binom{n}{11}+\ldots = \dfrac{1}{2}\left(2^{n-1}-2^{\frac{n}{2}}\sin\dfrac{n\pi}{4}\right)$.

X. 91. Calculate the following sums:

a) $\binom{n}{0}+\binom{n}{1}+\binom{n}{2}+\ldots+\binom{n}{n}$;

b) $\binom{n}{1}+2\binom{n}{2}+3\binom{n}{3}+\ldots+n\binom{n}{n}$;

c) $\binom{n}{0}+2\binom{n}{1}+3\binom{n}{2}+\ldots+(n+1)\binom{n}{n}$;

d) $\binom{n}{2}+2\binom{n}{3}+3\binom{n}{4}+\ldots+(n-1)\binom{n}{n}$;

e) $\binom{n}{0}+3\binom{n}{1}+5\binom{n}{2}+\ldots+(2n+1)\binom{n}{n}$;

f) $k\binom{n}{0}+(k+1)\binom{n}{1}+(k+2)\binom{n}{2}+\ldots+(k+n)\binom{n}{n}$;

g) $\binom{n}{0}-2\binom{n}{1}+3\binom{n}{2}+\ldots+(-1)^n(n+1)\binom{n}{n}$;

h) $3\binom{n}{1}+7\binom{n}{2}+11\binom{n}{3}+\ldots+(4n-1)\binom{n}{n}$;

i) $\binom{n}{1}-2\binom{n}{2}+3\binom{n}{3}-\ldots+(-1)^{n-1}n\binom{n}{n}$;

j) $\dfrac{\binom{n}{0}}{1}+\dfrac{\binom{n}{1}}{2}+\dfrac{\binom{n}{2}}{3}+\ldots+\dfrac{\binom{n}{n}}{n+1}$;

k) $\dfrac{\binom{n}{0}}{2}+\dfrac{\binom{n}{1}}{3}+\dfrac{\binom{n}{2}}{4}+\ldots+\dfrac{\binom{n}{n}}{n+2}$;

l) $\dfrac{\binom{n}{0}}{1\cdot 2}+\dfrac{\binom{n}{1}}{2\cdot 3}+\dfrac{\binom{n}{2}}{3\cdot 4}+\ldots+\dfrac{\binom{n}{n}}{(n+1)(n+2)}$;

m) $\dfrac{\binom{n}{0}}{1} - \dfrac{\binom{n}{1}}{2} + \dfrac{\binom{n}{2}}{3} - \ldots + (-1)^n \dfrac{\binom{n}{n}}{n+1}$.

X. 92. Calculate the following sums:

a) $\binom{n}{0} - \binom{n}{1} + \binom{n}{2} - \ldots + (-1)^n \binom{n}{m}$;

b) $\binom{n}{k} + \binom{n+1}{k} + \binom{n+2}{k} + \ldots + \binom{n+m}{k}$.

X. 93. Calculate the following sums:

a) $\binom{n}{0}^2 + \binom{n}{1}^2 + \binom{n}{2}^2 + \ldots + \binom{n}{n}^2$;

b) $\binom{n}{0}^2 - \binom{n}{1}^2 + \binom{n}{2}^2 + \ldots + (-1)^n \cdot \binom{n}{n}^2$;

c) $\binom{2n}{0} - \binom{2n-1}{1} + \binom{2n-2}{2} - \ldots + (-1)^n \binom{n}{n}$;

d) $\binom{2n}{n} + 2\binom{2n-1}{n} + 4\binom{2n-2}{n} + \ldots + 2^n \binom{n}{n}$;

e) $\binom{n}{1}^2 + 2\binom{n}{2}^2 + 3\binom{n}{3}^2 + \ldots + n\binom{n}{n}^2$.

X. 94. Prove that $\displaystyle\sum_{k=0}^{p} \binom{p}{k} P(m,k) P(n-p, m-k) = P(n,m)$.

X. 95. Prove that the equality $\binom{3n}{n} = \displaystyle\sum_{k=0}^{n} \binom{2n}{k}\binom{n}{k}$ holds

for any $n \geq 1$.

X. 96. Prove the identity $\left(\displaystyle\sum_{k=0}^{n} 2^k \binom{2n}{2k} \right)^2 - 2 \cdot \left(\displaystyle\sum_{k=0}^{n-1} 2^k \binom{2n}{2k+1} \right)^2 = 1$.

X. 97. Calculate the sums

$S_1 = 1 + \binom{n}{1}\cos x + \binom{n}{2}\cos 2x + \ldots + \binom{n}{n}\cos nx$ and

$S_2 = \binom{n}{1}\sin x + \binom{n}{2}\sin 2x + \ldots + \binom{n}{n}\sin nx$.

ANSWERS and SOLUTIONS

Chapter I

NATURAL AND INTEGER NUMBERS

I. 1. a) >; b) <; c) =; d) <; e) =; f) <; g) =; h) =; i) =; j) =.
I. 2. a) 4; **b)** 52; **c)** 15; **d)** 28; **e)** 112; **f)** 42; **g)** 89; **h)** 33;
i) 589; **j)** 33.
I. 3.

x	18	59	76
y	22	12	15
$x+y$	40	71	91

a	b	c	$a+b$	$a+c$	$a+b+c$
32	14	19	46	51	65
14	15	13	29	27	39

I. 4. a) 1; **b)** 5; **c)** 5; **d)** 5; **e)** 2; **f)** $b = 40$; **g)** $x = 1$;
h) $b = 52$; **i)** $a = 50$; **j)** $a = 40$.
I. 5. a)

$43 + 18 = 61$	$55 - 25 - 8 = 22$
$35 + 29 = 64$	$84 - 56 - 8 = 20$
$83 - 27 = 56$	$38 + 29 + 17 = 84$
$84 - 36 = 48$	$46 + 9 - 16 = 39$

b) $2 + 13 + 7 + 18 = (13 + 7) + (2 + 18) = 40.$ **c)**

$n = 51$	$a = 36$
$d = 4$	$a = 24$
$x = 11$	$x = 11$

d) 71; **e)** 120; **f)** 75;

g) 0, 5, 10, <u>15</u>, <u>20</u>, <u>25</u>, <u>30</u>, <u>35</u>, <u>40</u>, <u>45</u>, 50;

0, 3, 6, <u>9</u>, <u>12</u>, <u>15</u>, <u>18</u>, <u>21</u>, <u>24</u>, <u>27</u>, 30;

0, 10, 20, <u>30</u>, <u>40</u>, <u>50</u>, <u>60</u>, <u>70</u>, <u>80</u>, <u>90</u>, 100; **h)** i) 13,806. ii) 582.
iii) 571. iv) 197.

i) 7 times. **j)** $a = 42$; $a = 6$; $a = 9$. **I. 6.**

a	b	c	d	c-b	$3a$-b	$3c$-d+1	a+$3b$-c	$3b$+$2c$-d+2
7	21	63	189	42	0	1	7	2
10	16	16	48	0	14	1	42	34
7	15	15	30	0	6	16	37	47
5	15	45	135	30	0	1	5	2
4	19	36	144	17	− 7	−35	25	−13

I. 7. a) 0; **b)** 0; **c)** 217; **d)** 3; **e)** 1; **f)** 0; **g)** 0; **h)** 0; **i)** 0; **j)** 0.

I. 8. a) −5; **b)** 1; **c)** 1; **d)** −8; **e)** 8; **f)** −8; **g)** − 3.

I. 9. a) −49; **b)** −65; **c)** 189; **d)** 51; **e)** 1; **f)** −3; **g)** 16.

I. 10. a) 1; **b)** 72; **c)** −1; **d)** 0; **e)** 4; **f)** 8; **g)** −16; **h)** −10.

I. 11. a) −12; **b)** −63; **c)** −58; **d)** 17.

I. 12.

a	b	c	a-b	b-c	$(a$-$b)$+$2c$	$(2a$-$2b)$+$(b$-$c)$
26	15	14	11	1	39	23
26	25	24	1	1	49	3
125	100	100	25	0	225	50
45	40	30	5	10	65	20
60	16	6	30	10	42	70
8	7	1	1	6	3	8

I. 13. a) 1; **b)** −46; **c)** 40; **d)** −39; **e)** 65. **I. 14. a)** 6; **b)** −12; **c)** 6;
d) −31; **e)** 35; **f)** −7. **I. 15. a)** −144; **b)** 48; **c)** −192; **d)** −36; **e)** 7;

f) –5; **g)** –3; **h)** 4. **I. 16. a)** –30; **b)** 9; **c)** 23; **d)** –1; **e)** –4; **f)** 1;
g) –1; **h)** –6; **i)** 73; **j)** –17; **I. 17. a)** 60; **b)** 120; **c)** –32; **d)** 39;
e) –130; **f)** –72; **g)** 14. **I. 18. a)** $(80 + 9 - 9) \div (-40) = -2$;
b) $50 \cdot (59 - 59) = 0$; **c)** $22 \div (17 - 6) = 2$; **d)** $3 \cdot (18 + 11) = 87$;
e) $7 \cdot 17 = 119$; **f)** $10 + 53 = 63$; **g)** 1; **h)** –218; **i)** 5; **j)** $12 - 7 = 5$;
k) $-3 - 4 + 7 = 0$; **l)** $-12 \div (-6) - 7 = 2 - 7 = -5$.
I. 19. a) 5; **b)** $-3 \cdot [72 - 65 - (34 - 45)] = -3 \cdot 18 = -54$; **c)** 92;
d) $-9 \cdot [-5 - (12 - 9)] \cdot (-4) = -9 \cdot (-8) \cdot (-4) = -288$.
I. 20. a) –11; **b)** 1; **c)** –3; **d)** –124; **e)** 30; **f)** 18;

I. 21.

a	b	c	d	a²c-3a²b+a+b	a²+3ab²-abc
6	18	54	162	24	36
3	9	27	81	12	9
11	33	99	297	44	121
9	27	81	243	36	81
17	51	153	459	68	289

I. 22. a) –8; **b)** –5; **c)** 225; **d)** –1; **e)** 1; **f)** 17; **g)** 1; **h)** 1; **i)** 0;
j) -144; **k)** 190; **l)** $-29 \cdot 10^{23}$; **m)** 2^{81}; **n)** 3^{8}; **o)** 0; **p)** -1.
I. 23. a) -2^{12}; **b)** 3^{12}; **c)** 2^{5}; **d)** -1. **I. 24. a)** 1; **b)** 3 ;
c) -2^{4}; **d)** 3. **I. 25. a)** –4998; **b)** –1; **c)** 1.
I. 26. a) 3^{6}; **b)** –3; **c)** 1; **d)** 1; **e)** 7; **f)** -2^{111}.
I. 27. a) $8 \cdot \left\{ 10 + 1 \cdot \left[2^4 \cdot 2^4 \div 2^4 - 5 \cdot (27 - 25) \right] \div 3 + 2 \right\} =$
$= 8 \cdot \left[10 + (16 - 10) \div 3 + 2 \right] = 8 \cdot (10 + 6 \div 3 + 2) = 8 \cdot (10 + 2 + 2) = 112$;
b) $\left(5^2 + 3^2 \right) \div 2 = (25 + 9) \div 2 = 34 \div 2 = 17$;
c) $\left(2 \cdot 3^2 \right) + 1 = 18 + 1 = 19$; **d)** $\left(3^2 + 2^4 + 5^9 \right) \div \left(3^2 + 2^4 + 5^9 \right) = 1$.
I. 28. a) 0; **b)** $(-1)^{n+1}$; **c)** $(-1)^{n} + 2$; **d)** –50.
I. 29. a) i) $1 - 8 + 9 = 2$; **ii)** $(25 + 5 + 1) \div (27 + 4) = 31 \div 31 = 1$;

b) $\left(1000 \div 125 + 2^6 \div 2^4 - 27\right) \cdot 2 + 1 + 1 = \left(8 + 4 - 27\right) \cdot 2 + 2 = -28$;

c) $968 - \{182 - 5 \cdot [8 \cdot 5 \div (9 \cdot 5 \div 15 + 5) + 2 \cdot (13 - 12)]\} \cdot 5 =$
$= 968 - \{182 - 5 \cdot [40 \div (45 \div 15 + 5) + 2]\} \cdot 5 =$
$= 968 - \{182 - 5 \cdot [40 \div (3 + 5) + 2]\} \cdot 5 = 968 - [182 - 5 \cdot (5 + 2)] \cdot 5 =$
$= 968 - (182 - 5 \cdot 7) \cdot 5 = 968 - 147 \cdot 5 = 968 - 735 = 233$;

d) i) $4^2(15 - 14) = 4^2 = 16$, **ii)** $11^2(20 - 18 - 1) = 11^2 = 121$,

iii) $11^2(30 - 28) = 11^2 \cdot 2 = 121 \cdot 2 = 242$, **iv)** $2 + 2 = 4$,

v) $3^2 + 1 = 9 + 1 = 10$; **vi)** 24; **e)** 100; **f)** 326. **I. 30. a)** $-9x^5$;

b) $12m^6$; **c)** $-8x^7$; **d)** $40a^4b^4$; **e)** $14x^6y^3$; **f)** $72x^4y$;

g) $-12a^3b^4c$; **h)** $10m^4n^6$; **i)** $-20a^8b^7$.

I. 31. c). I. 32. a) $8a$; **b)** $3x^2y$; **c)** $-5mn$; **d)** $4m^2 - 3$;

e) $-2x^3 + 4x^3y$; **f)** $-rs^2 + 6rs$; **g)** $-7a^2b + 2ab$ **h)** x^2yz;

i) $-2x^2y + 5xy^2$.**I. 33.**

a	b	c	d	$6a^3c \div a - 18a^3b \div a + b$	$3a^3c^2 \div b - a^3cd \div b + b \div a$
5	15	45	135	15	3
3	9	27	81	9	3
1	3	9	27	3	3
9	27	81	243	27	3
17	51	153	459	51	3

I. 35. $7 \mid (5x + y) \Rightarrow 7 \mid [7x - (5x + y)] \Rightarrow 7 \mid (2x - y)$, $x, y \in \mathbf{Z}$.
I. 36. a) $a = 3$; **b)** $a = 4$; **c)** $a = 3$; **d)** $a = 7$; **e)** $a = 9$; **f)** $a = 14$;
g) $a = 77$; **h)** $a = 38$; **i)** $a = 40$; **j)** $x = 17$. **I. 37. a)** $x = 7$; **b)** $y = 9$;
c) $z = 5$; **d)** $x = -4$; **e)** $b = -3$; **f)** $x = 10$; **g)** $x = 39$; **h)** $x = -5$;
i) $y = 0$; **j)** $z = 19$; **k)** $b = 0$; **l)** $x = 15$; **m)** $x = 14$; **n)** $x = -11$.
I. 38. a) $x = 12$; **b)** $y = -4$; **c)** $a = -4$; **d)** $x = -16$; **e)** $y = 5$;
f) $z = 3$; **g)** $x = -8$; **h)** $t = -11$; **i)** $x = -1$.
I. 39. a) $x = 3$; **b)** $y = 22$; **c)** $z = -6$; **d)** $t = -18$; **e)** $x = -7$;

f) $y = -4$; **g)** $x = -4$; **h)** $x = -4$; **i)** $y = -45$. **I. 40. a)** $x = -8$;
b) $x = 3$; **c)** $y = -3$; **d)** $x = -5$; **e)** $t = 2$; **f)** $x = 7$; **g)** $a = -5$;
h) $x = -3$; **i)** $y = -15$; **j)** $x = -1$; **k)** $a = 12$; **l)** $t = -4$; **m)** $y = 3$;
n) $x = -5$; **o)** $z = 12$; **p)** $x = -7$; **q)** $x = 0$; **r)** $x = 1$; **s)** $x = 3$; **t)** $x = 6$;
u) $x = -2$; **v)** $x = -17$; **w)** $y = 4$; **x)** $x = -2$; **y)** $z = 5$; **z)** $a = -4$.
I. 41. a) $x = 7$; **b)** $x = 12$; **c)** $y = 20$; **d)** $a = 17$; **e)** $x = 12$;
f) $x = 13$; **g)** $x = -9$; **h)** $y = 4$; **i)** $x = 7$; **j)** $x = -5$; **k)** $y = 9$; **l)** $a = 5$;
m) $x = -6$; **n)** $y = 3$; **o)** $t = 3$; **p)** $x = -12$; **q)** $x = 9$; **r)** $x = 3$;
s) $x = -2$; **t)** $x = -3$; **u)** $x = 1$; **v)** $x = 1$; **w)** $x = 1$; **x)** $x = -1$.
I. 42. a) $x > 3$ or $x \in (3, \infty)$; **b)** $x \le 7$ or $x \in (-\infty, 7]$;
c) $x < 4$ or $x \in (-\infty, 4)$; **d)** $x \le -7$ or $x \in (-\infty, -7]$;
e) $x < 4$ or $x \in (-\infty, 4)$; **f)** $x \le -8$ or $x \in (-\infty, -8]$;
g) $x > -5$ or $x \in (-5, \infty)$; **h)** $x \le 5$ or $x \in (-\infty, 5]$;
i) $x \le 3$ or $x \in (-\infty, 3]$; **j)** $x < 18$ or $x \in (-\infty, 18)$;
k) $x \le 9$ or $x \in (-\infty, 9]$.
I. 43. a) $x > 5$ or $x \in (5, \infty)$; **b)** $x > 9$ or $x \in (9, \infty)$;
c) $x > -5$ or $x \in (-5, \infty)$; **d)** $x > 1$ or $x \in (1, \infty)$;
e) f) $x \le 3$ or $x \in (-\infty, 3]$; **f)** $x \le 6$ or $x \in (-\infty, 6]$;
g) $x < 1$ or $x \in (-\infty, 1)$; **h)** $x > -2$ or $x \in (-2, \infty)$;
i) $x \le -1$ or $x \in (-\infty, -1]$.
I. 44. a) $x < -2$ or $x \in (-\infty, -2)$; **b)** $x < -$ or $x \in (-\infty, -9)$;
c) $x \ge -14$ or $x \in [-14, \infty)$; **d)** $y \ge 11$ or $y \in [11, \infty)$;
e) f) $a \le -3$ or $a \in (-\infty, -3]$; **f)** $y > 13$ or $y \in (13, \infty)$.
I. 45. a) $x \in (-3, 4]$; **b)** $x \in [2, 4)$; **c)** $x \in [-4, -1)$; **d)** $x \in (1, 6]$;
e) $x \in (-1, 4]$; **f)** $x \in [-2, 4)$; **g)** $x \in (-2, 2]$; **h)** $x \in [-1, 3]$;
i) $x \in [1, 7]$; **j)** $x \in [-1, 1)$.

TEST I. 1

1) 1,4,7,10,13,16 and 3,7,11,15,19,23; **2) a)** 865 030;
b) 7 005 806; **3) a)** 2000; **b)** 35 000; **c)** 2886; **d)** 57; **e)** 79; **f)** 83;
4) 92; **5)** 47; **6) a)** $35.19; **b)** $11.73; **7)** $5.78; **8)** $5.78;
9) Matthew will walk 400m, 1600m, 4800 m and Alex
425m, 1700m, 5100 m, **10)** $7.33.

TEST I. 2

1) $a = 31$; **2)** $a = 80$; **3)** $a = 12$; **4)** $a = 56$; **5)** $a = 19$;

6) $a = 2$; **7)** $a = 14$; **8)** $a = 64$; **9)** $a = 41$; **10)** $a = 36$.

TEST I. 3

1) $100+(-1+101)+(-2+102)+(-3+103)+(-4+104)-5=500-a$
$100+100+100+100+100-5=500-a$. Therefore $a = 5$.
2) Let x be my age. $(4x+24) \div 2 - 2x = 2x+12-2x=12$.
Therefore my age is 12.
3) $2^7 - 2^6 - 2^5 - 2^4 = 2^4(2^3 - 2^2 - 2 - 1) = 2^4$
4) If you subtract 6 from one of the numbers it would be the same as if you subtract 1 from each of the six numbers, so that means their average is 1 less than the 12. Therefore the answer is 11.
5) The perimeter of the hexagon can't be 5 because if one side is 1, for example, the minimum perimeter would be 6, and 77.
6) $1^2 + 2^2 + 3^2 + ... + 18^2 + 19^2 + 20^2 = 2870$.
So then $1^2 + 2^2 + 3^2 + ... + 18^2 + 19^2 = 2870 - 20^2 = 2870 - 400 = 2470$.
7) $[(2+8)+(4+6)+10] \div [10+(8+2)+(6+4)]=1$.
8) A cube has 12 edges because there are 4 top edges, 4 bottom edges, and 4 vertical edges.
9) The amount he spent is $13 \times 7¢ = 91¢$. The amount he has left is $\$2.50 - 91¢ = \$1.59$. With 14¢ for a gumball he can buy 11 gumballs because $\$1.59 \div 14¢ = 11.35$ so he will have 5¢ left over.
10) $270 \div 30 = 9$. Therefore it will take her 9 days to read the whole book.

TEST I. 4

1) 273; **2)** i) $x = 153$, ii) $x = 4$, iii) $x = 9$, iv) $x = 40$; v) $x = 9$;
3) 6 ; **4)** $a = 36$; **5)** 250 ; **6)** 70 ; **7)** $a = 190, b = 38$;
8) $a = 212, b = 53$; **9)** $a = 503, b = 10$; **10)** \$30.

TEST I. 5

1) a) 6; **b)** 9; **c)** 0.8, d) 9; **2)** $a = 13$; **3)** $x=185.24$; **4)** 7;
5) $a = 45, b = 70, c = 25$; **6)** 10 pigs and 30 hens; **7)** a = 165 ;
8) 424.5; **9)** $a = 30, c = 15, b = 75$; **10)** 2950.

TEST I. 6

1) a) $a = 6$; **b)** $a = 9$; **c)** $a = 8$; **d)** $a = 9$; **2)** $a=12$; **3)** $x = 3$;
4) 7; **5)** $c = 55$; **6)** 153,846; **7)** 4,885; **8)** $a = 135, b = 89$;
9) 793; **10)** $x = 30, y = 156, z = 26$.

TEST I. 7

1) a) 2^9; **b)** 3^{10}; **c)** a^{15}; **d)** 7^{14}; **e)** 5^5; **f)** 2^2; **g)** 3^8;
i) 5^{12}; **2) a)** 2^8; **b)** 3^2; **c)** 2^{12}; **d)** 35^8; **3) a)** $2^{10} > 8^3$;
b) $32^{20} = 16^{25}$; **c)** $2^{15} < 3^{30}$; **d)** $3^{100} > 2^{150}$; **e)** $5^{30} < 3^{45}$;
4) a) 5; **b)** 2^5; **c)** 1; **d)** 32; **e)** 1; **f)** 4; **g)** 1;
5) a) $x = 4$, **b)** $x = 2^3$; **c)** $x = 3^2$; **d)** $x = 6$; **6) a)** $x = 4$;
b) $x = 66$; **c)** $x = 65$; **7)** $a > b$;
8) $a = 6^{2n+2}$, $b = 3^{2n+4} \cdot 2^{4n+4}$, $a < b$;
9) a) $n = 45$; **b)** $31 \cdot 3^n \cdot 5^{2n}$; **10)** $x = 4$.

TEST I. 8

1) a) $1 - 8 + 9 = 2$; **b)** $(25 + 5 + 1) \div (27 + 4) = 31 \div 31 = 1$;
2) $(1000 \div 125 + 2^6 \div 2^4 - 27) \cdot 2 + 1 + 1 = (8 + 4 - 27) \cdot 2 + 2 = -28$;
3) $968 - \{182 - 5 \cdot [8 \cdot 5 \div (9 \cdot 5 \div 15 + 5) + 2 \cdot (13 - 12)]\} \cdot 5 =$
$= 968 - \{182 - 5 \cdot [40 \div (45 \div 15 + 5) + 2]\} \cdot 5 =$
$= 968 - \{182 - 5 \cdot [40 \div (3 + 5) + 2]\} \cdot 5 = 968 - [182 - 5 \cdot (5 + 2)] \cdot 5 =$
$= 968 - (182 - 5 \cdot 7) \cdot 5 = 968 - 147 \cdot 5 = 968 - 735 = 233$
4) a) $4^2 (15 - 14) = 4^2 = 16$, **b)** $11^2 (20 - 18 - 1) = 11^2 = 121$,
c) $11^2 (30 - 28) = 11^2 \cdot 2 = 121 \cdot 2 = 242$, **d)** 4, **e)** $3^2 + 1 = 10$;
f) $2^7 - 2^6 - 2^5 - 2^4 + 2^3 = 2^3 (2^4 - 2^3 - 2^2 - 2 + 1) = 2^3 \cdot 3 = 24$;
5) $\{[(49 - 14) + 32 \div 4] \div (27 + 16) + 9\} \cdot 11 - [(4 + 9) \cdot 5 - 55] =$
$= [(35 + 8) \div 43 + 9] \cdot 11 - (13 \cdot 5 - 55) = (43 \div 43 + 9) \cdot 11 - (65 - 55) =$
$= 10 \cdot 11 - 10 = 110 - 10 = 100$;
6) $\{[(625 - 25) \div 25 + (32 + 64) \cdot 4] - 81\} - 1 =$
$= [(600 \div 25 + 96 \cdot 4) - 81] - 1 = [(24 + 384) - 81] - 1 = 326$;
7) $15 \cdot (3x - 14 + 9) + (2x + 25 \cdot 2 - 4 \cdot 9) \cdot 5 = 5 \cdot 8^2 \div 2$
$15 \cdot (3x - 5) + (2x + 50 - 36) \cdot 5 = 5 \cdot 64 \div 2$,
$15 \cdot (3x - 5) + (2x + 14) \cdot 5 = 160$, $3 \cdot (3x - 5) + (2x + 14) = 32$
$9x - 15 + 2x + 14 = 32$, $9x + 2x = 32 - 14 + 15$ $x = 3$;
8) $[(5x - 20 + 2^{10} \div 2^9 - 3^{10} \div 3^9) \cdot 5 - 2^{10} \div 2^9] \div 3 = 1$,

$[(5x-20+2-3)\cdot5-2]\div3=1$, $[(5x-21)\cdot5-2]\div3=1$,
$(25x-105-2)\div3=1$,
$25x-107=3$, $25x=3+107$, $25x=110$, $x=4.4$;
9) a) 5300; b) 49; c) 280; 10) $a=1$, $b=2^n$, $c=2^{2n}$, $n=1$.

TEST I. 9

1) a, $a+1$, $a+2$, $3(a+1)$; 2) a) 10; b) 50; 3) $a=6254$, $b=10$;
4) $(x-2)^{2x-4}=8^{16}$, $x=10$; 5) $A=6ab$; 6) $x=0, y=2^{1950}$, $(2\cdot3)^3$;
7) 891, 781, 561, 451, 341, 231, 121; 8) $E=0$; 9) $a=1$;
10) The last digit is 4.

TEST I. 10

1) 520; 2) 7; 3) $a=22$, $b=112$, $c=280$;
4) $(7\cdot8+4\cdot11)-50=14$; 5) 39; 6) 89; 7) 45; 8) 25;
9) 396.5; 10) 410.

TEST I. 11

1) a) 10; b) 25; c) 1; d) 8; e) 2; f) $6\frac{1}{9}$; g) 23; h) 0.15;

2) a) $x=6.98$; b) $x=20.47$; c) $x=0.4$; d) $x<4.2$;
e) $x=16.86$; f) $x\le20.5$; g) $x=4$ .3) a) 10^2; b) $(3^7)^2$;
c) 1000^2; d) $(2^4)^2$; 4) a) $a=4$, $b=-4$;
b) $x=28$; c) $-20,200$; 5) a) 11^2; b) $(2^5)^2$; c) 2010^2; d) $(2^8)^2$.
6) a) $10a+a+10b+b=11\cdot(a+b)$;
b) 18, 27, 36, 45, 54, 63, 72, 81; 7) $x+y=9$; 8) a) $\underbrace{100...00}_{37\ zeros}455$;

b) 15; c) $\overline{...5}-\overline{...5}$; d) odd –odd; 9) $N=6^{n-1}\cdot720$, the reminder
is 0; 10) $(\overline{a1b}+\overline{a2b}+...+\overline{a9b})\div(\overline{a5b})=9$.

TEST I. 12

2) $a=5$; 3) $a=3$; 4) $N=10^n\cdot x+2x=3x+10^n\cdot x-x=$
$=3x+x(10^n-1)=3x+3x\cdot3x=3x(3x+1)$, $N=\underbrace{333...3}_{n\ times}\cdot\underbrace{333...4}_{n\ times}$;

5) From (1) $a+b+c=d$,

$\overline{abc}+\overline{bca}+\overline{cab}=111(a+b+c)=111d=\overline{ddd}$; **6)** $a=3^2$; **7)** $a=$

$3111x$; **8)** $\overline{abc}=187$;

9) $\overline{abc2}=1000a+100b+10c+2$,

$3\cdot\overline{abc}=300a+30b+3c$, $\overline{4abc}=4000+100a+10\,b+c$,

$1200a+120b+12c=5232$, $100a+10b+c=436$,

$\overline{abc}=436$, $a=4$; $b=3$; $c=6$;

10) $20\cdot a+2\cdot b=2(10a+b)=2\cdot\overline{ab}\Rightarrow\overline{abb}^{\overline{ab}}=\left(\overline{ab}^2\right)^{\overline{ab}}\Rightarrow$

$\overline{abb}=\overline{ab}^2\Rightarrow10\overline{ab}+b=\overline{ab}^2$. Therefore b is a divisor of

$\overline{ab}\Rightarrow b=0$ and $a=1,\overline{ab}=10$.

TEST I. 13

1) $A=2002^2$; **2)** $A=\left(3^{n+2}\cdot2^{2n+2}\right)^2$;

3) $n=2^{2010}\left(2^3-2^2-2^1-2^0\right)=\left(2^{1005}\right)^2$;

4) $2^{6n+10}+2^{6n+10}+2^{6n+9}=5\cdot2^{3n+24}\Rightarrow2^{6n+9}(2+2+1)=5\cdot2^{3n+24}$,

$n=5$, the last digit is 2; **5)** $\left(2^0+2^1+2^2\right)\left(2^0+2^3+2^6+...+2^{96}\right)$;

6) $\left(3^1+3^2+3^3+3^4+3^5\right)\left(3^0+3^6+...+3^{2005}\right)$; **7)** $N=\overline{...7}$;

8) The last digit is 2; **9)** $a=2^{2n}\cdot3^{2n+2}(4\cdot9+2\cdot3-1)=6^{2n}\cdot369$;

10) $x=2^n$, $y=3^{n-1}$, $z=3$; $xyz=6^n$.

TEST I. 14

1) a) 123400; **b)** $a=1,\ b=0,\ c=1$;

2) a) $9^{60}=27^{40}$, **b)** $7^{1275}>5^{1275}$;

3) $a=2^{102}$, $b=2^{101}+2^{99}\cdot x$, $x=4$;

4) a) $x=\left(2^{1500}\right)^2\cdot5^2$, $y=5^2\cdot\left(3^{1000}\right)^2$;

b) $x=\left(2^{2999}\cdot5\right)^2\cdot10$; **c)** $x<y$; **5)** $x=121(a+b)$;

6) $x=110(a+b)$; **7)** $x=211(a+b)$;

8) $x\cdot1000+y\cdot100+z\cdot10+0+x\cdot100+y\cdot10+z=2002$,

$100x+10y+z=182$, $z=2$ and $y=8$;

9) $\overline{abc}\cdot10^{2007}+...+\overline{abc}\cdot10^3+\overline{abc}=\left(10^{2007}+...+\cdot10^3+10^0\right)\overline{abc}=$

$$= \left(10^3 + 10^0\right)\left(10^{2004} + ... + \cdot 10^3 + 10^0\right)\overline{abc}\,; \quad \textbf{10)}\ x = 222(a+b).$$

TEST I. 15

1) $x = \underbrace{1000000...00}_{2006\ times} + \underbrace{900...00}_{2002\ times} - 2 = \underbrace{1000900...00}_{2003\ times} - 2 = 10008\ \underbrace{9...9}_{2001\ times}\ 8$

Then the sum of the digits is $9 \times 2001 + 16 + 1 = 18{,}026$.

2) Let $a - 1$, a, $a + 1$ be the three consecutive numbers. We have

$$\left(a-1\right)^2 + a^2 + \left(a+1\right)^2 = 2\underbrace{99...998}_{n-1\ times}2\underbrace{00...00}_{n-2\ times}29.\ \text{Expanding and}$$

collecting terms we get $3a^2 = 2\underbrace{99...998}_{n-1\ times}2\underbrace{00...00}_{n-2\ times}27$ and

$a^2 = \underbrace{99...999}_{n-1\ times}4\underbrace{00...00}_{n-2\ times}09$. On the other hand

$$a^2 = 9 \cdot 10^{2n} + ... + 9 \cdot 10^{n+2} + 9 \cdot 10^{n+1} + 4 \cdot 10^n + 9 = \left(10^{n+1} - 3\right)^2.$$

Finally $a = 10^{n+1} - 3$. **4)** $x = 4$.

5) $x \cdot 1000 + y \cdot 100 + z \cdot 10 + 0 + x \cdot 100 + y \cdot 10 + z = 2002$

$x(1000 + 100) + y(100 + 10) + z(10 + 1) = 2002$

$1100x + 110y + 11z = 2002, \quad 100x + 10y + z = 182,$

$z = 2$, $y = 8$, $x = 1$.

6) a) $x = 3$, **b)** $x = 4.4$. **7)** $\overline{abc} = 148$.

8) Denoting $x = \underbrace{111...11}_{n\ times}$, we have

$$N = x \cdot 10^{n-1} + 2x = x \cdot \left(10^{n-1} - 1\right) + 3x = 3x \cdot x + 3x = 3x\left(3x + 1\right) =$$

$$= \underbrace{333...33}_{n\ times} \cdot \underbrace{333...34}_{n\ times}.\ \textbf{9)}\ n = 1.\ \textbf{10)}\ \text{The last digit is 5.}$$

Chapter II

RATIONAL NUMBERS AND FRACTIONS

II. 1. a) 0.75; **b)** 0.6; **c)** 0.73; **d)** $0.8\bar{3}$; **e)** $0.\bar{6}$; **f)** 0.35;

g) $0.9\overline{714285}$; **h)** $0.8\bar{6}$; **i)** 0.12. **II. 2. a)** $\frac{7}{10}$; **b)** $\frac{231}{100}$;

c) $\frac{1}{100}$; **d)** $\frac{13}{1000}$; **e)** $\frac{34}{99}$; **f)** $\frac{7}{10}$; **g)** $\frac{7}{3}$; **h)** $\frac{142}{99}$; **i)** $\frac{14}{15}$;

j) $\frac{17}{30}$; **k)** $\frac{32}{15}$; **l)** $\frac{17}{12}$. **II. 3. a)** $\frac{1}{3}$; **b)** $-\frac{7}{11}$; **c)** $\frac{11}{909}$;

d) $-\frac{5}{7}$; **e)** $\frac{1}{2}$; **f)** $\frac{1}{7}$; **g)** $\frac{2}{5}$; **h)** $\frac{1}{4}$; **i)** $\frac{1}{2}$; **j)** $\frac{108}{107}$.

II. 4. a) $-\frac{3}{4} = -0.75$; **b)** $-\frac{3}{4} < -\frac{2}{4}$; **c)** $-\frac{2}{5} < \frac{1}{7}$; **d)** $0.47 = \frac{47}{100}$;

e) $1.\bar{4} = 1\frac{4}{9}$; **f)** $-0.35 < -0.34$; **g)** $-\frac{3}{8} < -\frac{1}{3}$; **h)** $-\frac{2}{3} < -\frac{7}{8}$;

i) $\frac{2}{3} < \frac{3}{4}$. **II. 5. a)** -2, $-\frac{4}{3}$, $-\frac{1}{3}$, 0, $\frac{2}{3}$, $\frac{4}{3}$, 2;

b) $-1\frac{1}{5}, -\frac{4}{5}, -\frac{2}{5}, 0, \frac{2}{5}, 1\frac{1}{5}, 1\frac{3}{5}$; **c)** $-2\frac{1}{4}, -\frac{1}{4}, \frac{1}{4}, \frac{1}{2}, \frac{3}{4}$;

d) $\frac{2}{9}, \frac{1}{4}, \frac{5}{7}, \frac{3}{3}, \frac{9}{5}, \frac{8}{2}$; **e)** $\frac{4}{7}, \frac{3}{5}, 2\frac{5}{7}, \frac{4}{1}, 1\frac{9}{3}, 5\frac{3}{2}$.

II. 6. a) $-\frac{1}{4}$; **b)** $\frac{1}{3}$; **c)** $-\frac{5}{7}$; **d)** $\frac{1}{7}$; **e)** 0; **f)** 1.6; **g)** 1.21;

h) 1.26; **i)** –0.4; **j)** 0.16; **k)** – 0.7; **l)** – 3.22; **m)** 5.25;

n) – 1.19; **o)** 6.481. **II. 7. a)** $\frac{3}{14}$; **b)** $\frac{1}{14}$; **c)** $-\frac{3}{14}$; **d)** $-\frac{11}{10}$;

e) $-\frac{4}{15}$; **f)** $-\frac{14}{15}$; **g)** $\frac{1}{10}$; **h)** $-\frac{1}{4}$; **i)** $-\frac{11}{12}$; **j)** $-\frac{11}{12}$; **k)** $\frac{5}{6}$;

l) –11.68; **m)** 1.96; **n)** 0.92; **o)** – 0.1; **p)** – 0.01.

II. 8. a) $\frac{3}{7}$; **b)** 1; **c)** $-\frac{11}{14}$; **d)** $\frac{14}{45}$; **e)** $\frac{5}{24}$; **f)** $\frac{4}{3}$; **g)** $-\frac{23}{20}$;

h) $-\frac{3}{4}$; **i)** $\frac{11}{12}$; **j)** $\frac{3}{32}$. **II. 9. a)** $\frac{4}{15}$; **b)** $\frac{1}{35}$; **c)** $-3\frac{3}{20}$; **d)** $\frac{37}{72}$;

~ 227 ~

e) $\frac{11}{12}$; f) $\frac{23}{36}$; g) $-\frac{13}{144}$; h) $\frac{4}{3}$; i) $-8\frac{23}{30}$; j) 3; k) $-\frac{103}{20}$;

l) $-1\frac{7}{12}$; m) $-\frac{13}{24}$. **II. 10.** a) 2; b) -4; c) 1; d) -0.3;

e) 3.75; f) 1.5; g) $\frac{1}{3}$; h) 0.21; i) $-\frac{1}{15}$. **II. 11.** a) $\frac{2}{3}$; b) $\frac{12}{125}$;

c) $\frac{16}{49}$; d) $\frac{2}{5}$; e) $4\frac{1}{2}$; f) -1; g) 12; h) -7; i) -70; j) -5.1;

k) -2.6; l) -32; m) $\frac{8}{3}$; n) $\frac{1}{2}$. **II. 12.** a) $\frac{17}{21}$; b) $\frac{3}{10}$;

c) $-\frac{7}{8}$; d) $\frac{7}{10}$; e) $1\frac{2}{5}$; f) $-\frac{3}{2}$; g) $3\frac{1}{10}$; h) $\frac{1}{4}$;

i) $\frac{1}{12}$; j) 0; k) $-\frac{4}{5}$; l) $\frac{2}{27}$; m) $-\frac{23}{140}$; n) $\frac{5}{12}$; o) $\frac{1}{12}$;

p) $-\frac{2}{63}$; q) $\frac{5}{6}$; r) $2\frac{2}{15}$; s) -3; t) -1.

II. 13. a) $\frac{5}{6}$; b) -6; c) 1; d) $-5\frac{1}{17}$; e) -6; f) 4; g) 2; h) $\frac{5}{8}$.

II. 14. a) $2\frac{8}{11}$; b) $-\frac{17}{36}$; c) $-\frac{5}{6}$; d) $\frac{4}{5}$; e) $\frac{5}{6}$; f) $1\frac{4}{7}$;

g) $\frac{9}{4}$; h) $103\frac{1}{3}$; i) $3\frac{3}{25}$. **II. 15.** a) 0; b) 5; c) -20; d) 1; e) 3;

f) $1\frac{15}{16}$; g) 30; h) 1; i) 1; j) 24. **II. 16.** a)

$$\frac{\left(\dfrac{43}{10}\cdot\dfrac{12}{43}-\dfrac{1}{5}\right)\cdot 55}{\dfrac{31}{12}-\dfrac{1}{4}\cdot\dfrac{28}{3}\cdot\dfrac{5}{7}}\cdot\frac{1}{10}+\frac{76}{50}+\frac{362}{25}=\frac{\left(\dfrac{6}{5}-\dfrac{1}{5}\right)\cdot 55}{\dfrac{31}{12}-\dfrac{20}{12}}\cdot\frac{1}{10}+8=22\,;$$

b) $25\frac{1}{2}$; c) $7\frac{1}{5}$; d) $1\frac{1}{16}$; e) 1; f) 2; g) 9; h) $\frac{54}{55}$;

i) $A=\frac{75}{2}$; $B=16\frac{1}{2}$; $C=3$; $x=\frac{7500}{11}$. **II. 17.** a) $\frac{27}{125}$; b) $-\frac{8}{27}$;

c) $\frac{9}{25}$; d) $\frac{2401}{81}$; e) 1; f) $-\frac{3}{22}$; g) 1; h) $\frac{23}{11}$; i) $-\frac{1331}{1000}$;

j) $-\dfrac{16}{81}$; **k)** $-\dfrac{169}{100}$; **l)** $\dfrac{343}{1000}$; **m)** $-\dfrac{512}{125}$; **n)** $-\dfrac{2197}{1728}$.

II. 18. a) $\dfrac{1}{9}$; **b)** $\dfrac{1}{125}$; **c)** $\dfrac{1}{16}$; **d)** $-\dfrac{1}{216}$; **e)** $-\dfrac{1}{5}$; **f)** -4; **g)** $-\dfrac{5}{2}$;

h) $\dfrac{16}{9}$; **i)** $-\dfrac{27}{343}$; **j)** $-\dfrac{3}{2}$; **k)** -125; **l)** $\dfrac{9}{4}$; **m)** $-\dfrac{729}{125}$.

II. 19. a) 2^5, 2^4, 2^3, 2^2, 2^1, 2^0, $\dfrac{1}{2^1}$, $\dfrac{1}{2^2}$, $\dfrac{1}{2^3}$, $\dfrac{1}{2^4}$, $\dfrac{1}{2^5}$;

b) $\dfrac{1}{5^4}$, $\dfrac{1}{5^3}$, $\dfrac{1}{5^2}$, $\dfrac{1}{5^1}$, 5^0, 5^1, 5^2, 5^3, 5^4;

c) $\dfrac{1}{10^4}$, $\dfrac{1}{10^3}$, $\dfrac{1}{10^2}$, $\dfrac{1}{10^1}$, 10^0, 10^1, 10^2, 10^3, 10^4.

II. 20. a) 10^{-2}; **b)** 7^{-3}; **c)** 10^{-4}; **d)** 3^{-3}; **e)** 2^{-4}; **f)** 11^{-2};

g) 3^{-6}; **h)** 6^{-3}; **i)** 5^{-5}.

II. 21. a) 2^9; **b)** 7^6; **c)** -8^8; **d)** -3^5; **e)** 4^3;

f) $\dfrac{1}{7}$; **g)** $-\dfrac{1}{3^5}$; **h)** $\left(\dfrac{2}{5}\right)^8$; **i)** $\left(-\dfrac{2}{5}\right)^5$; **j)** $\left(\dfrac{3}{7}\right)^4$; **k)** 2^3; **l)** -3;

m) $\left(-\dfrac{2}{3}\right)^5$; **n)** -3; **o)** $\left(\dfrac{3}{7}\right)^3$; **p)** $\left(\dfrac{5}{11}\right)^6$; **q)** $\left(\dfrac{2}{3}\right)^4$; **r)** 2.3; **s)** 1.

II. 22. a) 5^4; **b)** 2^6; **c)** $\dfrac{1}{2^8}$; **d)** $\dfrac{1}{2^6}$; **e)** $\left(\dfrac{2}{5}\right)^4$; **f)** $-\dfrac{5}{6}$; **g)** $\left(\dfrac{3}{4}\right)^3$;

h) $-\dfrac{2}{3}$; **i)** $\dfrac{5}{3}$; **j)** $-\dfrac{4}{5}$; **k)** $\left(\dfrac{19}{10}\right)^3$; **l)** $\left(\dfrac{4}{9}\right)^2$; **m)** $\left(\dfrac{10}{38}\right)^8$;

n) $\left(\dfrac{13}{10}\right)^7$; **o)** $\left(\dfrac{8}{10}\right)^4$. **II. 23. a)** -6^{16}; **b)** 4^8; **c)** 3^{12}; **d)** -2^{12};

e) $\left(\dfrac{2}{5}\right)^6$; **f)** $\left(-\dfrac{3}{4}\right)^6$; **g)** $\left(-\dfrac{3}{4}\right)^6$; **h)** $\left(\dfrac{2}{3}\right)^8$; **i)** $\left(\dfrac{3}{4}\right)^{30}$; **j)** $\left(\dfrac{1}{2}\right)^4$;

k) $\left(\dfrac{9}{5}\right)^6$; **l)** $\left(\dfrac{1}{3}\right)^8$; **m)** $\left(-\dfrac{1}{2}\right)^{81}$; **n)** $\left(\dfrac{3}{2}\right)^6$; **o)** 2^6;

p) -2^6; **q)** $\left(\dfrac{5}{2}\right)^{12}$; **r)** $\left(\dfrac{3}{4}\right)^4$; **s)** $\left(\dfrac{5}{3}\right)^6$; **t)** $\left(\dfrac{2}{3}\right)^{24}$; **u)** 1; **v)** 1.

II. 24. **a)** $-\dfrac{3}{2}$; **b)** $\left(\dfrac{2}{5}\right)^3$; **c)** 0.3; **d)** -2.

II. 25. **a)** 3^4; **b)** -2^{10}; **c)** $-\dfrac{5^2}{2^3}$; **d)** 3^3; **e)** $-\dfrac{1}{2^{11}}$; **f)** -3^9;

g) 3^{18}; **h)** $\dfrac{1}{5}$; **i)** 2^{32}; **j)** $\dfrac{1}{3^6}$; **k)** 6^3; **l)** 3^{-7}; **m)** $-\dfrac{1}{2^7}$; **n)** -1;

o) $-\dfrac{2}{9}$; **p)** $\left(\dfrac{2}{3}\right)^9$; **q)** 6^5. **II. 26.** **a)** $\dfrac{25}{49}$; **b)** $\dfrac{1}{1000}$; **c)** $\dfrac{1}{2}$;

d) $\dfrac{1}{2}$; **e)** 9; **f)** 6^5; **g)** 3^{11}; **h)** 2^{20}; **i)** 5^8; **j)** $\dfrac{1}{2^3}$; **k)** $\dfrac{1}{6^8}$.

II. 27. **a)** 5; **b)** $-\dfrac{21}{2^4}$; **c)** $-\dfrac{1}{3^4}$; **d)** $\dfrac{19}{2^2}$; **e)** 7; **f)** 2; **g)** $-2\dfrac{1}{4}$;

h) 0; **i)** $2\dfrac{2}{3}$; **j)** $\dfrac{16}{121}$; **k)** $-3\dfrac{3}{8}$; **l)** $2\dfrac{17}{30}$; **m)** $1\dfrac{1}{4}$; **n)** $-\dfrac{9}{4}$; **o)** $1\dfrac{1}{9}$.

II. 28. **a)** $\dfrac{3}{7}$; **b)** $\left(\dfrac{3}{2}\right)^{10}$; **c)** 1; **d)** 1; **e)** $-\dfrac{2}{27}$; **f)** -1;

g) $3\dfrac{1}{2}$; **h)** 20; **i)** -24. **II. 29.** **a)** $\dfrac{1}{12}$; **b)** $\dfrac{1}{15}$; **c)** $x=\dfrac{1}{4}$; **d)** $-\dfrac{5}{12}$;

e) -15; **f)** 10; **g)** $1\dfrac{1}{7}$; **h)** -6; **i)** -20; **j)** -7; **k)** 3; **l)** -3;

m) 2; **n)** $\dfrac{3}{5}$. **II. 30.** **a)** $1\dfrac{1}{2}$; **b)** $1\dfrac{1}{3}$; **c)** $-\dfrac{2}{5}$; **d)** $3\dfrac{2}{3}$; **e)** $4\dfrac{1}{2}$;

f) $\dfrac{1}{14}$; **g)** $-\dfrac{5}{6}$; **h)** $8\dfrac{18}{49}$; **i)** $10\dfrac{2}{7}$; **j)** $1\dfrac{1}{8}$; **k)** -1; **l)** $-\dfrac{5}{6}$;

m) $\dfrac{15}{28}$; **n)** $-\dfrac{10}{21}$; **o)** -2; **p)** 3; **q)** -9; **r)** -39; **s)** -2; **t)** $-\dfrac{2}{5}$.

II. 31. **a)** -1; **b)** -8; **c)** 22; **d)** 30; **e)** $-\dfrac{1}{3}$; **f)** $\dfrac{1}{8}$; **g)** 12; **h)** 40;

i) 5; **j)** $\dfrac{2}{3}$; **k)** $1\dfrac{1}{2}$; **l)** $\dfrac{430}{473}$; **m)** -50; **n)** $1\dfrac{1}{8}$; **o)** $11\dfrac{1}{2}$; **p)** -32;

q) 13; **r)** impossible; **s)** 12.

II. 32. **a)** $x=-3$; **b)** $x=-\dfrac{1}{8}$; **c)** $x=2\dfrac{2}{5}$; **d)** $x=-5$;

e) $x = 5\dfrac{1}{2}$; f) $x = 4$; g) $x = \dfrac{2}{3}$; h) $x = -5\dfrac{1}{3}$.

II. 33. a) $x = 1$; **b)** $x = 1$; **c)** $x = 0.5$; **d)** $x = 2$; **e)** $x = 2$;

f) $x = 0.5$; g) $x = 5$. **II. 34. a)** $x = 10$; **b)** $x = 10$; **c)** $x = 5\dfrac{2}{11}$;

d) $x = 6\dfrac{1}{5}$; e) $x = -2\dfrac{1}{10}$; f) $x = 3900$; g) $x = 2\dfrac{4}{13}$; h) $x = 1\dfrac{182}{807}$;

i) $x = 26$; j) $x = 2010$. **II. 35. a)** $4\dfrac{1}{5}$; **b)** $\dfrac{5}{7}$; **c)** $-\dfrac{7}{15}$;

d) $1\dfrac{11}{30}$; e) $-\dfrac{3}{4}$; f) $\dfrac{5}{3}$; g) $-\dfrac{2}{7}$; h) $-1\dfrac{7}{9}$; i) $2\dfrac{1}{23}$; j) $2\dfrac{17}{20}$;

k) $\dfrac{5}{6}$; l) -18; m) $2\dfrac{29}{67}$; n) -126; o) $-\dfrac{4}{9}$; p) $\dfrac{8}{11}$;

q) 10; r) $\dfrac{13}{50}$; s) $\dfrac{10}{49}$; t) 8; u) $-\dfrac{5}{51}$; v) $3\dfrac{3}{2}$; w) -7;

x) $2\dfrac{19}{24}$; y) $8\dfrac{2}{11}$; z) $\dfrac{50}{57}$. **II. 36. a)** -3; **b)** $-3\dfrac{4}{7}$; **c)** 5; **d)** $-\dfrac{17}{24}$;

e) $\dfrac{44}{3}$; f) $\dfrac{28}{27}$; g) $3\dfrac{14}{29}$; h) -48; i) $2\dfrac{23}{90}$; j) $-5\dfrac{51}{219}$.

II. 37. a) $a = 1$; **b)** $x = 2\dfrac{33}{71}$; **c)** $a = 70$; **d)** $x = \dfrac{4}{25}$; **e)** $x = 30$;

f) $x = -\dfrac{20}{23}$; g) $x = 45$; h) $x = 105$; i) $x = -60$; j) $x = 4\dfrac{1}{2}$.

II. 38. a) $x = 1$; **b)** $x = 1$; **c)** $x = -2$; **d)** $x = 4$; **e)** $x = 2$.

II. 39. a) $-\dfrac{1}{9}$; **b)** $-\dfrac{7}{20}$; **c)** $\dfrac{5}{29}$; **d)** $\dfrac{16}{3}$; **e)** 1; **f)** $x = -\dfrac{1}{2}$.

II. 40. a) $x_1 = -\dfrac{1}{3}$, $x_2 = \dfrac{3}{4}$; **b)** $x_1 = -\dfrac{21}{4}$, $x_2 = -\dfrac{3}{2}$;

c) $x_1 = -\dfrac{24}{35}$, $x_2 = \dfrac{4}{3}$; d) $x_1 = \dfrac{91}{6}$, $x_2 = \dfrac{9}{14}$;

e) $x_1 = \dfrac{49}{8}$, $x_2 = -\dfrac{36}{25}$. **II. 41. a)** -4; **b)** $-\dfrac{1}{6}$; **c)** $\dfrac{2}{9}$; **d)** $-\dfrac{30}{13}$;

e) $\dfrac{31}{7}$; f) $-\dfrac{18}{23}$; g) $\dfrac{5}{2}$;

h) $-\dfrac{10}{19}$; **i)** $\dfrac{1}{4}$; **j)** $\dfrac{22}{5}$; **k)** 0; **l)** $-\dfrac{6}{19}$. **II. 42. a)** -5; **b)** $\dfrac{125}{216}$;

c) 4; **d)** 516; **e)** $-\dfrac{5^4}{3^6}$; **f)** $\dfrac{b}{c}=-\dfrac{1}{20}$, $\dfrac{c}{b}=-20$.

II. 43. a) $a=4\dfrac{3}{5}$; **b)** $a=\dfrac{5}{2}$; **c)** $m=-\dfrac{3}{2}$; **d)** $m=\dfrac{3}{14}$; **e)** $m=-\dfrac{15}{34}$.

II. 44. a) $x=-1$, $m\in\mathbf{R}\setminus\left\{\dfrac{1}{2}\right\}$; **b)** $x=\dfrac{a-3}{a-2}$, $a\in\mathbf{R}\setminus\{2\}$;

c) $x=\dfrac{12}{a-4}$, $a\in\mathbf{R}\setminus\{4\}$; **d)** $x=\dfrac{1+2m}{1-3m}$, $m\in\mathbf{R}\setminus\left\{\dfrac{1}{3}\right\}$;

e) $x=\dfrac{3a-4}{4a}$, $a\in\mathbf{R}\setminus\{0\}$; **f)** $x=\dfrac{3a-2}{a+5}$, $a\in\mathbf{R}\setminus\{-5\}$;

g) $x=\dfrac{1}{a-5}$, $a\in\mathbf{R}\setminus\{5\}$; **h)** $x=\dfrac{4a-3}{a-8}$, $a\in\mathbf{R}\setminus\{8\}$;

i) $x=\dfrac{6a}{a-1}$, $a\in\mathbf{R}\setminus\{1\}$.

II. 45. a) 1; **b)** Denote $n!=1\cdot2\cdot3\cdot...\cdot n$, then

$$\left(\dfrac{1}{1!}-\dfrac{1}{2!}\right)+\left(\dfrac{1}{2!}-\dfrac{1}{3!}\right)+...+\left(\dfrac{1}{x!}-\dfrac{1}{(x+1)!}\right)=\dfrac{2010!-1}{2010!}, \quad x=2009;$$ **c)** 1;

II. 46. 8-foot and 4-foot. **II. 47.** 24-foot and 12-foot.
II. 48. 25 \$140 suits and 15 \$210 suits.
II. 49. Let $a-2$, $a-1$, a, $a+1$, and $a+2$ be the five positive consecutive numbers.
$a(a-1)(a-2)=a(a+1)(a+2)-216$, $a=6$. The numbers are 4, 5, 6, 7, and 8. **II. 50.** The first gear has 39 teeth and the second gear has 57 teeth. **I. 51.** Let x and y be the numbers. $x=y-27.8$ and $x+y=66.7$. Therefore $x=19.45$ and $y=47.25$.
II. 52. Let m and n be the two numbers. Then $m=3n$ and $m+n=\dfrac{34}{9}$. Therefore $m=\dfrac{17}{6}$ and $n=\dfrac{17}{18}$.

II. 53. $5.8a=a-4$, $a=\dfrac{5}{6}$.

II. 54. The length is 418 yards and the width is 202 yards.

II. 55. $x^2 + 83 = (x+3)(x+5)$, $x = \dfrac{17}{2}$.

II. 56. The length is 20 feet and the width is 11 feet.

II. 57. When the angles of the triangle are $2x$, $2x$, and $5x$, $x = 20$. When the angles of the triangle are $5x$, $5x$, and $2x$, $x = 15$.

II. 58. $\dfrac{3}{4}$. **II. 59.** Let x be the length and y be the width.

$x + y = 4\dfrac{1}{2}$ and $x = 2y + 1.2$. Then $x = \dfrac{17}{5}$ and $y = \dfrac{11}{10}$.

II. 60. Let A, B, and C be the three angles, then $\angle A = 72°$. $B = 3C$ and $B + C = 108$. Therefore $\angle B = 81°$ and $\angle C = 27°$.

II. 61. Let x and y be the price for 1 kg of apples and 1 kg of oranges, respectively. $3x + 2y = \dfrac{101}{20}$ and $x = y + \dfrac{3}{10}$, $x = \dfrac{89}{100}$ and $y = \dfrac{119}{100}$. 1 kg of apples costs $0.89 and 1 kg of oranges costs $1.19.

II. 62. Let x and y be the math and physics textbooks, respectively, then $10x + 15y = 2000$ and $x + y = 163$. Therefore $x = 89$ and $y = 74$.

II. 63. Let x and y be the two numbers. $x + y = 150$ and $x = 4y + 15$. Then $x = 123$ and $y = 27$.

II. 64. Let a and b be the length and width, respectively. $a - b = 13$ and $a + b = 63$. The length is 38 cm and the width is 25 cm.

II. 65. a) 1.7 and $\dfrac{10}{17}$; **b)** $\dfrac{70}{51}$ and $\dfrac{51}{70}$; **c)** $\dfrac{6}{5}$ and $\dfrac{5}{6}$;

d) $\dfrac{10}{9}$ and $\dfrac{9}{10}$. **II. 66. a)** $-\dfrac{9}{5}$; **b)** $\dfrac{65}{3}$. **II. 67. a)** 2π; **b)** 8π.

II. 68. a) $\dfrac{3}{4}$; **b)** $\dfrac{3}{4}$.

II. 69. a) $\dfrac{2}{3}$; **b)** $\dfrac{12}{7}$; **c)** $\dfrac{5}{9}$; **d)** $\dfrac{1}{3}$; **e)** $\dfrac{1}{2}$; **f)** $\dfrac{3}{5}a^2$; **g)** $\dfrac{1}{6}$; **h)** $\dfrac{3}{8}$;

i) $\dfrac{1}{3}$; **j)** $\dfrac{75}{2}$; **k)** $\dfrac{3}{8}$; **l)** $\dfrac{5}{6}$.

II. 70. a) $-\dfrac{43}{15}$; **b)** $-\dfrac{14}{53}$; **c)** $\dfrac{25}{158}$; **d)** $-\dfrac{2}{23}$; **e)** $-1\dfrac{2}{69}$.

II. 71. a) $x = 20$; **b)** $x = \dfrac{117}{40}$; **c)** $x = 8\dfrac{4}{5}$; **d)** $a = 8$;

e) $z = \dfrac{19}{2}$; **f)** $z = \dfrac{10}{63}$; **g)** $a = 2\dfrac{1}{2}$; **h)** $n = 2\dfrac{1}{2}$; **i)** $x = 7$;

j) $x = 7\dfrac{1}{7}$; **k)** $x = 2$; **l)** $x = 52$; **m)** $x = 27\dfrac{1}{2}$; **n)** $x = \dfrac{91}{1200}$.

II. 72. a) $x = 15$, $y = 35$; **b)** $x = \dfrac{91}{6}$, $y = \dfrac{49}{6}$;

c) $x = -\dfrac{20}{7}$, $y = -\dfrac{90}{7}$; **d)** $x = -25\dfrac{32}{43}$, $y = -10\dfrac{21}{43}$;

e) $x = -16\dfrac{7}{13}$, $y = -14\dfrac{5}{26}$; **f)** $x = \dfrac{1}{7}$, $y = 1\dfrac{13}{21}$.

II. 73. a) 2 ; **b)** 30; **c)** 22 ; **d)** 30; **e)** $\dfrac{20}{29}$; **f)** $-\dfrac{15}{2}$;

g) $-\dfrac{3}{14}$; **h)** 13; **i)** 4; **j)** Impossible; **k)** $\dfrac{17}{4}$; **l)** 48; **m)** $\dfrac{26}{15}$.

II. 74. a) $\dfrac{3}{8}$; **b)** $\dfrac{22}{15}$; **c)** $\dfrac{1}{2}$; **d)** $-\dfrac{4}{7}$; **e)** $\dfrac{18}{35}$; **f)** $-\dfrac{1}{9}$; **g)** $-\dfrac{1}{45}$.

II. 75. a) $\dfrac{3}{4}$; **b)** $\dfrac{1}{4}$; **c)** $\dfrac{3}{25}$; **d)** $\dfrac{1}{250}$; **e)** $\dfrac{1}{2}$; **f)** $\dfrac{7}{1000}$; **g)** $\dfrac{23}{1000}$;

h) $\dfrac{2}{125}$. **II. 76. a)** 0.05; **b)** 0.03; **c)** 0.34; **d)** 0.17; **e)** 0.0033;

f) 0.0415; **g)** 0.0025; **h)** 0.0145. **II. 77. a)** 100; **b)** 6; **c)** 16.5;

d) 0.54; **e)** 2.25; **f)** 78; **g)** 99; **h)** 40.6; **i)** $10.8\overline{3}$.

II. 78. a) $16.\overline{6}\,\%$; **b)** 80%; **c)** 75%; **d)** 64%;

e) 48%; **f)** $32.\overline{432}\,\%$; **g)** 5000%; **h)** 500%; **i)** 80%; **j)** $66.\overline{6}\,\%$;

k) 64%.

II. 79. a) $\dfrac{20}{100} \cdot x = 15$, $x = 75$; **b)** $x = 26.\overline{6}$; **c)** 22.5; **d)** $x = 106.\overline{6}$;

e) $x = 200$; **f)** $x = 600$; **g)** $x = 500$; **h)** $63.\overline{3}$; **i)** $x = 72$.

II. 80. $x = 18$, $y = 24$, and $z = 30$. **II. 81.** $x = 4, y = 6, z = 10$,

and $t = 14$. **II. 82.** $x = 6, y = 4$, and $z = 8$.

II. 83. $x = 8, y = 12$, and $z = 15$.

II. 84. $x = 2, y = 3$, and $z = 4$. **II. 85.** $x = 20, y = 12$, and $z = 24$.

II. 86. $x = 5$, $y = 3$, and $z = 4$. **II. 87.** $x = \dfrac{8}{9}$, $y = \dfrac{4}{3}$, and $z = \dfrac{16}{9}$.

II. 88. $x = 6$, $y = 8$, and $z = 12$.

II. 89. $a = 4$, $b = 6$, $c = 8$, and $d = 10$.

II. 90. a) $\dfrac{2}{3}$; **b)** $\dfrac{2}{3}$; **c)** $\dfrac{4}{9}$. **II. 91. a)** $\dfrac{2}{3}$; **b)** $1\dfrac{2}{11}$; **c)** $1\dfrac{2}{5}$.

II. 93. a) $a = 10^n \cdot 2 + 1 = \underbrace{200...001}_{n-1\ zeros}$, the sum of the digits is three

which means a is divisible by three. $b = 10^n \cdot 5 + 1 = \underbrace{500...001}_{n-1\ zeros}$,

$c = 10^n \cdot 8 + 7 = \underbrace{800...007}_{n-1\ zeros}$, $d = 10^{n+1} \cdot 25 - 1 = \underbrace{2499...999}_{n+1\ zeros}$.

b) $\dfrac{7d + c}{b - a} = \dfrac{10^{n+1} \cdot 175 - 7 + 10^n \cdot 8 + 7}{10^n \cdot 5 + 1 - 10^n \cdot 2 - 1} = \dfrac{10^n(174 + 8)}{10^n(5 - 2)} = \dfrac{183}{3} = 61.$

c) No, c is not a perfect square because the last digit is seven and a perfect square does not have the last digit seven.

II. 94. a) $156 + 13^{101}$; **b)** -6. **II. 95.** The number is 111.

TEST II. 1

1) The coins he might have in his pocket is $\dfrac{1}{20} = \dfrac{5}{100} = 0.05$ dollars;

5 cents. **2) a)** $\dfrac{3}{5} = \dfrac{6}{10} = 06$; **b)** $\dfrac{3}{2} = \dfrac{15}{10} = 1.5$;

c) $2\dfrac{1}{5} = 2\dfrac{2}{10} = 2.2$; **d)** $5\dfrac{4}{5} = 5\dfrac{8}{10} = 5.8$.

3) a) $\dfrac{1}{4} = \dfrac{25}{100} = 0.25$; **b)** $\dfrac{3}{10} = \dfrac{30}{100} = 0.3$;

c) $\dfrac{2}{50} = \dfrac{4}{100} = 0.04$; **d)** $\dfrac{2}{25} = \dfrac{8}{100} = 0.8$.

4) a) $7\dfrac{1}{2} = 7\dfrac{50}{100} = 7.5$; **b)** $6\dfrac{2}{5} = 6\dfrac{40}{100} = 6.4$;

c) $3\dfrac{11}{25} = 3\dfrac{44}{100} = 3.44$; **d)** $11\dfrac{1}{100} = 11.01$.

5) **a)** $\dfrac{75}{100}=\dfrac{3}{4}$; **b)** $0.08 < 0.8$; **c)** $0.6=\dfrac{60}{100}$; **d)** $\dfrac{16}{10}>\dfrac{13}{10}$.

6) **a)** $0.9=\dfrac{9}{10}=\dfrac{90}{100}$; **b)** $0.40=\dfrac{4}{10}=\dfrac{40}{100}$;

c) $0.75=\dfrac{75}{100}=\dfrac{15}{20}$; **d)** $0.5=\dfrac{5}{10}=\dfrac{50}{100}$.

7) **a)** $77\div8=\dfrac{77}{8}=9\dfrac{5}{8}$; **b)** $84\div9=\dfrac{84}{9}=9\dfrac{3}{9}$;

c) $45\div7=\dfrac{45}{7}=6\dfrac{3}{7}$; **d)** $79\div6=\dfrac{79}{6}=13\dfrac{1}{6}$.

8) $7\div5=\dfrac{7}{5}=\dfrac{140}{100}=1.4$ of a pizza.

9) The area of the yard is $28.54\times31 = 884.74 \ m^2$.

10) $0.94 \ \text{m}\times6 = 5.64 \ \text{m}$.

TEST II. 2

a) $1\dfrac{1}{5}$; **b)** $\dfrac{24}{125}$; **c)** $\dfrac{4}{51}$; **d)** 12; **e)** 12 ; **f)** 7 ; **g)** 1.3 ;

h) $1\dfrac{4}{5}$; **i)** -12; **j)** $\dfrac{2}{3}$.

TEST II. 3

1) $x = 4$; **2)** $x=-11$; **3)** $x = 2$; **4)** $x = 1/5$; **5)** $x = -2$; **6)** $x = -1$;
7) $x = -6$; **8)** $x = -7/19$; **9)** $x = 7/2$; **10)** $x = 1$.

TEST II. 4

1) $x = 2$, $y = 14$, $\dfrac{x+y}{2} = 8$; **2)** $x = 9\dfrac{1}{8}$; **3)** $x = 63$, $y = 70$, $z = 84$;

4) **a)** 1, **b)** 2; **5)** $a = 1$; **6)** **a)** $x = 7$; **b)** $x = 6$; **c)** $x = 2$;

7) **a)** 6.1; **b)** 30; **c)** 1; **8)** $x = \dfrac{29}{9}$; **9)** 1,536;

10) $a = 10$, $b = \dfrac{81}{10}$, $c = \dfrac{439}{560}$.

1) $x = \dfrac{1.5 \cdot 0.5}{0.2} = \dfrac{0.75}{0.2} = \dfrac{7.5}{2} = 3.75$. **2) a)** $x = 6$; **b)** $x = 20/9$;

c) $x = 0.75$; **d)** $x = 1.2$.

3) We have $\dfrac{x}{0.2} = \dfrac{y}{0.05} = \dfrac{x+y}{0.2+0.05} = \dfrac{10}{0.25} = 40$. From

$\dfrac{x}{0.2} = 40 \Rightarrow x = 8$, and $\dfrac{y}{0.05} = 40 \Rightarrow y = 2$. **4)** We have

$\dfrac{x}{3} = \dfrac{y}{4} = \dfrac{z}{5} = \dfrac{x+y+z}{3+4+5} = \dfrac{24}{12} = 2$. Therefore $x = 6, y = 8, z = 10$.

5) $x = 20, z = 24, y = 8$. **6)** $x = 25, y = 10, z = 20$.

7) $a = 6, b = 8, c = 10$.

8) Let be $\dfrac{x}{12} = \dfrac{y}{15} = \dfrac{z}{30} = k$. Then $x = 12k, \ y = 15k, \ z = 30k$, and

$xyz = (12k)(15k)(30k) = 5400k^3$, so

$5400k^3 = 1600 \Rightarrow k^3 = \left(\dfrac{2}{3}\right)^3 \Rightarrow k = \dfrac{2}{3}$. Hence $x = 12 \cdot \dfrac{2}{3} = 8$,

$y = 15 \cdot \dfrac{2}{3} = 10$, $z = 30 \cdot \dfrac{2}{3} = 20$. **9)** $x = 33, y = 42, z = 51$.

TEST II. 6

1) a) 50; **b)** 2; **2) a)** 32; **b)** 1.15; **3)** $a = -11/3, \ b = 110.8$,

$x = 328.4/11$; **4)** $\dfrac{1}{256}$; **5)** $\dfrac{1}{7} \cdot \left[\dfrac{1}{5} \cdot \left(\dfrac{1}{3} \cdot (x-2) - 4\right) + 6\right] + 8 = 3$,

$\dfrac{1}{7} \cdot \left[\dfrac{1}{5} \cdot \left(\dfrac{1}{3} \cdot (x-2) - 4\right) + 6\right] = -5$, $\dfrac{1}{5} \cdot \left(\dfrac{1}{3} \cdot (x-2) - 4\right) = -41$,

$\dfrac{1}{3} \cdot (x-2) - 4 = -205$, $x = -601$. **6)** $\dfrac{2}{669}$.

7) $\left(\dfrac{1}{\sqrt{2}} - \dfrac{1}{1}\right) + \left(\dfrac{1}{\sqrt{3}} - \dfrac{1}{\sqrt{2}}\right) + \ldots + \left(\dfrac{1}{\sqrt{2010}} - \dfrac{1}{\sqrt{2009}}\right) = -1 + \dfrac{1}{\sqrt{2010}}$.

8) $x\left(\dfrac{1}{2}+\dfrac{1}{3}+...+\dfrac{1}{n}\right)=\dfrac{1}{2}+\dfrac{1}{3}+...+\dfrac{1}{n}\Rightarrow x=1$.

9) $AB = BD = 6$, $BC = BD - CD = 6 - 4 = 2$ and

$AM = AB + \dfrac{BC}{2} = 6 + 1 = 7$.

10) $AB + CD + DF = AG - (BC + FG) = 57 - 24 = 33$cm,

$AB = 11$cm, $BF = BC + CD + DF = 12 + 11 + 11 = 34$cm,

$CG = CD + DF + FG = 11+11+12 = 34$cm. If E is the midpoint of

CG, then $AE = AC + \dfrac{CG}{2} = 23 + 17 = 40$cm.

TEST II. 7

1) $x = 2$, $y = 84$, $\dfrac{x+y}{2} = 43$; **2)** $x = 9\dfrac{1}{8}$; **3)** $x = 63$, $y = 70$,

$z = 84$; **4) a)** 1, **b)** 2; **5)** $a = 1$; **6) a)** $x = 7$; **b)** $x = 6$;

c) $x = \dfrac{42}{19}$; **7) a)** 1; **b)** 50; **c)** $\dfrac{1}{100}$; **8)** $x = -\dfrac{1}{3}$;

9) 48, 32; **10)** $a = 10$, $b = \dfrac{81}{10}$, $c = \dfrac{439}{560}$.

TEST II. 8

1) a) 1. **b)** 50. **c)** 1. **2)** $x = 1$. **3)** $\dfrac{5}{8}$. **4)** $x = 5.5$. **5)** $x = 1\dfrac{1}{6}$. **6)** 1.

7) a) 1; **b)** 2. **8)** $a = 1$. **9) a)** 0; **b)** $x = 4$; **c)** $x = 1, 2, 3$. **10)** 7.5.

TEST II. 9

1) a) Mary is 10 years old, **b)** When Mary is 20 years old her mother is 40 years old. **2) a)** 850; **b)** $a = 300$.

3) b) $S = \dfrac{24}{25}$, **d)** $n = 100$. **4) a)** $\dfrac{a}{b} = -52$;

b) $n = \left(7 + 7^2 + 7^3\right)\left(1 + 7^3 + ... + 7^{2007}\right)$. **5)** $N = 2^{6n} \cdot 64$.

6) $\dfrac{a}{4} = \dfrac{b}{6} = \dfrac{c}{8} = \dfrac{a+c}{12}$, $\dfrac{a+c}{12} = \dfrac{b}{6}$, $a+c = 2b$. $a = \dfrac{504}{13}$, $b = \dfrac{756}{13}$,

$c = \dfrac{1008}{13}$. **7)** $\dfrac{49}{145}$. **8)** $a = 15$; $b = 24$; $c = 36$.

9) a) 1400; **b)** 36; **c)** $k = 9800$. **10)** 55.

1) a) $\dfrac{10+x}{99}+\dfrac{20+x}{99}+...+\dfrac{90+x}{99}=\dfrac{26x}{11}$, $\dfrac{450+9x}{99}=\dfrac{26x}{11}$, $x=2$;

b) $x=1$. **2)** $a=6^{2n+2}\cdot16$. **3)** $S_1=6^{n+1}\cdot5$, $S_2=15^n\cdot25$,

$S_3=231^n\cdot2000$, $n\in N$, $S_4=2^{3n}\cdot3^{4n}\cdot1118$.

4) $\dfrac{\left(3^n+4\right)\cdot3^n+17\cdot\left(3^n+4\right)}{2\cdot\left(3^n+4\right)}=\dfrac{3^n+17}{2}$, 3^n+17 is an even number.

5) $A=\dfrac{3^{2n}\cdot2^{2n}\cdot5^{n+1}\cdot7-2^{2n+1}\cdot3^{2n+1}\cdot5^n}{3^n\cdot2^n\cdot5^{n-1}\cdot19+2^{n+1}\cdot5^n\cdot3^n}=$

$=\dfrac{3^{2n}\cdot2^{2n}\cdot5^n\left(5\cdot7-2\cdot3\right)}{3^n\cdot2^n\cdot5^{n-1}\left(19+2\cdot5\right)}=6^n\cdot5$. **6) a)** $\dfrac{7^x\cdot2}{6^x}$; **b)** $\dfrac{11^x\cdot10}{13^x}$.

7) a) $N=666$, $a=1$, $b=2$, $c=3$. **b)** $\dfrac{111}{185}\left(a+b+c\right)$;

8) $6\cdot\overline{ac}=\overline{abc}$, $6\left(10a+c\right)=100a+10b+c$, $\overline{abc}=108$.

9) $a+b+c=3$, $111, 102, 120, 210, 201$.

9) $\left(a,b\right)=\left\{\left(1,6\right);\ \left(2,5\right);\ \left(3,4\right)\right\}$.

11) $a+\dfrac{\overline{bcd}}{999}+b+\dfrac{\overline{cda}}{999}+c+\dfrac{\overline{dab}}{999}+d+\dfrac{\overline{abc}}{999}=10\Rightarrow$

$3d+\dfrac{111\left(a+b+c+d\right)}{999}=10\Rightarrow d=3$.

Chapter III

ALGEBRAIC EXPRESSIONS AND LINEAR FUNCTIONS

III. 1. a) $\dfrac{xb^2 - 7a^2b}{2}$; **b)** $5nm^2 - 12an$; **c)** $0.18ax$; **d)** $\dfrac{13x - 10x^2}{4}$;

e) $\dfrac{-5bxy - 6x^2y}{4}$; **f)** $\dfrac{48x}{5}$; **g)** $\dfrac{320b - 9a}{20}$; **h)** $-ax^3$; **i)** $\dfrac{79by}{10}$;

j) $\dfrac{13a^2b + 21ab^2}{6}$. **III. 2. a)** $\dfrac{3ab^2}{5}$; **b)** $4(m - n)^2$; **c)** $4(a + b)^3$;

d) $8 - x^2$; **e)** $-(5x^3 + x^2 + 5)$; **f)** $11b + 6c - 6a$; **g)** $-(x^2 + 1)$;

h) $a + b + c$; **i)** $\dfrac{8ab}{7} - \dfrac{bc}{10} - \dfrac{7ac}{15}$; **j)** $\dfrac{7a^2x}{6} - \dfrac{ax^2}{12} + \dfrac{25x^2}{24}$.

III. 3. a) $9a^6$; **b)** $14x^{10}$; **c)** $6x^6$; **d)** $-x^9$; **e)** $-6a^6b^3$;
f) $5a^2b^5$; **g)** $3x^4y^3z^4$.

III. 4. a) $x^2 + 2x + 1$; **b)** $x^2 - 2x + 1$; **c)** $x^2 - 4x + 4$;
d) $x^2 + 6x + 9$; **e)** $4x^2 + 4x + 1$; **f)** $9x^2 - 6x + 1$; **g)** $4x^2 - 8x + 4$;
h) $16x^2 - 16x + 4$; **i)** $25x^2 + 30x + 9$; **j)** $16x^2 - 24x + 9$;
k) $9x^2 - 6x + 1$; **l)** $x^2 + 6x + 9$; **m)** $36x^2 - 12x + 1$;
n) $0.25x^2 + 4x + 16$; **o)** $a^4 + 2a^2 + 1$; **p)** $25u^2 - 20u + 4$;
q) $a^2 + 14a + 49$; **r)** $4b^2 - 12b + 9$; **s)** $9x^2 + 6x + 1$;
t) $4x^4 - 4x^3 + x^2$; **u)** $x^2 + 10x^3 + 25x^4$;
v) $a^4 - 4a^2x + 4x^2$; **x)** $t^4 - 10t^2y^3 + 25y^6$.

III. 5. a) $\dfrac{x^2}{4} - \dfrac{xy}{3} + \dfrac{y^2}{9}$; **b)** $x^2 - \dfrac{2x}{3} + \dfrac{1}{9}$; **c)** $\dfrac{16}{9} - \dfrac{2b}{3} + \dfrac{b^2}{16}$;

d) $\dfrac{9a^2}{4} - \dfrac{3ab}{5} + \dfrac{4b^2}{25}$; **e)** $4r^2 - \dfrac{16r}{3} + \dfrac{16}{9}$; **f)** $\dfrac{4a^2}{25} + \dfrac{a}{2} + \dfrac{25}{16}$;

g) $\dfrac{9a^2}{64} + \dfrac{a}{2} + \dfrac{4}{9}$; **h)** $\dfrac{a^4}{36} + \dfrac{2a^3}{9} + \dfrac{4a^2}{9}$;

i) $\dfrac{9x^8y^2}{4} + 2x^7y^3 + \dfrac{4x^6y^4}{9}$; **j)** $\dfrac{4a^{10}b^2}{49} + \dfrac{6a^8b^3}{35} + \dfrac{9a^6b^4}{100}$;

k) $9x^4y^2 - 12x^2y^2 + 4y^2$; **l)** $4y^2 - 2x^3y^3 + 0.25x^6y^4$;

m) $4x^4y^4 - 0.8x^5y^3 + 0.04x^6y^2$; **n)** $2 + 2a\sqrt{2} + a^2$;

o) $3a^2 + 2ab\sqrt{6} + 2b^2$; **p)** $x^2 - 2x\sqrt{3} + 3$; **q)** $2x^2 - 2xy + \dfrac{y^2}{2}$;

r) $\dfrac{4x^2}{3} + 4x + 3$; **s)** $\dfrac{9x^2}{2} + 6x + 2$; **t)** $4y^2 - 4y\sqrt{3} + 3$.

III. 6. a) $x^2 - 4$; **b)** $4x^2 - 9$; **c)** $9x^2 - 1$; **d)** $9x^2 - 25$;

e) $x^2 - 0.16$; **f)** $-4x^2 + 9$; **g)** $0.04x^2 - 1$; **h)** $9x^2 - 16$; **i)** $\dfrac{x^2}{9} - \dfrac{1}{4}$;

j) $\dfrac{a^2}{4} - \dfrac{4b^2}{9}$; **k)** $9x^2 - \dfrac{1}{9}$; **l)** $\dfrac{y^2}{9} - \dfrac{9}{4}$; **m)** $\dfrac{a^2}{9} - 9a^4$;

n) $\dfrac{25x^2}{9} - \dfrac{4y^2}{25}$; **o)** $\dfrac{16x^2}{9} - \dfrac{9y^2}{4}$; **p)** $-x^2 + 3$; **q)** $5x^2 - 1$.

III. 7. a) $-3x^2 - 4x - 1$; **b)** $-2x^2 + 15x - 9$; **c)** $-6x^2 - 14x - 17$;
d) $-x^4 + 20x^2 - 4$; **e)** $8x^3$; **f)** $18x + 54$; **g)** $8x - 25$;
h) $-7x + 57$; **i)** $84x + 163$.

III.8. a) $x^2 + y^2 + z^2 + 2xy + 2yz + 2xz$ **b)** $x^2 + y^2 + 4 + 4x + 4y$;
c) $x^2 + y^2 + 0.01 + 0.2x + 0.2y$; **d)** $a^4 + 2a^3 + 3a^2 + 2a + 1$;
e) $x^4 + y^4 + z^4 + 2x^2y^2 + 2y^2z^2 + 2x^2z^2$; **f)** $6 + 2\sqrt{2} + 2\sqrt{3} + 2\sqrt{6}$;
g) $a^2 + b^2 + 2ab - 2b - 2a + 1$; **h)** $x^2 + 4y^2 + 4xy + 4y - 2x + 1$;
i) $x^2 + 4y^2 + 4z^2 - 4xy + 8zy - 4xz$; **j)** $4x^2 - 4x - 4x\sqrt{2} + 2\sqrt{2} + 3$.

III. 9. a) $(a + b)(x - y)$; **b)** $(x - y)(a - b)$; **c)** $(x + y)(m - n)$;
d) $(a - b)(x + 2)$; **e)** $(x + z)(a + y)$; **f)** $(a + d)(b - c)$;
g) $(m + y)(x - n)$; **h)** $(2 - 5a)(2x - 3)$; **i)** $(2a + b)(x - y)(x + y)$;
j) $(y + \sqrt{3})(a - \sqrt{2})$; **k)** $(x - y)(a^2 + b)$; **l)** $(x - 1)(x^2 + 1)$;
m) $(x + 1)^2$; **n)** $(x + 1)(x^2 - 2)$; **o)** $(4x - 3)(x - 3)$;
p) $35x - 14$; **q)** $(x + 2\sqrt{3})(x + 1)$; **r)** $(x + \sqrt{3})(4 - 3x)$.

III. 10. a) $\dfrac{x}{y}$; **b)** $\dfrac{15x^2}{y^2}$; **c)** $\dfrac{8}{y}$; **d)** $-\dfrac{5}{2x}$; **e)** $\dfrac{x}{y^2}$; **f)** $\dfrac{m}{x}$; **g)** 1; **h)** $\dfrac{b}{a}$;

i) -11; **j)** $\dfrac{3}{2}$. **III. 11. 1) a)** $\dfrac{5x}{3}$; **b)** $\dfrac{4y + x}{y}$; **c)** $\dfrac{24x + 11}{12}$;

d) $\dfrac{6x+5}{10}$; **e)** 1; **f)** $\dfrac{6x+7}{15}$; **g)** $\dfrac{12x-1}{6}$; **h)** $\dfrac{17x-3}{4}$; **i)** $\dfrac{17}{2x}$;

j) $\dfrac{14x^2-4x-5}{14x^2}$; **k)** $\dfrac{1}{12(x+1)}$; **l)** $-\dfrac{1}{15(x-2)}$. **III. 12. a)** $\dfrac{y}{x}$;

b) $\dfrac{5}{x+4}$; **c)** $\dfrac{7}{9(x-3y)}$; **d)** $\dfrac{40x^2-40}{5x+5}$; **e)** $\dfrac{1}{2x-3}$; **f)** $\dfrac{x^3-y^3}{x^6}$;

g) $\dfrac{x-1}{x+7}$; **h)** $\dfrac{x^2+2xy+4y^2}{x+2y}$; **i)** $\dfrac{x-2y-2}{2y-x-1}$; **j)** $2x-2ax+2a^2$.

III. 13. a) $\dfrac{y}{4xz}$; **b)** $\dfrac{a}{y^2}$; **c)** $\dfrac{4}{3bx^2}$; **d)** $\dfrac{x-4}{x+3}$; **e)** $\dfrac{x}{2}$; **f)** $\dfrac{x-2}{x+2}$;

g) $\dfrac{x}{2}$; **h)** $3x-1$; **i)** $\dfrac{1}{5x-1}$; **j)** $\dfrac{x-3}{x+3}$; **k)** $a-5$; **l)** $\dfrac{6x-5}{6x+5}$;

m) $\dfrac{2(x+4)}{x-4}$; **n)** $\dfrac{2x+3}{4(x-1)}$; **o)** $\dfrac{3x+2}{2(2x+1)}$; **p)** $\dfrac{a^2+2}{a^2-2}$; **q)** $\dfrac{x+4}{x-4}$;

r) $\dfrac{x+2}{x-2}$; **s)** $\dfrac{x^2-5x+3}{x^2-5x-1}$; **t)** $\dfrac{x^2-4x-1}{x^2-4x-3}$; **u)** $\dfrac{2x^2-3x-2}{2x^2-3x-1}$.

III. 14. a) $\dfrac{2}{x}$; **b)** $\dfrac{3}{x-2}$; **c)** $\dfrac{x}{x+2}$; **d)** 1; **e)** $\dfrac{2}{(x^2-1)(x-1)}$;

f) $\dfrac{3a^2-x^2+2a}{6a^3x^2}$; **g)** $\dfrac{6}{x^4-1}$; **h)** $\dfrac{6}{x-1}$; **i)** $\dfrac{2(x+1)}{x(x-3)}$;

j) $\dfrac{2(x+9)}{(x-2)(x+5)}$; **k)** $\dfrac{2(x^2+1)}{x^2-1}$; **l)** $\dfrac{x-3}{x}$; **m)** $\dfrac{2(3x-4)}{(x-1)(x+1)}$;

n) $\dfrac{1}{x+1}$; **III. 15. a)** $\dfrac{1-3x}{(x-1)(x-2)}$; **b)** $\dfrac{1-x^3-x}{x^3-1}$; **c)** $\dfrac{1}{x}$; **d)** -1;

e) $\dfrac{4-19x^2}{4(x-2)}$; **f)** $\dfrac{3x^2+2y^2+8xy}{2x+y}$; **g)** $\dfrac{3x^2+2}{2x}$; **h)** 1.

III. 16. a)
$$\left[\dfrac{x^2+2xy+y^2+2y^2-(x^2+xy+y^2)+x^2-y^2}{x^3-y^3}\right]\div\dfrac{x-y}{xy}=$$
$$=\dfrac{x^2+2xy+y^2+2y^2-x^2-xy-y^2+x^2-y^2}{(x-y)(x^2+xy+y^2)}\cdot\dfrac{xy}{x-y}=$$

$$= \frac{x^2 + xy + y^2}{x^2 + xy + y^2} \cdot \frac{1}{xy} = \frac{1}{xy}$$

b) $\left\{ \dfrac{x^2 - y^2}{(x+y)^2} + \left(\dfrac{x^2 - y^2}{(x+y)^2} \right)^3 + 2\left[\dfrac{x^2 - y^2}{(x+y)^2} \right]^2 \right\} \div \dfrac{x^2 - y^2}{(x+y)^4} =$

$$= \frac{x^2 - y^2}{(x+y)^2} \left\{ 1 + \left[\frac{x^2 - y^2}{(x+y)^2} \right]^2 + 2\frac{x^2 - y^2}{(x+y)^2} \right\} \cdot \frac{(x+y)^4}{x^2 - y^2} =$$

$$= \left[1 + \frac{x^2 - y^2}{(x+y)^2} \right]^2 (x+y)^2 = \left[\frac{2x(x+y)}{(x+y)^2} \right]^2 (x+y)^2 = 4x^2$$

;

c) $x - 1$; **d)** $-\dfrac{m+n}{m^2 n^2}$; **e)** $\dfrac{x-y}{2xy}$; **f)** $-\dfrac{1}{4x}$; **g)** $\dfrac{1}{2a}$; **h)** $\dfrac{y}{x+y}$;

i) $\dfrac{2}{x}$; **j)** $\dfrac{b-2}{2b}$. **III. 17. a)** $E(x) = 1$; **b)** $x \in \{0, 1, 3, 4\}$;

III. 18. a) $x \in \{-2, -1, 0, 1\}$; **II)** $x = -2$, which is not a solution.

III. 19. a) $x \in \mathbf{R} \setminus \{-5, 5\}$; **b)** $\dfrac{9}{x+5}$; **c)** $x = 4$.

III. 20. a) $x \in \mathbf{R} \setminus \{-2, 0, 2\}$; **b)** $-\dfrac{x(x-2)}{x+2}$; **c)** $x \in \{0, 1\}$;

d) $x = -6$. **III. 21. a)** $x \in \mathbf{R} \setminus \{-1, 1\}$; **b)** $-\dfrac{x+1}{2(x-1)}$; **c)** $x = 2$.

III. 22. a) $x \in \mathbf{R} \setminus \left\{ -2, -\dfrac{2}{3}, 2 \right\}$; **c)** $x = 1$. **III. 23. a)** $11x - 48$;

b) $4\dfrac{4}{11}$; **c)** 9. **III. 24. b)** $x \in \{-11, -8, -7, -6, -4, -3, -2\}$.

III. 25. a) $x \in \mathbf{R} \setminus \{-1, 1\}$; **b)** $\dfrac{x^4 + 1}{x^2 - 1}$;

III. 26. a) $x \in \mathbf{R} \setminus \{-1, 1\}$; **c)** $E(\sqrt{2}) = 4(3 - 2\sqrt{2})$.

III. 27. a) $E_1(x): x \in \mathbf{R} \setminus \{-1, 1\}$ and

$E_2(x): x \in \mathbf{R} \setminus \{-1, 0, 1\}$; **b)** $E_1(x) = -2x$, $E_2(x) = \dfrac{x-1}{x+1}$; **c)** -1;

III. 28. a) $x \in \mathbf{R} \setminus \left\{ -\dfrac{1}{2}, \ -\dfrac{1}{3}, \ \dfrac{1}{3}, \ \dfrac{1}{2} \right\}$; **b)** $x \in \{0, \ 1\}$;

III. 29. a) $x \in \mathbf{R} \setminus \left\{ -3, \ \dfrac{1}{2}, \ 3 \right\}$; **c)** $x \in \{-5, -4, -2, -1\}$.

III. 30. a) $x \in \mathbf{R} \setminus \{-1, \ 0, \ 1\}$,

$$E = \left[\left(\frac{x}{x+1} - \frac{x^2}{(x+1)^2} \right) \div \left(\frac{x}{(x-1)(x+1)} - \frac{1}{x+1} \right) \right] \frac{x+1}{x} =$$

$$= \left[\left(\frac{x(x+1) - x^2}{(x+1)^2} \div \frac{x - x + 1}{(x-1)(x+1)} \right) \right] \frac{x+1}{x} = \left(\frac{x^2 + x - x^2}{x+1} \cdot \frac{x-1}{1} \right) \frac{x+1}{x} =$$

$$= \frac{x(x-1)}{x} = x - 1;$$ **b)** $x \in \mathbf{R} \setminus \{-2, \ -1, \ 0, \ 1, \ 2\}$, $E = 2x$;

c) $x \in \mathbf{R} \setminus \{-4, \ -1, \ 1\}$; $E(x) = 1 - x$; **d)** $\dfrac{1}{x+2}$; **e)** -1;

f) $x - y + 3$; **g)** $x + 5$; **h)** $\dfrac{x-1}{x-3} = x - 1, x = 4$; **i)** $\dfrac{1}{y^2 + y + 1}$;

j) $x \notin \left\{ -\dfrac{3}{2}, -\dfrac{7}{5}, -1, \ 1, \ 0 \right\}$, 1.

III. 31. a) $x = 3.5$; **b)** $x = \dfrac{1}{4}$; **c)** no solution; **d)** $x \in \mathbf{R} \setminus \{3\}$;

e) no solution; **f)** no solution; **g)** $x = 2$; **h)** $x \in \mathbf{R} \setminus \{-1\}$;

i) $x = 3$; **j)** $x = -5$; **k)** $x = \dfrac{a(a+b)}{b}$; **l)** $x = \dfrac{2}{3}$;

m) $x = -\dfrac{10}{21}$; **n)** $x = -3$; **o)** $x = \dfrac{5}{27}$; **p)** $x = 24$;

q) $x = \dfrac{1}{8}$; **r)** $x = 2\dfrac{a+b}{a-b}$; **s)** $x = 2$; **t)** $x = 2$.

III. 32. a) $x = 1$; **b)** $x = 1$; **c)** $x = 2$; **d)** $x = -1$; **e)** $x = 2$;

f) $x = \dfrac{ab}{a+b}$; **g)** $x = \dfrac{a}{a+1}$; **h)** $x = -a^4$; **i)** $x = \dfrac{2a^2}{3}$;

j) $x = \dfrac{a+1}{4}$; **k)** $x = 1$; **l)** $x = 4$; **m)** $x = \dfrac{16}{17}$. **III. 33. a)** $x = 2$;

b) $x = 6$; **c)** $x = \dfrac{3\sqrt{3}}{5}$; **d)** $x = \sqrt{2}$; **e)** $x = \sqrt{3} - 1$;

f) $x = -\dfrac{19}{36}$; **g)** $x = 14 - 2\sqrt{5}$; **h)** $x = 4\sqrt{2}$; **i)** $x = -\dfrac{1}{20} - \sqrt{2}$;

j) $x = 1$; **k)** $x = \dfrac{\sqrt{2}-1}{2}$; **l)** $x = \dfrac{10 + 3\sqrt{2}}{8}$; **m)** $x = 5$;

n) $x = \dfrac{4\left(1 - \sqrt{3}\right)}{9}$; **o)** $x = 3$; **p)** $z = -\dfrac{2\left(5 + 9\sqrt{2}\right)}{21}$; **q)** $x = 3$;

r) $x = 0$; **s)** $x = 16$; **t)** $x = 4$; **u)** $x = 0$; **v) i)** $a = -1$; **ii)** $a = 11$;
iii) $a = -4$; **iv)** $a = 3$; **w) i)** $m = 6$; **ii)** $m = 3$; **iii)** $m = 2$;
iv) $m = -6$.

III. 34. a) $x_1 = -8$, $x_2 = 4$; **b)** $x_1 = -2$, $x_2 = 8$; **c)** $x_1 = -3$,

$x_2 = \dfrac{5}{3}$; **d)** $x_1 = -4$, $x_2 = 6$; **e)** $x = \pm 3$; **f)** Impossible;

g) $x_1 = -\dfrac{11}{6}$, $x_2 = \dfrac{5}{6}$; **h)** $x_1 = -\dfrac{2}{3}$, $x_2 = 0$; **i)** $x = 3$; **j)** $x = \pm m$;

k) $x_1 = a - 1$, $x_2 = -a - 1$; **l)** Impossible; **m)** $x = \pm 1$, $a \neq 0$;

n) $x \in \left(-\infty,\ 0\right]$; **o)** $x = \pm 1$, $m \neq -\dfrac{1}{2}$; **q)** $x_1 = m - 1$, $x_2 = 1 - m$,

$m \neq 0$; **r)** $x = \pm 1$, $m \neq -2$; **s)** $x = \pm 1$, $m \neq 0$; **t)** $x_1 = 1$, $m \neq -2$,

$x_2 = \dfrac{m+2}{2-m}$, $m \neq 2$; **u)** If $a < 0$, $x_1 = \dfrac{2}{a}$, if $a > 0$, $x_2 = \dfrac{2}{3a}$,

and if $a = 0$, there is no solution; **v)** $x = -1$;

w) No solutions; **x)** $x = -0.5$; **y)** $x_1 = -\dfrac{8}{3}$, $x_2 = -\dfrac{4}{5}$; **z)**

$$|x| = \begin{cases} -x & if \ \ x \in \left(-\infty,\ 0\right) \\ x & if \ \ x \in \left[0,\ \infty\right) \end{cases}, \ |x + 2| = \begin{cases} -x - 2 & if \ \ x \in \left(-\infty,\ -2\right) \\ x + 2 & if \ \ x \in \left[-2,\ \infty\right) \end{cases} ; \ \textbf{i)}$$

$x \in \left(-\infty \ -2\right)$. The equation becomes $-x - x - 2 = x + 5$.

Thus $x = -\dfrac{7}{3}$; **ii)** $x \in \left[-2,\ 0\right)$. The equation becomes

$-x + x + 2 = x + 5$. Thus $x = -3$ (Impossible);

iii) $x \in \left[0,\ \infty\right)$. The equation becomes $x + x + 2 = x + 5$. Thus

$x = 3$. Therefore $x = -\dfrac{7}{3}$ and $x = 3$ are the solutions of the equation.

III. 35. a) $x = \dfrac{3}{2+m}, m \neq -2$; **b)** $x = 1, m \neq -1$;

c) $x = \dfrac{4}{m-3}, m \neq 3$; **d)** $x = \dfrac{m-1}{3}$; **e)** $x = \dfrac{3ab}{a^2+b^2}, a^2+b^2 \neq 0$;

f) $x = \dfrac{m-3}{m^2+m-2}, m \neq 1, m \neq -2$; **g)** $x = \dfrac{1}{m+1}, m \neq -1, \ m \neq 3$;

h) $x = \dfrac{1}{m^2-3n-4}, m \neq -1, m \neq 4$; **i)** $x = \dfrac{4m+m^2}{m-4}, m \neq 4$;

j) $x = \dfrac{2m}{m^2+2m}, m \neq -2, m \neq 0$; **k)** $x = -1-m, m \neq 1$;

l) $x = \dfrac{3a-7}{2a-1}, m \neq \dfrac{1}{2}$; **m)** $x = \dfrac{-1-2m}{m+2}, m \neq -2$; **n)** $x = -\dfrac{a}{m}, m \neq 0$;

o) $x = \dfrac{2am}{a+m}, m \neq -a$; **p)** $x = \dfrac{5}{m}, m \neq 0$; **q)** $x = \dfrac{1}{a-m}, m \neq a$;

r) $x = a+m, m \neq a$; **s)** $x = \dfrac{m\sqrt{2}}{4\sqrt{2}-4m}, m \neq \sqrt{2}$;

t) $x = \dfrac{2-3m}{m-1}, m \neq 1$; **u)** $x = \dfrac{4a^2-3m^2}{2a-m}, m \neq 2a$. **III. 36. a)** $x_1 = 0$,

$x_2 = -2$; **b)** $x = 3$; **c)** $x = \pm\sqrt{3}$; **d)** $x_1 = 0$, $x_2 = -\dfrac{5}{3}$; **e)** $x = 2$,

$x \neq 0$; **f)** $x_1 = \dfrac{5}{2}$, $x_2 = -\dfrac{5}{2}$; **g)** $x_1 = -17$, $x_2 = 15$; **h)** $x_1 = \dfrac{1}{5}$,

$x_2 = 5$; **i)** $x_1 = 1$, $x_2 = 2$, $x_3 = -3$; **j)** $x_1 = -2$, $x_2 = \dfrac{9}{4}$;

k) $x = -4$; **l)** $x_1 = 1$, $x_2 = -3$; **m)** $x_1 = -\dfrac{1}{8}$, $x_2 = 1$; **n)** $x = 1$;

o) $x_1 = 3$, $x_2 = -\dfrac{3}{2}$; **p)** $x = -3$; **q)** No solution; **r)** $x = 1$, $x \neq \pm3$.

III. 37. a) $(-\infty, 2]$, **b)** $\left(\dfrac{1}{3}, \ \infty\right)$, **c)** $\left(-\infty, \ -\dfrac{1}{7}\right]$,

d) $\left(-\infty, \ -1\dfrac{1}{2}\right]$; **e)** $[0, \ \infty)$, **f)** $\left(-\infty, \ 3\right]$; **g)** $\left(-\infty, \ 1\right]$;

h) $\left(-\infty, \ -2\right)$, **i)** $\left(-\infty, -2\right)$; **j)** $\left(\dfrac{2}{3}, \ \infty\right)$; **k)** $\left(-\infty, \ 1\right)$;

l) $\left[8\dfrac{1}{2}, \ \infty\right)$; **m)** $[0, \ \infty)$; **n)** $\left(-\infty, \ 5\right)$; **o)** $\left(-\infty, \ -\dfrac{1}{6}\right]$;

p) $\left(-\infty, \ -3\dfrac{3}{7}\right]$. **III. 38. a)** $(1, \ -1)$; **b)** $(2, \ -1)$; **c)** $(-1, \ -1)$

; **d)** $(2, \ 1)$; **e)** $(3, \ 1)$; **f)** $(1, \ 1)$; **g)** $(-1, \ -1)$; **h)** $(-5, \ -2)$;
i) $(7, \ 6)$; **j)** $(6, \ 10)$. **III. 39. a)** $(1, \ -2)$; **b)** $(1, \ 0)$;
c) $(2, \ 3)$; **d)** $(2, \ -1)$; **e)** $\left(\sqrt{2}, \ -\sqrt{3}\right)$; **f)** $\left(\sqrt{5}-\sqrt{3}, \ \sqrt{5}+\sqrt{3}\right)$;

g) $\left(\dfrac{1}{3}, \ -\dfrac{1}{3}\right)$; **h)** $\left(\dfrac{1}{2}, \ -\dfrac{1}{2}\right)$; **i)** $(28, \ 82)$; **j)** $(1, \ -3)$.

III. 40. a) $(6, \ -6)$; **b)** $(2, \ 1)$; **c)** $(4, \ -1)$; **d)** $(1, \ -1)$;

e) $\dfrac{x}{a}=3-\dfrac{y}{b}$, substitute in the second equation:

$$\dfrac{1}{3}\left(3-\dfrac{y}{b}\right)+\dfrac{y}{4b}=\dfrac{5}{6}, \ 1-\dfrac{y}{3b}+\dfrac{y}{4b}=\dfrac{5}{6}, \ \dfrac{-4y+3y}{12b}=\dfrac{5}{6}-1,$$

$\dfrac{-y}{12b}=\dfrac{-1}{6}\Rightarrow 12b=6y\Rightarrow 2b=y$ and similarly $x=a$; **f)** Multiply

the second equation by $\dfrac{1}{a-b}$ and add the equations

$$x\left(\dfrac{1}{a+b}+\dfrac{1}{a-b}\right)=2a+\dfrac{4ab}{a-b}\Rightarrow x\dfrac{2a}{(a+b)(a-b)}=\dfrac{2a^2+2ab}{a-b},$$

finally $x=(a+b)^2$. Replace x in the second equation, to find y:
$(a+b)^2-y=4ab$, and $y=(a-b)^2$;

g) $\dfrac{1}{x+y-1}=a$ and $\dfrac{1}{x-y+2}=b$ then the system becomes

$\begin{cases}2a+3b=5\\b-2a=-1\end{cases}$, $x=\dfrac{1}{2}$, $y=\dfrac{3}{2}$; **h)** Denote

$$\frac{1}{k}=\frac{3}{x+y}=\frac{4}{x+z}=\frac{5}{y+z}, \quad x+y=3k, \ x+z=4k, \ y+z=5k$$

and replacing in the third equation of the system:

$$(3k)(4k)(5k)=60 \Rightarrow k^{3}=1 \Rightarrow k=1. \text{ Therefore } \begin{cases} x+y=3 \\ x+z=4, \\ y+z=5 \end{cases}$$

add these three equations: $2(x+y+z)=12 \Rightarrow x+y+z=6$, $x=1$, $y=2$, $z=3$; **i)** $y=7$, $x=5$.

III. 41. Let a, b, c be the sides of the triangle, then $a+b=45$, $b+c=52$, $c+a=48$. $a=\dfrac{41}{2}$, $b=\dfrac{49}{2}$, $c=\dfrac{55}{2}$.

III. 42. b)

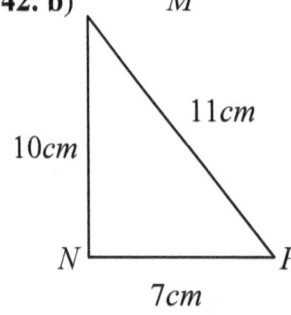

$$\frac{MN}{AC}=\frac{NP}{AB}=\frac{MP}{BC} \text{ and } AB+BC+AC=70. \text{ Applying the}$$

proportions properties, we get

$$\frac{MN}{AC}=\frac{NP}{AB}=\frac{MP}{BC}=\frac{MN+NP+MP}{AC+AB+BC}=\frac{7+10+11}{70}=\frac{28}{70}=\frac{2}{5}.$$

III. 43. $a+b+c=222$, $b=10c$ and $a=10b=100c$, $a=200$, $c=2$, $b=20$;

III. 44. a) $x=\dfrac{19}{2}$, $y=\dfrac{11}{2}$; **b)** $x=1$, $y=1$; **c)** $x=1$, $y=1$, $z=1$;

d) $x=1$, $y=5$, $z=1$;

e) $\begin{cases} (x+y-z)(x+y+z)=9 \\ (y+z-x)(y+z+x)=45 \\ (z+x-y)(z+x+y)=27 \end{cases}$. Adding side by side, factor out the

common factor $(x+y+z)$ we get

$(x+y+z)(x+y+z)=81 \Rightarrow (x+y+z)=\pm9$. **I)** $(x+y+z)=9$, then

$x=2$, $y=3$, $z=4$, **II)** $x+y+z=-9$, then $x=-2$, $y=-3$,

$z=-4$; **f)** $\dfrac{x}{y}=-1$; **g)** $A=0.\overline{2}$.

III. 45. 1) $a=-3$. **III. 46.** $f(x)=2x-3$.

III. 47.

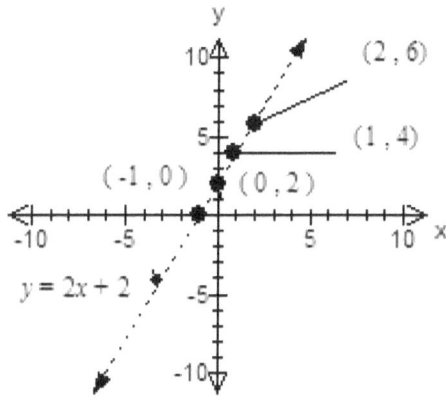

III. 48.

III. 49. $a=1$. **III. 50.** $\left(\dfrac{1}{3}, 1\dfrac{2}{3}\right)$, yes.

III. 51. $f(x)=-3x-2\sqrt{6}$.

III. 52. a) 1; **b)**
c) $M(3, 0)$ and $N(0, -3)$;

d) A; **e)** 4.5; **f)** $\dfrac{3\sqrt{2}}{2}$;

g) 45°; **h)** $x = 1$;
i) $x \in (-\infty,\ 4]$.

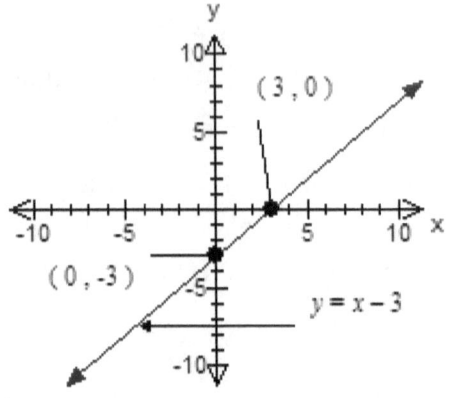

III. 53. $f(x) = -2x + 4$.
III. 54. $f(4) + f(1) = -3$.
A and B are on the graph.
III. 55. a) $(0, -8)$ and $(4, 0)$;
b)

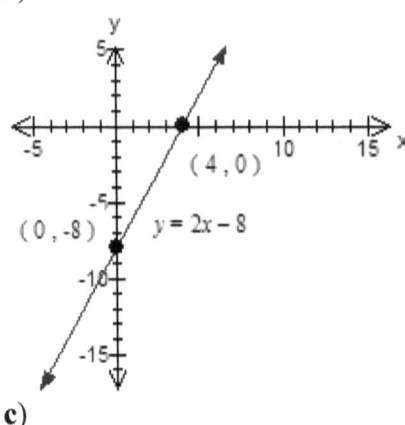

c)

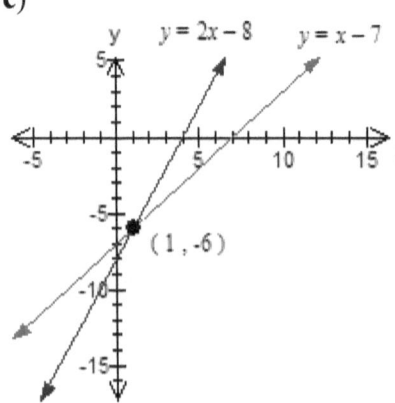

d) $4\left(4 + \sqrt{5}\right)$ and 16; **e)** $\dfrac{8\sqrt{5}}{5}$;

f) $2\left(3 + \sqrt{5}\right)$ and $2\sqrt{5}$; **g)** $h(x) = x - 6$;

h) $x \in \left(-\infty, \ -\dfrac{7}{3} \right]$; **i)** $(-1, 2)$; **k)** $C(11, 14)$.

III. 56. a) $M(-1, 1)$; **b)** $M(2, 0)$; **c)** $M(0, 1)$; **iv)** $M(3, 2)$.

III. 57. $m = -1$. **III. 58.** $m = \dfrac{3\sqrt{6} - 5\sqrt{2}}{2}$.

III. 59. a) $f(x) = 2 - x$; **b)** $f(x) = 2x - 3$; **c)** $f(x) = 5 - 0.5x$;
d) $f(x) = -4x - 2$; **e)** $f(x) = x - 7$; **f)** $f(x) = 5x + 1$. **III. 60. a)** no;
b) yes;
c) no; **d)** yes. **III. 61.** $m = 1, n = 2$. **III. 62.** $m = 3, n = -1$.
III. 63. $a = 1$.

III. 64. $A_{\triangle ABC} = 3$ and $P_{\triangle ABC} = 4 + 2\sqrt{2} + \sqrt{5}$.

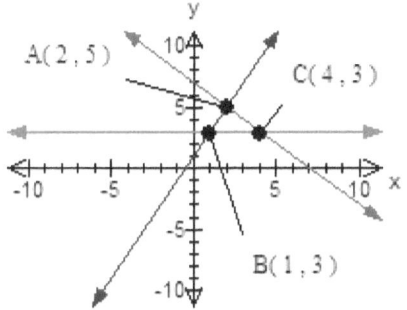

III. 65. a) $x = 2$; **b)** $f(-x) + f(x) = 10$. **III. 66. b)** Area $= 6$; **c)** $a = -1$;

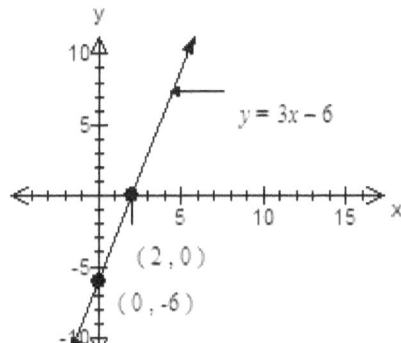

d) 105.

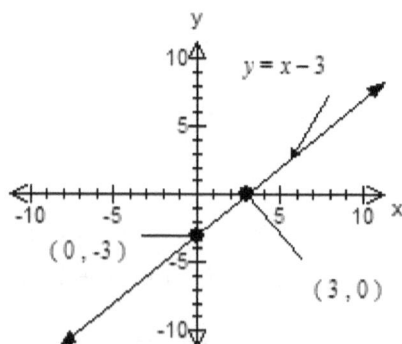

III. 67. a)

c) Area = 4.5; **d)** $\dfrac{3\sqrt{2}}{2}$; **e)** $E = 6$; **III. 68.** $f(x) = 2x + 7$.

III. 69. a) $a = 1$;

b) $b = 3$; **c)** $a < 1$; **d)** $M(2, 2)$. **III. 70. b)** $A(6, 6)$;

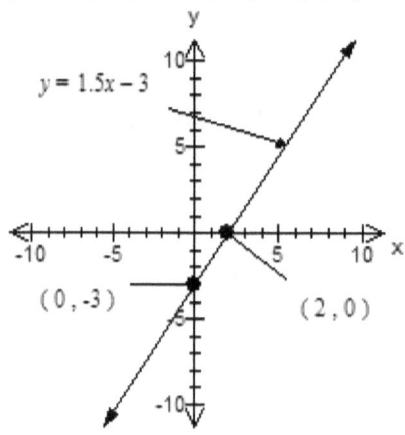

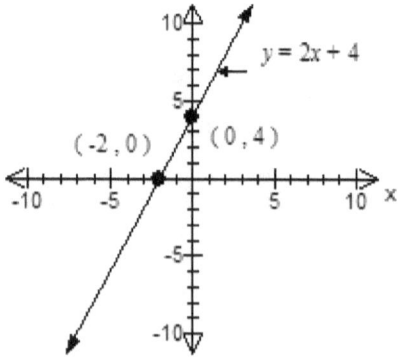

III. 71. a) $m = 1$; **b)**

c) Area = 4;

III. 72. b) $(2, 0)$ and $(0, -4)$; **c)** Area $= 4$; **d)** Q is the

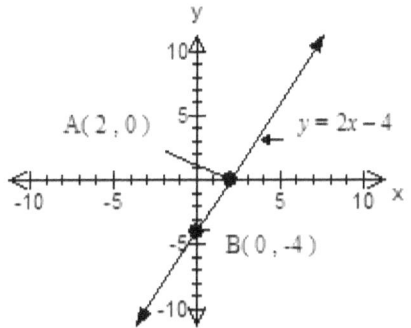

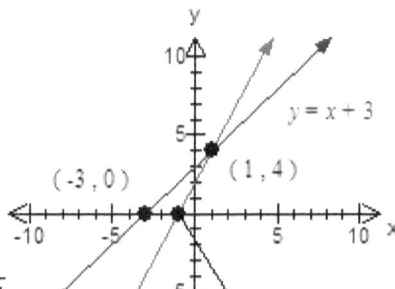

midpoint of the segment AB; **e)** $\dfrac{4\sqrt{5}}{5}$.

III. 73. b) Area $= 4$;

III. 74. a) $a = 1$; **b)** Geometric mean $= 1$;
c) x intercept $= (2.5, 0)$ and y intercept $= (0, 5)$.

III. 75. a) $\left[\dfrac{2x+1}{3}\right] \le \dfrac{2x+1}{3} < \left[\dfrac{2x+1}{3}\right] + 1$ or

$\dfrac{x-1}{3} \le \dfrac{2x+1}{3} < \dfrac{x-1}{3} + 1$ with the solution $-14 \le x < -8$. On the other

hand $\dfrac{x-4}{2} = k \in \mathbf{Z}$, then $x = 2k + 4$. Therefore $-14 \le 2k + 4 < -8$

which implies $k \in [-9, -6)$, but k is an integer number, thus
$k \in \{-9, -8, -7\}$. The solutions of the equation are -16, -12, and -10; **b)**
-5 and -2; **c)** 7, 7.5, 8, 8.5, 9, 9.5;

d) 5, $\dfrac{19}{3}$, $\dfrac{23}{3}$; **e)** Denote $\left[3x - \dfrac{2}{5}\right] = t$, $t \in \mathbf{Z}$. Then $t = 2x + 1$ or

$x = \dfrac{t-1}{2}$ and the initial equation becomes $\left[\dfrac{15t-19}{10}\right] = t$, therefore

$t \le \dfrac{15t-19}{10} < t + 1$. Since $t \in \mathbf{Z}$, we obtain $t_1 = 4$ and $t_2 = 5$. The

solutions of the equation are $x_1 = 1.5$ and $x_2 = 2$; **f)** $x_1 = \dfrac{3}{4}$, $x_2 = 1$;

g) $x_1 = -\dfrac{5}{2}$, $x_2 = -\dfrac{3}{2}$; **h)** $x \in \left\{ -\dfrac{1}{6}, 0, \dfrac{1}{6}, \dfrac{1}{3}, \dfrac{1}{2}, \dfrac{2}{3} \right\}$; **i)** 1, 3, 5.

III. 76. a) Denote $\left[x + \dfrac{2}{3} \right] = k$, $k \in \mathbf{Z}$, thus $\left[x + \dfrac{3}{5} \right] = 1 - k$. Then

$k - \dfrac{2}{3} \le x < k + \dfrac{2}{3}$ and respectively $-k + \dfrac{2}{5} \le x < -k + \dfrac{8}{5}$. The last

two inequalities become $-\dfrac{2}{3} < -x + k \le \dfrac{2}{3}$ and $\dfrac{2}{5} \le x + k < \dfrac{8}{5}$. Now

adding them side by side we get $-\dfrac{4}{30} < k < \dfrac{34}{30}$. Since k is an

integer number, then $k = 0$ or 1. Therefore

$x \in \left[-\dfrac{2}{3}, \dfrac{2}{3} \right) \cap \left[\dfrac{2}{5}, \dfrac{8}{5} \right) = \left[\dfrac{2}{5}, \dfrac{2}{3} \right)$ or

$x \in \left[\dfrac{1}{3}, \dfrac{5}{3} \right) \cap \left[-\dfrac{3}{5}, \dfrac{3}{5} \right) = \left[\dfrac{1}{3}, \dfrac{3}{5} \right)$ and finally $x \in \left[\dfrac{1}{3}, \dfrac{3}{5} \right)$;

b) $x \in \left[\dfrac{1}{3}, \dfrac{2}{3} \right)$.

Chapter IV

POWERS AND RADICALS

IV. 1. a) x^4; **b)** y^4; **c)** $4a^3$; **d)** $-3b^3$; **e)** $-2x^2y$;

f) $3a^2b^3$; **g)** 2^3a^3; **h)** a^6b^6; **i)** a^{15}; **k)** a^{21}; **j)** $-3^3x^{12}y^9$;

l) x^{11}; **m)** x^{n+1}; **n)** x^{2n}; **o)** a^{m+n+1}. **IV. 2. a)** $6y^5$;

b) $\left(3x^7y^5\right)$; **c)** $8y^3$; **d)** $16x^9$; **e)** $\dfrac{a}{b}$; **f)** $\dfrac{2}{b^4}$; **g)** $4r^7s$;

h) $\dfrac{8}{9}t^{10}$; **i)** $8y^3$; **j)** 12; **k)** $\dfrac{1}{x^3y^{12}}$; **l)** $\dfrac{d^7}{c^8}$;

m) $\dfrac{y^2z}{x^2}$; **n)** x^3; **o)** $x^3y^{12}z^{21}$; **p)** $\dfrac{s^4}{q^7r^6}$; **q)** $\dfrac{c^8}{3a^3}$; **r)** $\dfrac{1}{x^6y^4}$.

IV. 3. a) $x^{\frac{13}{15}}$; **b)** $-10a^{\frac{9}{4}}$; **c)** $2^4 \cdot b^{\frac{9}{10}}$; **d)** $\dfrac{1}{2^2 \cdot x^4}$; **e)** $\dfrac{1}{c^{\frac{2}{3}}d}$;

f) $2x^9y^{12}$; **g)** $y^{\frac{1}{2}}$; **h)** $\dfrac{1}{a^2}$; **i)** $\dfrac{2x^6}{y^2}$; **j)** $\dfrac{x^3}{y^9z^6}$; **k)** $\dfrac{x^{15}}{y^{10}}$; **l)** $\dfrac{16x^3}{yz^2}$;

m) a^{13}; **n)** $27a^{\frac{5}{4}}$; **o)** $12x^{\frac{5}{6}}$; **p)** $\dfrac{1}{3b^2}$; **q)** $x^{\frac{2}{3}} \cdot y^{\frac{5}{2}}$; **r)** 64; **s)** $y^{\frac{5}{4}}$.

IV. 4. a) $\sqrt{14}$; **b)** 2; **c)** 5; **d)** $\sqrt{39}-\sqrt{26}$; **e)** $2\sqrt{3}$;

f) $2\sqrt{2}-2$; **g)** $\sqrt{2}$; **h)** $17\sqrt{5}$; **i)** 0; **j)** $7\sqrt{2}$; **k)** $-12\sqrt{2}-3\sqrt{5}$;

l) $12-18\sqrt{2}+12\sqrt{3}$; **m)** 10; **n)** $4\sqrt{5}$; **o)** $42\sqrt{10}-174$; **p)** 1;

q) 4; **r)** 5; **s)** 33. **IV. 5. a)** $-\sqrt{3}$; **b)** $2\sqrt{5}$;

c) $3\sqrt{6}$; **d)** -72; **e)** $4\sqrt{3}+13\sqrt{2}$; **f)** 48; **g)** $-3\sqrt{5}$;

h) -1; **i)** 14; **j)** 172; **k)** -2; **l)** $\dfrac{3\sqrt{2}+\sqrt{3}}{3}$;

m) $\left(\sqrt{5}+\sqrt{2}\right)\cdot\left(\sqrt{2}+1\right)$; **n)** $\sqrt{2}+2$; **o)** 1; **p)** $6\dfrac{1}{5}$; **q)** $3-\sqrt{3}$;

r) $\sqrt{5}-\sqrt{2}$; **s)** $\sqrt{2}-1$; **t)** $\sqrt{2}-2\sqrt{3}$; **u)** $\dfrac{9}{10}$.

IV. 6. a) $\sqrt{54} = 3\sqrt{6}$ and $\sqrt[3]{-108} = -\sqrt[3]{2^2 \times 3^3} = -3\sqrt[3]{4}$;

b) $26 \times \sqrt[3]{2}$; **c)** $\dfrac{-23 \times \sqrt[3]{3}}{6}$. **IV. 7. a)** $\dfrac{2\sqrt{3}}{3}$; **b)** $\sqrt{3}$; **c)** $\dfrac{\sqrt{6}}{2}$;

d) $-3 + 2\sqrt{2}$; **e)** $\sqrt{6}$; **f)** $-3 + 2\sqrt{2}$; **g)** $7 + 2\sqrt{3}$;

h) $\sqrt{7} - \sqrt{3}$; **i)** $2\sqrt{2} + \sqrt{10} - \sqrt{5} - 1$; **j)** $\sqrt{3} - \sqrt{2}$;

k) $\sqrt[3]{9} - \sqrt[3]{21} + \sqrt[3]{49}$; **l)** $\sqrt{5} - 1$; **m)** $\sqrt{6} - \sqrt{2}$; **n)** 4 ; **o)** $-5 + 3\sqrt{2}$.

IV. 8. a) $2 - \sqrt{3}$; **b)** $\left.\dfrac{x(a+b)}{ab}\right|_{x=\sqrt{ab}} = \dfrac{a+b}{\sqrt{ab}}$. **IV. 9. a)**

$A = \left(\sqrt[3]{x} - 1\right)\left(\sqrt[3]{x^2} + \sqrt[3]{x} + 1\right)\left(\sqrt[3]{x} + 1\right)\left(\sqrt[3]{x^2} - \sqrt[3]{x} + 1\right) = (x-1)(x+1)$.

b) $B = \left(\sqrt{x^3} - 2 \cdot \sqrt[4]{y}\right)$; **c)** $D = 2 \cdot \sqrt[3]{xy}$. **IV. 12.** $(-1, 1)$.

IV. 13. a) $2x - 1 + \sqrt{(x-4)^2} = 2x - 1 + |x - 4| = \begin{cases} 3x - 5 & if \quad x > 4 \\ x + 3 & if \quad x \leq 4 \end{cases}$;

b) $x - 3 + \sqrt{(x-3)^2} = x - 3 + |x - 3| = \begin{cases} 2(x-3) & if \quad x > 3 \\ 0 & if \quad x \leq 3 \end{cases}$.

IV. 14. a) $x \in \left[\dfrac{1}{2}, \ \infty\right)$; **b)** $x \in \mathbf{R}$; **c)** $x \in \left(-\infty, \ -\dfrac{1}{3}\right]$; **d)** $x \in \mathbf{R}$;

e) $x \in \left(-\infty, \ \dfrac{1}{3}\right]$; **f)** $x \in \mathbf{R}$; **g)** $x \in (-\infty, \ 3] \cup [4, \ \infty)$;

h) $x \in [-2, \ 2]$; **i)** $x \in [-4, \ -1] \cup [2, \ \infty)$; **j)** $x \in [1, \ 3]$;

k) $x \in (-\infty, \ -1] \cup [1, \ \infty)$; **l)** $x \in [2, \ 5]$; **m)** $x \in (-\infty, \ 2]$.

IV. 15. a) $\sqrt{32} > \sqrt{27}$; **b)** $\sqrt[4]{405} > \sqrt[4]{216}$; **c)** $\sqrt{2205} > \sqrt{2016}$;

d) $\sqrt{\dfrac{48}{14}} > \sqrt{\dfrac{47}{14}}$; **e)** $\sqrt{\dfrac{4}{75}} = -\sqrt{\dfrac{24}{450}}$; **f)** $-\sqrt[3]{378} < -\sqrt[3]{144}$;

g) $\sqrt[3]{\dfrac{2}{11}} > \sqrt[3]{\dfrac{7}{325}}$.

IV. 16. $a = \dfrac{\left(\sqrt{7} - \sqrt{5}\right)^2}{\left(\sqrt{7} + \sqrt{5}\right)\left(\sqrt{7} - \sqrt{5}\right)} = \dfrac{12 - 2\sqrt{35}}{2} = 6 - \sqrt{35}$,

$b = \dfrac{12 + \sqrt{35}}{2} = 6 + \sqrt{35}$, $a + b = 12$, $a \cdot b = 1$ and

$a^2 + b^2 = (a+b)^2 - 2ab = 142$.

IV. 17.

$$a = \sqrt{\frac{\left(2+\sqrt{3}\right)^2}{2^2-3}} + \sqrt{\frac{\left(2-\sqrt{3}\right)^2}{2^2-3}} = \sqrt{\left(2+\sqrt{3}\right)^2} + \sqrt{\left(2-\sqrt{3}\right)^2} =$$

$$= 2+\sqrt{3}+2-\sqrt{3} = 4 \text{ and}$$

$b = 11+4\sqrt{7} - 2\sqrt{11+4\sqrt{7}}\sqrt{11-4\sqrt{7}} + 11 - 4\sqrt{7} = 22 - 3 = 19$.

IV. 18. a) $\sqrt[12]{216}$, $\sqrt[12]{625}$, $\sqrt[12]{628}$; **b)** $\sqrt{4}$, $\sqrt[4]{15}$, $\sqrt[3]{11}$;

c) $\sqrt[12]{777}$, $\sqrt[3]{6}$, $\sqrt{5}$; **d)** $\sqrt[12]{250}$, $\sqrt[6]{16}$, $2\cdot\sqrt[3]{3}$;

e) $\sqrt[15]{448}$, $\sqrt[10]{20}$, $\sqrt[5]{50}$; **f)** $\sqrt[3]{7}$, $\sqrt[12]{4000}$, $\sqrt[4]{17}$.

IV. 19. a) $\dfrac{7-4\sqrt{3}}{\sqrt[3]{26-15\sqrt{3}}} = \dfrac{\left(2-\sqrt{3}\right)^2}{\sqrt[3]{\left(2-\sqrt{3}\right)^3}} = \dfrac{\left(2-\sqrt{3}\right)^2}{2-\sqrt{3}} = 2-\sqrt{3}$.

IV. 21. $S_1 = -1 + \sqrt{n}$; $S_2 = 1 - \dfrac{1}{\sqrt{n+1}}$.

IV. 22. a) $a = \sqrt{x \times \sqrt[3]{\sqrt{x^3}}} = \sqrt{x \times \sqrt[6]{x^3}} = \sqrt{x \times \sqrt{x}}$ and $b = \dfrac{1}{\sqrt[3]{a}}$; **b)**

$$\sqrt[4]{\frac{x}{32}} \cdot \frac{\sqrt[4]{2}\left(\sqrt[4]{2^3\,x}+2\right)}{\sqrt[4]{x}\left(\sqrt[4]{x}-\sqrt[4]{2}\right)} \times \frac{\sqrt[4]{2}\left(\sqrt[4]{2^3}-\sqrt[4]{x^3}\right)}{\sqrt{x}+\sqrt[4]{2x}+\sqrt{2}} =$$

$$= \sqrt[4]{\frac{x}{32}} \times \frac{\sqrt[4]{2^2}\left(\sqrt[4]{2^3\,x}+2\right)\left(\sqrt[4]{2^3}-\sqrt[4]{x^3}\right)}{\sqrt[4]{x}\left(\sqrt[4]{x^3}-\sqrt[4]{2^3}\right)} = \sqrt[4]{\frac{x}{32}} \times \frac{-\sqrt[4]{2^2}\left(\sqrt[4]{2^3\,x}+2\right)}{\sqrt[4]{x}} =$$

$$= \frac{\sqrt[4]{x}}{\sqrt[4]{2^5}} \times \frac{-\sqrt[4]{2^5}\left(\sqrt[4]{x}+\sqrt[4]{2}\right)}{\sqrt[4]{x}} = -\sqrt[4]{x}-\sqrt[4]{2};$$

c) $a - b$; **d)** $x^2 \cdot \sqrt[4]{a}$; **e)** $\sqrt{ab}$; **f)** $\dfrac{1}{2\sqrt{ab}}$; **g)** $\dfrac{\sqrt{a}}{\sqrt{2a+1}-\sqrt{a}}$;

h) $\dfrac{x^2 - \sqrt{x^4 - a^4}}{a^2}$, $E = \dfrac{n}{m}$; **i)** $\left(\sqrt{x}-2\right)\left(\sqrt{x}-3\right)$; **j)** $2\cdot\sqrt[4]{\dfrac{y}{x^2}}$.

IV. 23. a) 1; **b)**

$$\left(\sqrt[6]{1+\frac{1}{\sqrt{x}}} \cdot \sqrt[3]{\sqrt{x}+1} - \sqrt[3]{\sqrt{x}-1} \cdot \sqrt[6]{1-\frac{1}{\sqrt{x}}}\right)^{-2} \frac{\sqrt[12]{x}}{\sqrt{x}+\sqrt{x-1}} =$$

$$= \left(\frac{\sqrt[6]{\left(\sqrt{x}+1\right)^3}}{\sqrt[12]{x}} - \frac{\sqrt[6]{\left(\sqrt{x}-1\right)^3}}{\sqrt[12]{x}} \right)^{-2} \frac{\sqrt[12]{x}}{\sqrt{x}+\sqrt{x-1}} =$$

$$= \left(\frac{\sqrt[12]{x}}{\sqrt{\sqrt{x}+1}-\sqrt{\sqrt{x}-1}} \right)^2 \frac{\sqrt[12]{x}}{\sqrt{x}+\sqrt{x-1}} = \left(\frac{\sqrt[12]{x}\left(\sqrt{\sqrt{x}+1}+\sqrt{\sqrt{x}-1}\right)}{\left(\sqrt{\sqrt{x}+1}\right)^2 - \left(\sqrt{\sqrt{x}-1}\right)^2} \right)^2 \frac{\sqrt[12]{x}}{\sqrt{x}+\sqrt{x-1}} =$$

$$= \frac{\sqrt[12]{x^2}\left(\sqrt{x}+1+2\sqrt{x-1}+\sqrt{x}-1\right)}{4} \frac{\sqrt[12]{x}}{\sqrt{x}+\sqrt{x-1}} = \frac{\sqrt[12]{x^3}\left(2\sqrt{x}+2\sqrt{x-1}\right)}{4\left(\sqrt{x}+\sqrt{x-1}\right)} = \frac{\sqrt[4]{x}}{2}$$

c) $1-\sqrt{x^2-1}$; d) $\dfrac{\sqrt{p^2-q^2}}{\sqrt{p}}$. **IV. 24.** a) $\dfrac{x}{2y-x}$;

b) $\dfrac{-16x\sqrt{x}}{(1-x)(1-x^2)}$; c) 1; d) $x\sqrt{2}$;

e) $\dfrac{1}{\sqrt[4]{x^2-1}}$; f) $\dfrac{1}{x^2}$; g) 2; h) $y(x+y)$; i) 0; j) $\dfrac{24\sqrt[4]{y}}{\sqrt{x}}$.

IV. 25. a) $\dfrac{2x^4}{x^8-16y^8}$, 3; b) ab, 1; c) $\dfrac{a-b}{2ab}$, $-\dfrac{7}{24}$;

d) $\sqrt{\left(a^3+b^3\right)^2-a}$, 0; e) $\dfrac{-\sqrt{ax}}{2x-a}$, 1; f) $-\dfrac{2\sqrt{x}}{3}$, -2;

g) $a\left(100-72\sqrt{ax}\right)$, 100; h) $\dfrac{1}{m-n}$, $\dfrac{\sqrt{5}}{5}$; i) $\dfrac{4x}{x-4}$, 20; j) $-x$, 1.

IV. 26. a) $-\sqrt{\dfrac{3x-2}{3x+2}}$; b) $x^3 \cdot \sqrt[4]{a}$; c) $\sqrt[3]{\dfrac{2n}{1+n}}$; d) $a\sqrt{\dfrac{a}{a+4b}}$;

e) 1; f) $\sqrt[4]{x}$; g) 1; h) $\pm\dfrac{1}{\sqrt{x}}$; i) $-\dfrac{2\sqrt{x}}{3}$; j) $\left|\sqrt[4]{a}-\sqrt[6]{b}\right|$.

IV. 27. a) $\sqrt{4\sqrt{2}+2\sqrt{6}} = \sqrt[4]{2}\sqrt{4+2\sqrt{3}} = \sqrt[4]{2}\left(\sqrt{3}+1\right)$; b) $3+2\sqrt{2}$;

c) $2-\sqrt{3}$; d) $-2+\sqrt{5}$.

IV. 28. $S = \left(\sqrt{2}+\sqrt{2^3}+\sqrt{2^5}+...+\sqrt{2^{2009}}\right)+$

$+\left(\sqrt{2^2}+\sqrt{2^4}+\sqrt{2^6}+...+\sqrt{2^{2010}}\right)=$

$$= \sqrt{2}\left(1 + 2 + 2^2 + \dots + 2^{1004}\right) + \left(2 + 2^2 + 2^3 + \dots + 2^{1005}\right) =$$
$$= \sqrt{2}\left(2^{1005} - 1\right) + \left(2^{1006} - 1\right). \textbf{ IV. 29. } S = \frac{n \cdot (2n+1)}{4},$$

$$\left(\sqrt{x_1 - 1} - \frac{1}{2}\right)^2 + \left(\sqrt{x_2 - 2} - \frac{1}{2}\right)^2 + \dots + \left(\sqrt{x_n - n} - \frac{1}{2}\right)^2 = 0,$$

$$x_1 = \frac{5}{4}, x_2 = \frac{9}{4}, \dots, x_n = \frac{4n+1}{4}. \text{ IV. 30. a) We have}$$

$\left(\sqrt{5} - 2\right)^n + \sqrt{m} = \sqrt{m+1}$, squaring in both sides and collecting the

terms we obtain $2\left(\sqrt{5} - 2\right)^n \sqrt{m} = 1 - \left(\sqrt{5} - 2\right)^{2n}$ or

$2\sqrt{m} = \left(\sqrt{5} + 2\right)^n - \left(\sqrt{5} - 2\right)^n$. Therefore

$$m = \frac{1}{4}\left[\left(\sqrt{5} + 2\right)^n - \left(\sqrt{5} - 2\right)^n\right]^2;$$

b) $m = \frac{1}{4}\left[\left(\sqrt{17} + 4\right)^n - \left(\sqrt{17} - 4\right)^n\right]^2 + 1;$

c) $m = \frac{1}{4}\left[\left(5 - \sqrt{27}\right)^n - \left(5\sqrt{2} + 7\right)^n\right]^2.$

IV. 31. a) $x - 2 \geq 0 \Rightarrow x \in [2, \infty), \sqrt{x-2} = t, t \geq 0$

$\sqrt{t^2 + 2t + 1} - \sqrt{t^2 - 2t + 1} = 1$, $|t+1| - |t-1| = 1$. I) $t \in (1, \infty)$

$t + 1 - t + 1 = 1$ impossible. II) $t \in [0,1)$ $t + 1 + t - 1 = 1 \Rightarrow t = \frac{1}{2}$,

$x = 2.25$; **b)** $x + 3 \geq 0, \sqrt{x+3} = t \geq 0,$

$\sqrt{t^2 + 4t + 4} + \sqrt{t^2 - 4t + 4} = 6 \Leftrightarrow t + 2 + |t - 2| = 6.$

I) $t \in [2, \infty), t + 2 + t - 2 = 6 \Leftrightarrow t = 3, \sqrt{x+3} = 3 \Rightarrow x = 6.$

II) $t \in [0,2), t + 2 - t + 2 = 6$ which is impossible; **c)** $x \in [-2, -1];$

d) $x \in [0, 7];$ **e)** $x \in [5, 10].$

IV. 32. a) $x \in [2, \infty), x - 2 + 2\sqrt{x-2}\sqrt{2x-3} + 2x - 3 = 25,$

$2\sqrt{(x-2)(2x-3)} = 30 - 3x$. $30 - 3x \geq 0 \Rightarrow x \in (-\infty, 10]$ then the

set of permissible values is $x \in [2, \infty) \cap (-\infty, 10] = [2, 10]$. Squaring

both sides of the last equation, we obtain

$4(2x^2 - 7x + 6) = (30 - 3x)^2$, $x_1 = 6$ and $x_2 = 146$ (not solution).

b) $x \in \left[\dfrac{1}{2}, 2\right]$, $x_1 = 1$, $x_2 = \dfrac{17}{9}$;

c) $x \in \left[-\dfrac{1}{3}, \dfrac{4}{3}\right]$; $x_1 = 0$, $x_2 = 1$; **d)** $x \in [1, 6]$; $x_1 = 2$, $x_2 = 5$;

e) $x \in [0, \infty)$, $x = 1$; **f)** $x \in [-2, \infty)$, $x = 2$; **g)** $x \in [0, \infty)$,

$x = \dfrac{1}{2}$; **h)** $x \in [3, \infty)$, $x_2 = 3$; **i)** $x \in [2, \infty)$; $x_1 = 6$, $x_2 = 18$;

j) $x \in (-\infty, -4] \cup [4, \infty)$, $x = 5$; **k)** $x \in \left(-\infty, \dfrac{3}{4}\right]$; $x = -\dfrac{5}{8}$.

l) $x \in \left[-\dfrac{1}{2}, 1\right]$, $x_1 = 0$, $x_2 = \dfrac{8}{9}$; **m)** $x \in \left[\dfrac{6}{5}, \infty\right)$, $x = \dfrac{11}{5}$.

IV. 33. a) $x = 2$; **b)** $x = -1$; **c)** $x_1 = 7$, $x_2 = 8$; **d)** $x = 0$;
e) $x = 4$; **f)** $x = 3$; **g)** $x = 4$; **h)** $x = 14$; **i)**) $x = -1$;
IV. 34. a) $x \in [-1, \infty)$, $\sqrt[6]{x+1} = t$, $2\sqrt{5t+4} = \sqrt{2t-1} + \sqrt{20t+5}$.
Squaring both sides we get
$20t + 16 = 2t - 1 + 20t + 5 + 2\sqrt{(2t-1)(20t+5)}$ and collecting terms,
we obtain $\sqrt{(2t-1)(20t+5)} = 6 - t$. First side of this equation is
positive, this means $6 - t \geq 0 \Rightarrow t \in (-\infty, 6]$, then the set of
permissible values is $x \in [-1, \infty) \cap (-\infty, 6] = [-1, 6]$. Squaring both
sides we obtain $39t^2 + 2t - 41 = 0$ whose roots are $t_1 = -\dfrac{41}{39}$

(impossible) and $t_2 = 1$. From

(1) $\sqrt[6]{x+1} = 1 \Rightarrow x = 0$ is the unique solution of the initial
equation; **b)** $x = 1$; **c)** $x = 2$.
IV. 35.

$$\dfrac{1}{(\sqrt{x}-2)(\sqrt{x}-3)} + \dfrac{1}{(\sqrt{x}-1)(\sqrt{x}-2)} + \dfrac{1}{\sqrt{x}(\sqrt{x}-1)} = \dfrac{1}{\sqrt{x}(\sqrt{x}-3)}.$$
$$\sqrt{x}(\sqrt{x}-1) + \sqrt{x}(\sqrt{x}-3) + (\sqrt{x}-2)(\sqrt{x}-3) = (\sqrt{x}-1)(\sqrt{x}-2).$$

Finally
$x = 1$ and $x = 4$. The equation doesn't have any solution.

IV. 36. a) $\sqrt[3]{x+1}+\sqrt[3]{x+2}=-\sqrt[3]{x+3}$. Cubing both sides of the last equation $x+1+x+2+3\sqrt[3]{x+1}\sqrt[3]{x+2}\left(\sqrt[3]{x+1}+\sqrt[3]{x+2}\right)=-x-3$

$3\cdot\sqrt[3]{x+1}\sqrt[3]{x+2}\left(-\sqrt[3]{x+3}\right)=-3x-6\Rightarrow\sqrt[3]{x+1}\sqrt[3]{x+2}\sqrt[3]{x+3}=x+2$

Cubing again both sides $(x+1)(x+2)(x+3)=(x+2)^3\Rightarrow x=-2$;

b) $\sqrt[3]{12+x^2}=a$ and $\sqrt[3]{12-x^2}=b$. Therefore $a+b=2\sqrt[3]{3}$ and

$a^3+b^3=24$. $x=\pm2\sqrt{3}$; **c)** $x=25$; **d)** $x=3$; **e)** $x=2$;

f) $x=-1$; **g)** $x_1=-2$, $x_2=-1.5$, $x_3=-1$;

h) $x_1=6$, $x_{2,3}=6\pm\dfrac{12}{7}\sqrt{21}$; **i)** $x\in[0,\infty)$. Dividing both

sides of the equation by $\sqrt[3]{x-4}=\sqrt[3]{\left(\sqrt{x}-2\right)\left(\sqrt{x}+2\right)}$ we obtain

$2\cdot\sqrt[3]{\dfrac{\left(\sqrt{x}-2\right)^2}{\left(\sqrt{x}-2\right)\left(\sqrt{x}+2\right)}}-\sqrt[3]{\dfrac{\left(\sqrt{x}+2\right)^2}{\left(\sqrt{x}-2\right)\left(\sqrt{x}+2\right)}}=3\Rightarrow$

$2\cdot\sqrt[3]{\dfrac{\sqrt{x}-2}{\sqrt{x}+2}}-\sqrt[3]{\dfrac{\sqrt{x}+2}{\sqrt{x}-2}}=3$, $\sqrt[3]{\dfrac{\sqrt{x}-2}{\sqrt{x}+2}}=t$, $2t-\dfrac{1}{t}=3$, then $t_1=1$

and $t_2=\dfrac{1}{2}$, $x=\left(\dfrac{18}{7}\right)^2$.

IV. 37. Denote $\sqrt[3]{x^2+2}=m$, $\sqrt[3]{4x^2+3x-2}=n$,

$\sqrt[3]{3x^2+x+5}=p$, $\sqrt[3]{2x^2+2x-5}=q$, and we have

$\begin{cases}m+n=p+q\\m^3+n^3=p^3+q^3\end{cases}\Rightarrow mn(m+n)=pq(p+q)$

$\begin{cases}m+n=p+q\\(m+n)^3-3mn(m+n)=(p+q)^3-3pq(p+q)\end{cases}\Rightarrow$

I) If $m+n=p+q=0\Rightarrow p=-q\Rightarrow 5x^2+3x=0\Rightarrow\begin{cases}x_1=0\\x_2=-\dfrac{3}{5}\end{cases}$

II) If $m+n=p+q\neq0\Rightarrow mn=pq\Rightarrow$

$\Rightarrow\dfrac{x^2+2}{2x^2+2x-5}=\dfrac{3x^2+x+5}{4x^2+3x-2}\Rightarrow(x^2+2x-7)\cdot(2x^2+x+3)=0$.

We have

1) $x^2 + 2x - 7 = 0 \Rightarrow x_{3,4} = -1 \pm 2\sqrt{2}$

2) $2x^2 - 3x + 4 = 0 \Rightarrow x_{5,6} \notin R$.

Therefore $x \in \left\{ -1 - 2\sqrt{2}, \ -\dfrac{3}{5}, \ 0, \ -1 + 2\sqrt{2} \right\}$.

IV. 38. **a)** $\sqrt[3]{3x - 1} = a$ and $\sqrt{3x + 7} = b$, $a + b = 6$ and
$a^3 - b^2 = -8$, $(a - 2)(a^2 + a + 14) = 0 \Rightarrow a = 2$, $x = 2$; **b)** $x_1 = -2$,
$x_2 = 7$, $x_3 = -1$; **c)** $x = 2$; **d)** $x = 23$; **e)** $x = 8$; **f)** $x_1 = 1$, $x_2 = 2$.

IV. 39. $x \neq 1$, $\sqrt[5]{x^3} + \sqrt[5]{x^2} + \sqrt[5]{x} - 14 = 0$, $\sqrt[5]{x} = t$,

$(t - 2)(t^2 + 3t + 7) = 0$, $x = 32$. **IV. 40.** Conditions $\begin{cases} 28 + 5x \geq 0 \\ 54 - 5x \geq 0 \end{cases} \Leftrightarrow$

$-\dfrac{28}{5} \leq x \leq \dfrac{54}{5}$.

Denote $\sqrt[4]{54 - 5x} = t \geq 0$ and $\sqrt[4]{28 + 5x} = z \geq 0$. We have

$\begin{cases} t + z = 4 \\ t^4 + z^4 = 82 \end{cases} \Leftrightarrow \begin{cases} t + z = 4 \\ \left[(t + z)^2 - 2tz \right]^2 - 2t^2 z^2 = 82 \end{cases} \Rightarrow$

$(16 - 2tz)^2 - 2t^2 z^2 = 82 \Rightarrow t^2 z^2 - 32tz + 87 = 0$. Therefore $tz = 3$ or

$tz = 29$. Now solve the systems $\begin{cases} t + z = 4 \\ tz = 3 \end{cases}$ and $\begin{cases} t + z = 4 \\ tz = 29 \end{cases}$. The

second system does not have real solutions; the first system is

equivalent to the systems $\begin{cases} t = 1 \\ z = 3 \end{cases}$ and $\begin{cases} t = 3 \\ z = 1 \end{cases}$. Since $x = \dfrac{t^4 - 28}{5}$,

the final solutions of the initial equation are

$x_1 = -\dfrac{27}{5}$ and $x_2 = \dfrac{53}{5}$.

IV. 41. **a)** $x = 3^{10}$; **b)** $x = 5$; **c)** $x_1 = 32$, $x_2 = 243$; **d)** $x = 65$;

e) $x = 8$; **f)** $x = 32$; **g)** $x = \pm 1$; **h)** $x = \pm\dfrac{24}{25}$; **i)** $x_1 = 2$, $x_2 = 9$;

$x_{3,4} = \dfrac{11}{2} \pm \dfrac{7\sqrt{141}}{12}$; **j)** $x > 0$, $\sqrt{x} = t > 0$,

$5t^3 + t - 42 = 0 \Rightarrow (t-2)(5t^2 + 10t + 21) = 0$, $x = 4$;

k) $x^{\frac{2}{15}} = t$, $5t^3 + 3t^2 - 8 = 0 \Rightarrow (t-1)(5t^2 + 8t + 8) = 0$, $x = \pm 1$.

IV. 42. a) $x \in (-1,1)$. Dividing both members of the equation by

$\sqrt[n]{1-x^2}$, $\sqrt[n]{\dfrac{1+x}{1-x}} + \sqrt[n]{\dfrac{1-x}{1+x}} = 4$, $\sqrt[n]{\dfrac{1+x}{1-x}} = t$, $t + \dfrac{1}{t} = 4$,

$t_{1,2} = 2 \pm \sqrt{3}$, $x = \dfrac{\left(2 \pm \sqrt{3}\right)^n - 1}{\left(2 \pm \sqrt{3}\right)^n + 1}$; **b)** $x = \dfrac{225}{289}$; **c)** $x = \dfrac{8}{27}$;

d) $x = 3$; **e)** $x = 0$; **f)** $x = 1$.

IV. 43. a) Domain is $x \in (-1, \infty)$. The equation becomes

$\sqrt{x^2 + 6x} + \sqrt{x^2 + 6x + 5} = 4 + \sqrt{21}$. Denote $x^2 + 6x = t$, then

$\sqrt{t} + \sqrt{t+5} = 4 + \sqrt{21}$. Squaring in the both sides of the equation

and collecting the terms, we get $\sqrt{t(t+5)} - 4\sqrt{21} = 16 - t$. Squaring

again and collecting the terms, we have $8\sqrt{21t(t+5)} = 80 + 37t$ and

again squaring in both sides, it yelds to $25t^2 - 800t + 6400 = 0$. The

root is $t = 16$ and therefore only $x = 2$ is the solution of the given

equation;

b) $x_1 = 10$, $x_2 = 32$; **c)** $x_1 = 5$, $x_2 = 24$.

IV. 44. $x \geq 0$ and $2 - \sqrt{2 + \sqrt{2 + x}} \geq 0 \Rightarrow x \in [0, 2]$. In this case

denoting $x = 2\cos t$, $t \in \left[0, \dfrac{\pi}{2}\right]$, we get

$\sqrt{2 - \sqrt{2 + \sqrt{2 + x}}} = \sqrt{2 - \sqrt{2 + \sqrt{2 + 2\cos t}}} = 2\sin\dfrac{t}{8}$ and similarly

$\sqrt{6 + 3\sqrt{2 + \sqrt{2 + x}}} = 2\sqrt{3}\cos\dfrac{t}{8}$. The initial equation becomes

$2\sqrt{3}\cos\dfrac{t}{8} + 2\sin\dfrac{t}{8} = 4\cos t \Rightarrow \dfrac{\sqrt{3}}{2}\cos\dfrac{t}{8} + \dfrac{1}{2}\sin\dfrac{t}{8} = \cos t \Rightarrow$

$\cos\left(\dfrac{\pi}{6} - \dfrac{t}{8}\right) = \cos t$ with the solution $t = \dfrac{4\pi}{27}$ or $x = 2\cos\dfrac{4\pi}{27}$.

IV. 45. a) $x \in (-\infty, \ -2]$; **b)** $x \in \left(\dfrac{3 + 2\sqrt{21}}{2}, \ \infty \right)$; **c)** $x \in (2, \ \infty)$;

d) $x \in (-\infty, \ 0] \cup (4.5, \ \infty)$; **e)** $x \in \left(\dfrac{-5 - \sqrt{5}}{2}, \ 0 \right]$;

f) $x \in \left(0, \ \dfrac{3 - \sqrt{13}}{2} \right] \cup (3, \ \infty)$; **g)** $x \in \left[-2, \ \dfrac{7}{4} \right)$;

h) $x \in [49, \ \infty)$; **i)** $x \in \left(-\infty, \ \dfrac{1}{3} \right) \cup (2, \ \infty)$; **j)** $x \in (11, \ \infty)$.

IV. 46. a) (4; 9) and (9; 4); **b)** (1; 9) and (9; 1); **c)** (1; 4) and (4; 1); **d)** (3; 2); **e)** (3; 1.5) and (6; 3);
f) (27; 8) and (8; 27); **g)** (1; 16) and (16; 1); **h)** (17/2; 5/3).

IV. 47.

$$(\sqrt{x_1 - 1} - 1)^2 + (\sqrt{x_2 - 2^2} - 2^2)^2 + \ldots + (\sqrt{x_n - n^2} - n^2)^2 = 0.$$

Chapter V

QUADRATIC FUNCTION AND APPLICATIONS

V. 1. **a)** $(x-7)(x+7)$; **b)** $(5x-1)(5x+1)$; **c)** $(9x-2)(9x+2)$;
d) $5(2x+11)$; **e)** $(16x-35)(-2x+7)$; **f)** $(x+3)(x+4)$;
g) $(x+3)(x-12)$; **h)** $(x-2)(x+8)$. **V. 2.** **a)** $f(x)=(x-1)^2+7$;
b) $f(x)=-(x-2.5)^2+8.25$; **c)** $f(x)=0.\overline{6}(x-3)^2-5$;
d) $f(x)=0.5(x-0.2)^2-0.28$; **e)** $f(x)=-3(x-0.1\overline{6})^2+0.18\overline{3}$;
f) $f(x)=5(x-0.2)^2-0.9\overline{7}$.

V. 3. **a)** $x=\dfrac{3}{4}$; **b)** $x=\dfrac{1}{6}$; **c)** $x=\dfrac{3}{8}$; **d)** $x=\dfrac{1}{5}$; **e)** $x=\dfrac{1}{8}$; **f)** $x=\dfrac{3}{4}$.

V. 4. **a)** $y_{min}=\dfrac{2}{3}$; **b)** $y_{max}=-\dfrac{25}{8}$;

c) $y_{min}=6$; **d)** $y_{max}=5$; **e)** $y_{max}=\dfrac{14}{5}$; **f)** $y_{min}=-4$.

V. 5.
a)

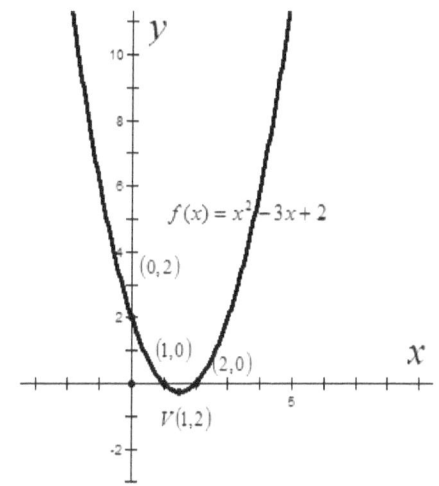

b)

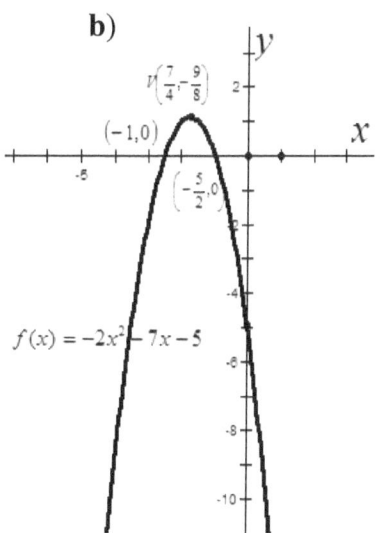

c)

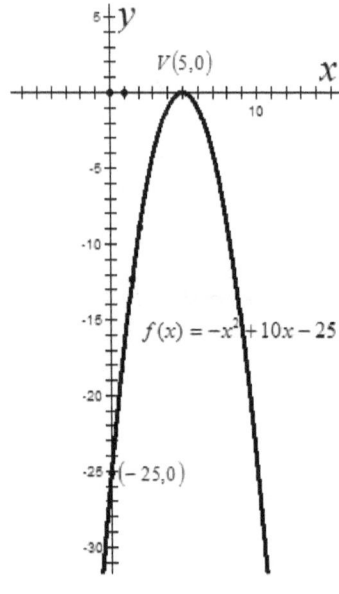

$V(5,0)$

$f(x) = -x^2 + 10x - 25$

$(-25,0)$

d)

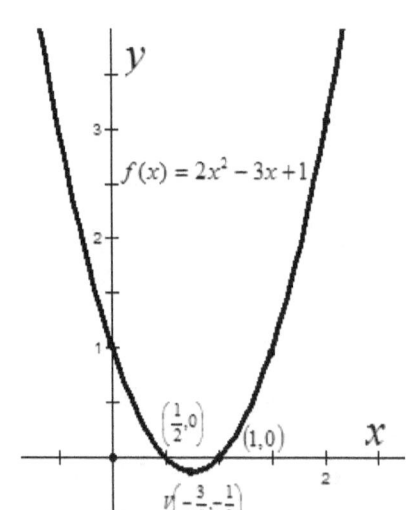

$f(x) = 2x^2 - 3x + 1$

$\left(\frac{1}{2},0\right)$ $(1,0)$

$V\left(-\frac{3}{4},-\frac{1}{8}\right)$

e)

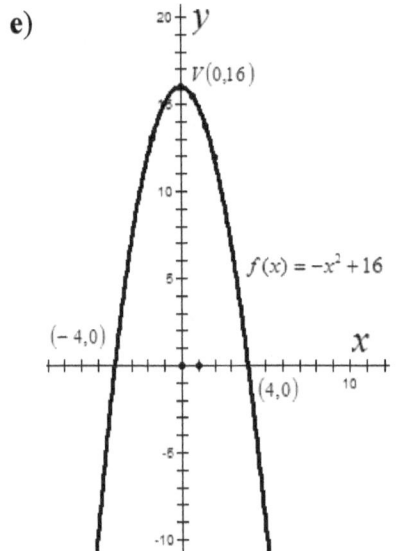

$V(0,16)$

$f(x) = -x^2 + 16$

$(-4,0)$

$(4,0)$

f)

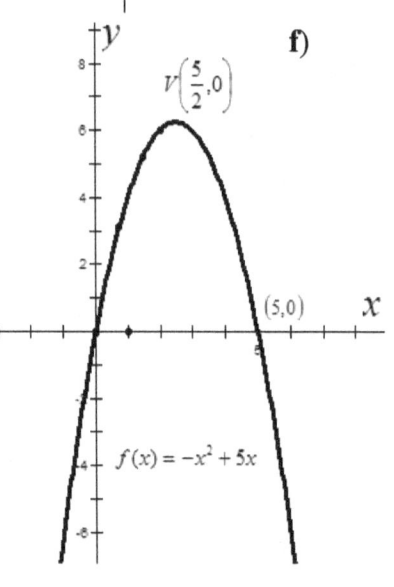

$V\left(\frac{5}{2},0\right)$

$(5,0)$

$f(x) = -x^2 + 5x$

V. 6. A function is inversible if it is a bijective function. A function is bijective if it is one to one (injective) and onto (surjective).
A function $f : A \to B$, $f(x) = y$ is *one to one* (*injective*) if for any $x_1 \neq x_2$, $x_1, x_2 \in A$ implies $f(x_1) \neq f(x_2)$.
A function $f : A \to B$, $f(x) = y$ is onto (surjective) if for any $y \in B$ there is a $x \in A$ such that $f(x) = y$.

a) *One to one* (*injective*): Let $x_1, x_2 \in (-\infty, 3]$, $x_1 < x_2 \leq 3$. We assume that

(1) $f(x_1) = f(x_2)$ then we have

$x_1^2 - 6x_1 + 8 = x_2^2 - 6x_2 + 8$ or $(x_1 - x_2)(x_1 + x_2 - 6) = 0$. Since $x_1 + x_2 < 6$ (acording to the condition $x_1 < x_2 \leq 3$) it implies $x_1 = x_2$, which is a contradiction.

Onto (surjective): Let any $y \in [-1, \infty)$, $y = x^2 - 6x + 8$, then $x^2 - 6x + 8 - y = 0$ (2) with the solutions $x_{1,2} = 3 \pm \sqrt{y+1}$. We have to show that the equation (2) has only one solution on the interval $(-\infty, 3]$.

Consider $x_1 = 3 + \sqrt{y+1}$. It is obvious that

$x_1 = 3 + \sqrt{y+1} \geq 3$, so x_1 can not be a solution for (2).

Finally $x_2 = 3 - \sqrt{y+1} \leq 3$ is a solution for (2). Indeed

$f(x_2) = (3 - \sqrt{y+1})^2 - 6(3 - \sqrt{y+1}) + 8 =$
$= 9 - 6\sqrt{y+1} + y - 1 - 9 + 6\sqrt{y+1} + 8 = y$.

Therefore

$f^{-1} : [-1, \infty) \to (-\infty, 3]$, $f^{-1}(x) = 3 - \sqrt{x+1}$

is the inverse of the given function.

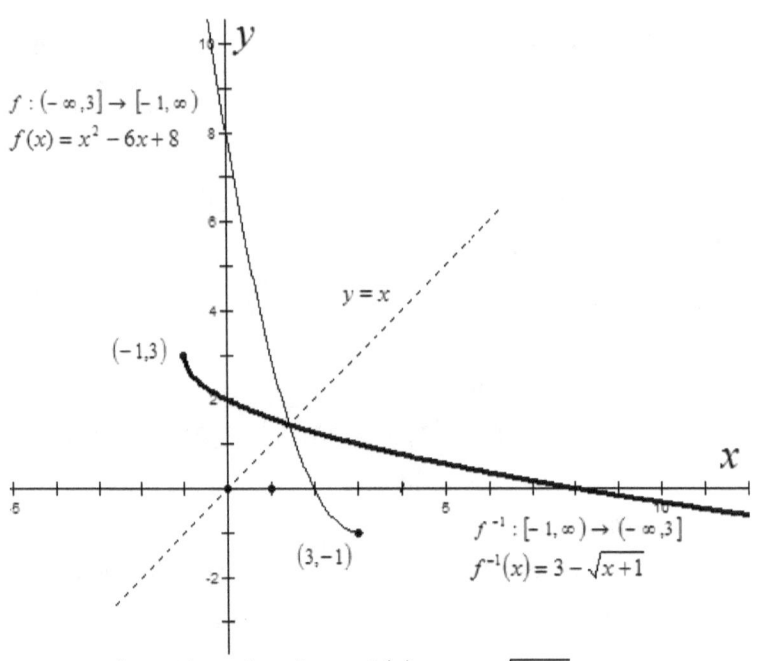

$f : (-\infty, 3] \rightarrow [-1, \infty)$

$f(x) = x^2 - 6x + 8$

$y = x$

$(-1,3)$

$(3,-1)$

$f^{-1} : [-1, \infty) \rightarrow (-\infty, 3]$

$f^{-1}(x) = 3 - \sqrt{x+1}$

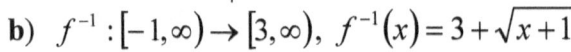

b) $f^{-1} : [-1, \infty) \rightarrow [3, \infty)$, $f^{-1}(x) = 3 + \sqrt{x+1}$

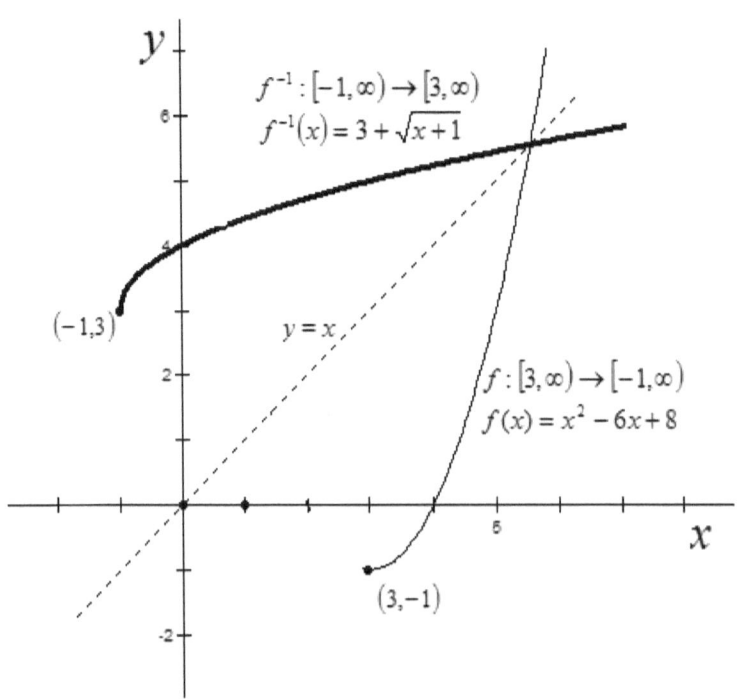

$f^{-1} : [-1, \infty) \rightarrow [3, \infty)$

$f^{-1}(x) = 3 + \sqrt{x+1}$

$(-1,3)$

$y = x$

$f : [3, \infty) \rightarrow [-1, \infty)$

$f(x) = x^2 - 6x + 8$

$(3,-1)$

c) $f:\left(-\infty,\dfrac{9}{4}\right]\to\left(-\infty,\dfrac{5}{2}\right]$, $f^{-1}(x)=\dfrac{5-\sqrt{9-4x}}{2}$;

d) $f:\left(-\infty,\dfrac{9}{4}\right]\to\left[\dfrac{5}{2},\infty\right)$, $f^{-1}(x)=\dfrac{5+\sqrt{9-4x}}{2}$.

V. 7. $AB=8$. **V. 8.** $a_1=-2$ and $a_2=3$.

V. 9. $a=1,\,c=-2$. **V. 10.** **a)** $f(x)=2x^2-3x+1$,

b) $f(x)=x^2-3x+2$; **c)** $f(x)=x^2-5x+4$;

d) $f(x)=-\dfrac{2}{3}x^2+\dfrac{4}{3}x+\dfrac{16}{3}$; **e)** $f(x)=-5x^2+10x-3$;

f) $f(x)=\dfrac{1}{9}x^2-\dfrac{10}{9}x+\dfrac{106}{9}$ or $f(x)=x^2-2x+10$.

V. 11. $f(2)=3$. **V. 12.** $f(x)=-x^2+6x-4$. **V. 13.** $A_2(-3,\ 5)$.

V. 14. $a=1$, $b=\dfrac{16}{3}$.

V. 15. The equation $3x^2+(a-10)x+b+11=0$ has the roots $x_1=-1$ and $x_2=4$. $a=-8.6$, $b=-27.6$.

V. 16. $A(x_1,\ 0)$, $B(x_2,\ 0)$,

$V\left(\dfrac{x_1+x_2}{2},\ f\left(\dfrac{x_1+x_2}{2}\right)\right)$, $A_{\Delta VAB}=\dfrac{1}{2}\left|x_1-x_2\right|\left|f\left(\dfrac{x_1+x_2}{2}\right)\right|$,

V. 17. $\dfrac{a_1+a_2+...,+a_n}{n}$. **V. 18.** $f(x)=-x^2+6x-4$ or

$f(x)=-9x^2+6x+4$. **V. 19.** **a)** $\dfrac{1}{7}$; **b)** $\dfrac{4}{3}$; **c)** -3, $\dfrac{1}{13}$; **d)** $14,\,-4$;

e) -1; **f)** $\dfrac{1}{3}$; **g)** -1; **h)** -1; $\dfrac{5}{2}$; **i)** $x\in\left(-\infty,\ -\dfrac{13}{4}\right]\cup(-1,\ 3)$.

V. 20. **a)** $2m$; **b)** $m-1$; **c)** $4m^2-2m+2$; **d)** $2m\left(4m^2-3m+3\right)$;

e) $\pm2\sqrt{m^2-m+1}$; **f)** $16m^4-16m^3+18m^2-4m+2$; **g)** $\dfrac{2m}{m-1}$;

h) $\dfrac{4m^2-2m+2}{m^2-2m+1}$; **i)** $\dfrac{2m\left(4m^2-3m+3\right)}{(m-1)^3}$; **j)** $\sqrt{2m+2\sqrt{m-1}}$;

k) $\sqrt[4]{x_1 + x_2 + 6\sqrt{x_1 x_2} + 4 \cdot \sqrt[4]{x_1 x_2}\left(\sqrt{x_1} + \sqrt{x_2}\right)}$. **V. 20.** $E = \dfrac{4}{3}$.

V. 22. $m = \dfrac{2}{3}$. **V. 23. a)** $x^2 - 9 = 0$; **b)** $x^2 + 2x = 0$;

c) $x^2 + 2x - 8 = 0$; **d)** $2x^2 - 2x - 1 = 0$; **e)** $3x^2 - 2x - 1 = 0$.

V. 24. a) $y_1 + y_2 = \dfrac{x_1 + x_2}{x_1 x_2} = \dfrac{-m}{2m - 1}$, $y_1 \cdot y_2 = \dfrac{1}{x_1 x_2} = \dfrac{1}{2m - 1}$,

$(2m - 1)y^2 - (3m - 2)y + 3m - 1 = 0$.

Another solution: use the substitution $x = \dfrac{1}{y}$ for the given

equation at x.

b) $(2m - 1)y^2 - (3m - 2)y + m = 0$; **c)** $(2m - 1)y^2 - m^2 y + m^2 = 0$;

d) $(2m - 1)y^2 - (m^2 - 4m + 2)y + 1 = 0$;

e) $(2m - 1)^2 y^2 - (m^2 - 4m + 2)y + 1 = 0$;

f) $(2m - 1)^3 y^2 - (m^3 - 3m^2 + 3m)y + 1 = 0$.

V. 25. $x^2 - 6x + 1 = t$, then

$$\dfrac{t + (1 - t)}{t} + \dfrac{(t + 1) + (1 - t)}{t + 1} + \dots + \dfrac{(t + 2009) + (1 - t)}{t + 2009} = 2010 \Rightarrow$$

$$\left(1 + \dfrac{1 - t}{t}\right) + \left(1 + \dfrac{1 - t}{t + 1}\right) + \left(1 + \dfrac{1 - t}{t + 2}\right) + \dots + \left(1 + \dfrac{1 - t}{t + 2009}\right) = 2010 \Rightarrow$$

$$(1 - t)\left(\dfrac{1}{t} + \dfrac{1}{t + 1} + \dfrac{1}{t + 2} + \dots + \dfrac{1}{t + 2009}\right) = 0 \Rightarrow t = 1 \text{ or}$$

$x^2 - 6x + 1 = 1 \Rightarrow x^2 - 6x = 0$. Therefore $x_1 = 0$ and $x_2 = 6$ are the roots of the initial equation.

V. 26. $2x(2 + 4 + \dots + 2010) - 2x(1 + 3 + \dots + 2011) =$

$= (1^2 - 0^2) + (3^2 - 2^2) + \dots + (2011^2 - 2010^2)$

$2x[(0 - 1) + (2 - 3) + \dots + (2010 - 2011)] =$

$= (1 + 0) + (3 + 2) + \dots + (2011 + 2010)$

$2x(-2010) = 1 + 2 + 3 + \dots + 2011 \Rightarrow 2x(-2010) = \dfrac{(1 + 2011)2011}{2}$.

Finally $x = -\dfrac{1006 \cdot 2011}{4020}$. **V. 27. a)** $x \in (-\infty, \ -6) \cup (-1, \ \infty)$;

b) $x \in [2,\ 3]$; **c)** $x \in \left[-1,\ \dfrac{11}{3}\right]$; **d)** $x \in \left(-\infty,\ -\dfrac{2}{5}\right] \cup [1,\ \infty)$; **e)** $x \in \mathbf{R}$;

f)

x	$-\infty$	$\left(-10-2\sqrt{22}\right)$	-2	$\left(-10+2\sqrt{22}\right)$	2	∞
$x^2+20x+12$	+	+0 –	–	–	0 +	+
x^2-4	+ +	+ +	0 –	–	–	0 +
$\dfrac{x^2+20x+12}{2(x^2-4)}$	+	+ 0 –	– – –|	+ 0	– – |	+

$x = \left[-10-2\sqrt{22},\ -2\right) \cup \left[-10+2\sqrt{22},\ 2\right)$;

g) $x \in (-\infty,\ -1) \cup \left(\dfrac{2}{3},\ 1\right] \cup [3,\ \infty)$;

h) $x \in (-3,\ -2) \cup \{0\}$;

i) $x \in (5,\ 6.5)$;

j)

x	$-\infty$	-1	1	2	9	∞
$-x+9$	+	+	+	+	+	0 – –
$x+1$	– –	0 +	+	+	+	+ +
$x-2$	– –	–	–	0 +	+	+
$x-1$	– –	–	0 +	+	+	+ +
$\dfrac{-x^2+8x+9}{x^2-3x+2}$	– –	0 + + |	– – |	+	0 – –

$x \in [-1,\ 1) \cup (2,\ 9]$; **k)** $x \in (-\infty,\ 0) \cup (1,\ \infty)$;

l) $x \in [-5,\ -3) \cup [-1,\ 1)$; **m)** $x \in \left[0,\ \dfrac{8}{5}\right] \cup \left[\dfrac{5}{2},\ \infty\right)$;

n) $x \in \left(-\infty,\ \dfrac{11-\sqrt{105}}{2}\right) \cup \left(1,\ \dfrac{11+\sqrt{105}}{2}\right)$;

o) $x \in (-\infty,\ -3) \cup (-2,\ -1)$; **p)** $x \in [-3,\ -2) \cup (3,\ 3.5]$.

V. 28. a) $x \in (-\infty, \ -1) \cup (4, \ \infty)$; **b)** $x \in \left(3, \ \dfrac{7}{2}\right]$; **c)** $x \in (0, \ \infty)$;

d) $x = 1$; **e)** $x \in \left(-\infty, \ -2 - \sqrt{10}\right] \cup \left(-2, \ -2 + \sqrt{6}\right] \cup (2, \ \infty)$;

V. 29 a) $[-1, 3]$; **b)** $[-1, 3)$; **c)** $[-1, 8]$; **d)** $[3, \infty)$; **e)** $[0, +\infty)$;

f) $[-1, \infty)$; **g)** $(-1, +\infty)$.

V. 30. a) $y \in \left[\dfrac{3}{4}, \ +\infty\right)$; **b)** $\left[-\dfrac{49}{4}, \ +\infty\right)$;

c) $[-5, \infty)$; **d)** Denote $m = \dfrac{x^2 - x - 3}{x^2 + x + 2}$,

$(1 - m)x^2 - (m + 1)x - 2m - 3 = 0$, $\Delta \geq 0$, $m \in \left[\dfrac{-1 - \sqrt{92}}{7}, \ \dfrac{-1 + \sqrt{92}}{7}\right]$

Therefore the range is $\left[\dfrac{-1 - \sqrt{92}}{7}, \ \dfrac{-1 + \sqrt{92}}{7}\right]$. **e)** $\left[\dfrac{1 - \sqrt{3}}{2}, \ \dfrac{1 + \sqrt{3}}{2}\right]$;

f) $y \in (-\infty, \ -1]$. **V. 31.** $a = \pm 4\sqrt{3}$, $b = -1$. **V. 32.**

$x \in \left[0.25, \ 1 + 2\sqrt{2}\right]$. **V. 33.** $m \in [0, \ +\infty)$, $x = 3 \pm \sqrt{3}$.

V. 34. $m \in [2, \infty) \cup \{1\}$. **V. 35. b)** $x \in \left(3 - 2\sqrt{3}, \ 0\right) \cup \left(3 + 2\sqrt{3}, \ \infty\right)$.

V. 36. $m \in [3, \ +\infty)$. **V. 37.** $x \in \left(-1, \ \dfrac{3 - \sqrt{17}}{2}\right] \cup \left[\dfrac{3 + \sqrt{17}}{2}, \ +\infty\right)$;

V. 38. a) $x_1 = 2x_2$, $m_1 = 3$, $m_2 = -\dfrac{15}{8}$;

b) $x \in \left(3 - \sqrt{4}, \ 0\right) \cup \left(3 + \sqrt{14}, \ \infty\right)$. **V. 39.** $\Delta < 0$, $m \in (1, \ 2)$.

V. 40. a) $x_v = m + 2$, $y_v = -m^2 - 3m - 2$, $y = -x^2 + x$;

b) $\Delta > 0$, $m \in (-\infty, -2) \cup (-1, \infty)$. **V. 41.** $y = -x^2 + x - 1$.

V. 42. a) $x_v = \dfrac{-(m + 1)}{m}$, $y_v = \dfrac{1}{m}$, $x_v = -1 + y_v$; **b)** $AB = |\, x_1 - x_2 \,|$,

$FV = \dfrac{1}{|m|}$; **c)** $m(x^2 + 2x + 1) + 2x + 2 - y = 0$, $(-1, 0)$.

V. 43. a) $f_{-1}(x) = x^2 - 3$, $x_{\min} = 0$, f_{-1} decreases on $(-\infty, \ 0]$

and increases on $[0, \ \infty)$. **b)** $x_v = y_v$, $m = -6 \pm \sqrt{13}$. **c)** $a > 0$,

$\Delta < 0$, $m \in \varnothing$. **V. 44.** $\Delta > 0$, $12m + 13 > 0$.

V. 45. $\Delta = 4(m^2 + m + 1) > 0$. **V. 46.** $m = -1$. **V. 47.** $(-1, 7)$, $(1, 3)$.

V. 48. $m \in \left(\dfrac{1}{7}, \infty\right)$. **V. 49.** $a = m - 1 < 0$ and $y_{\min} = \dfrac{4m + 5}{1 - m} < 0$,

$m \in \left(-\infty, -\dfrac{5}{4}\right)$. **V. 50. a)** $y_{min} < 0$, $y_{min} = -(m + 1) < 0$,

$m \in (-1, \infty)$; **b)** $\Delta = 0$, $m = -1$; **c)** $x_{min} = m - 1$, $y_{min} = -(m + 1)$

and the line is $y = -x - 2$.

V. 51. $a > 0$, $\Delta < 0$, $m \in (2, \infty)$. **V. 52.** $\Delta < 0$, $a < 0$,

$m \in \left(1, \dfrac{5}{3}\right)$. **V. 53.** $a < 0$, $\Delta < 0$, $m \in (-\infty, -2)$.

V. 54. a) $(x_1 - 1)^2 + (x_2 - 1)^2 = (x_1 + x_2)^2 - 2x_1 x_2 - 4(x_1 + x_2) + 8$,

where $x_1 + x_2 = m + 1$ and $2x_1 x_2 = m^2 - 2m + 3$ (Vietas formulas);

b) Denote $\dfrac{x_1}{x_2} = t$, then $t^2 - 4t + 1 = \dfrac{x_1^2 - 4x_1 x_2 + x_2^2}{x_2^2} =$

$= \dfrac{(x_1 + x_2)^2 - 6x_1 x_2}{x_2^2} = \dfrac{m^2 + 2m + 1 - 3m^2 + 6m - 9}{x_2^2} = \dfrac{-2(m - 2)^2}{x_2^2} \le 0$

Therefore $t^2 - 4t + 1 \le 0$, which implies $t \in \left[2 - \sqrt{3}, \ 2 + \sqrt{3}\right]$;

c) From $s = x_1 + x_2 = m + 1$ and $p = x_1 x_2 = \dfrac{m^2 - 2m + 3}{2}$, eliminate

m and get $2p = s^2 - 4s + 6$.

V. 55. a) $x_1 + x_2 = m + 3$, $2x_1 x_2 = m^2 - 4m + 24$;

b) $(x_1 - 5)^2 = 5 - (x_2 - 5)^2 \le 5$, thus $|x_1 - 5| \le \sqrt{5}$

and similarly for x_2; **c)**

$|x_1 - x_2| = \sqrt{(x_1 - x_2)^2} = \sqrt{(x_1 + x_2)^2 - 4x_1 x_2} = \sqrt{10 - (m - 7)^2} \le \sqrt{10}$

d) Since $x_1 + x_2 = m + 3$ and $\Delta = 4\left[10 - (m - 7)^2\right]$, from

$10 - (m - 7)^2 \ge 0$, we get $7 - \sqrt{10} \le m \le 7 + \sqrt{10}$, therefore

$-4 \le x_1 + x_2 \le 4$;

e) $x_1^2 + x_2^2 = (x_1 + x_2)^2 - 2x_1 x_2 = 10m - 15$ and since $m \le 7 + \sqrt{10}$

then $x_1^2 + x_2^2 \le 55 + 10\sqrt{10}$;

f) Denote $\dfrac{x_1}{x_2} = t$,

$$t^2 - \frac{5}{2}t + 1 = \frac{2x_1^2 - 5x_1x_2 + 2x_2^2}{2x_2^2} = \frac{-5(m-6)^2}{4x_2^2} \le 0.$$

Therefore $2t^2 - 5t + 2 \le 0$, which implies $t \in [0.5,\ 2]$;

g) $\Delta = 4\left[10 - (m-7)^2\right] \le 0$, $m \in \left(-\infty\ \ 7 - \sqrt{10}\right] \cup \left[7 + \sqrt{10},\ \infty\right)$;

h) Find the smallest possible value of $f_{\min} = \dfrac{1}{2}(m^2 - 14m + 39)$.

The minimum of $f_{\min}$ is holding for $m = 7$, therefore $f_{\min}(7) = -1.5$;

i) From $s = x_1 + x_2 = m + 3$ and $2p = 2x_1x_2 = m^2 - 4m + 24$,
eliminate m and get $2p = s^2 - 10s + 45$.

j) $f(5)f(6) < 0$, $\left(m^2 - 14m + 44\right)\left(m^2 - 16m + 60\right) < 0$.

V. 56. a) $(3, 1)$, and $(-3, -2)$; **b)** $(1, 2)$ and $\left(\dfrac{11}{4},\ \dfrac{9}{8}\right)$;

c) $(3, 2)$, and $\left(-\dfrac{5}{2},\ \dfrac{19}{4}\right)$; **d)** $(1, 3)$, and $\left(\dfrac{29}{2},\ -\dfrac{3}{2}\right)$;

e) $(-1, 2)$, and $\left(-\dfrac{13}{4},\ \dfrac{1}{2}\right)$; **f)** $(2, 1)$ and $\left(-\dfrac{4}{3},\ \dfrac{23}{3}\right)$; **g)** $y \ne 0$,

Denote $\dfrac{x}{y} = t$, from the first equation we get $t^3 + t^2 - 36 = 0 \Rightarrow$

$(t - 3)(t^2 + 4t + 12) = 0$, then $t = 3$. In the second equation denote

$xy = z$ it yelds to $z^2 + z = 12$ with the

solutions $z_1 = 3$ and $z_2 = -4$. Therefore the initial system is reduced

to the systems $\begin{cases} \dfrac{x}{y} = 3 \\ xy = 3 \end{cases}$ and $\begin{cases} \dfrac{x}{y} = 3 \\ xy = -4 \end{cases}$. Only the first system has real

solutions, which are $(3, 1)$, and $(-3, -1)$.

V. 57. a) Multiply the first equation by -6 and add it to the second

equation. We get $-5x^2 + xy + 18y^2 = 0$. Denote $\dfrac{x}{y} = t \Rightarrow$

$5t^2 - t - 18 = 0$. The initial system is equivalent with two systems.

I): $\begin{cases} x = 2y \\ x^2 + xy = 6 \end{cases}$ with the solutions $x_1 = 2$, $y_1 = 1$ and $x_2 = -2$,

$y_2 = -1$. II): $\begin{cases} x = -\dfrac{9}{5}y \\ x^2 + xy = 6 \end{cases}$ with the solutions $x_3 = -\dfrac{3\sqrt{6}}{2}$, $y_3 = \dfrac{5}{\sqrt{6}}$

and $x_4 = \dfrac{3\sqrt{6}}{2}$, $y_4 = -\dfrac{5}{\sqrt{6}}$;

b) $(3, 2)$ and $(-3, -2)$; c) $(-1, 2), (1, -2)$ and $\left(\dfrac{5\sqrt{3}}{3}, \dfrac{4\sqrt{3}}{3} \right)$,

$\left(-\dfrac{5\sqrt{3}}{3}, -\dfrac{4\sqrt{3}}{3} \right)$; d) $(1, 1)$ and $(-1, -1)$; e) $(-1, 1)$ and $(1, -1)$;

f) $(1, 1), (-1, -1)$ and $\left(-\dfrac{5}{\sqrt{37}}, \dfrac{4}{\sqrt{37}} \right), \left(\dfrac{5}{\sqrt{37}}, -\dfrac{4}{\sqrt{37}} \right)$.

V. 58. a) Denote $\begin{cases} x + y = s \\ xy = p \end{cases}$, the system becomes $\begin{cases} s^2 - 2p = 5 \\ p = 2 \end{cases}$,

with the solutions $\begin{cases} s = 3 \\ p = 2 \end{cases}$ and $\begin{cases} s = -3 \\ p = 2 \end{cases}$ or $\begin{cases} x + y = 3 \\ xy = 2 \end{cases}$ and

$\begin{cases} x + y = -3 \\ xy = 2 \end{cases}$. Solving the last two systems we get the solutions:

$x_1 = 1$, $y_1 = 2$, $x_2 = 2$, $y_2 = 1$ and $x_3 = -1$, $y_3 = -2$ and

$x_4 = -2$, $y_4 = -1$; b) $(0, 0)$ and $(2, 2)$; c) $(1, 2)$ and $(2, 1)$;

d) $(1, 5)$, $(5, 1)$, $(3, 2)$, and $(2, 3)$; e) $(2, 3)$; $(-2, -3)$; $(3, 2)$, and

$(-3, -2)$; f) $\left(\dfrac{1}{2}, \dfrac{1}{3} \right)$, $\left(\dfrac{1}{3}, \dfrac{1}{2} \right)$; g) $(-2, 3)$, $(3, -2)$;

h) $(2, 2)$; $(-2, -2)$, $(-2, 2)$, and $(2, -2)$; i) $(1, 3)$; $(-1, -3)$, $(3, 1)$, and $(-3, -1)$. V. 59. a) $(1.6, 1.3)$; $(1.4, 1.5)$;

b) $(3, 1)$, and $(-1, -3)$; c) $(-3, 1)$, and $(1, -3)$;

d) $(3, 5)$, and $(-3, -5)$;

e) $(4, 1)$, and $(-4, -1)$; f) $(3, 2)$, and $(-3, -2)$;

g) $(a, 2a)$, and $(2a, a)$; **h)** $(-1, 2)$ and $\left(\dfrac{17 \cdot \sqrt[3]{4}}{4}, \ \dfrac{19 \cdot \sqrt[3]{4}}{4} \right)$;

i) $(2, -5, -4)$, $(-2, -1, 0)$; **j)** $(1, 3, 4)$, $(-1, -3, -4)$;
k) $(1, -2, 3)$, $(1, -3, 2)$, $(2, -1, 3)$, $(2, -3, 1)$, $(3, -1, 2)$, $(3, -2, 1)$.

V. 60. Taking $x = a^6$, $y = b^6$, $z = c^6$, the inequality becomes
$\left(a^2 + b^2 + c^2\right)^3 > \left(a^3 + b^3 + c^3\right)^2$. Expanding and collecting terms, we

obtain $3\left(b^2 + c^2\right)a^4 - 2\left(b^3 + c^3\right)a^3 + 3\left(b^2 + c^2\right)^2 a^2 +$
$$+ b^2 c^2 \left[\left(2b^2 + 2c^2\right) + (b - c)^2 \right] > 0 \text{ on the other hand}$$

$3\left(b^2 + c^2\right)a^2 - 2\left(b^3 + c^3\right)a + 3\left(b^2 + c^2\right)^2 > 0$ is true because $\Delta_a < 0$.

V. 61. a) Denote by x the price at which each CD will be sold.
Number of Cd-s sold $= 4{,}000 - 400$(number of \$1 increase) $= 4{,}000$
$- 400(x - 5)$. Profit per CD is $x - 2$. Total profit $P(x) = (\#$ CD-s
sold)(profit per CD)$= (6{,}000 - 400x)(x - 2) =$
$= -400x^2 + 6{,}800x - 12{,}000$.
b) Profit is maximized for \$8.50 per CD amd maximum monthly
profit is $P(8.5) = \$16{,}900$.
V. 62. The fare is reduced to $8 - 0.25x$ dollars per car, then the
number of cars is $300 + 10x$. The total daily revenue is
$R(x) = (8 - 0.25x)(300 + 10x)$ or
$R(x) = -2.5x^2 + 5x + 2400$. The maximum for this function is
2402.50 and it is attained for $x = 1$. The fare is \$7.75 per car.
V. 63. Let x be the number trees which maximize the yield per acre.
The number of bushes produced by each tree is $30 - 0.5(x - 40)$. The

total yield in bushels per acre is $f(x) = x\left(50 - \dfrac{x}{2}\right)$. The maximum of

f is obtained for $x = 50$.
V. 64. $h_{\max} = 31.25$. **V. 65.** Let x the width and y the length. We
have $2x + y = 100$. $A = xy = x(100 - 2x)$. The function
$A(x) = -x^2 + 100x$, $x \in [0, 50]$ has a maximum when $x = 25$, and
finally $y = 50$. **V. 66.** $AM = $ x, the area of the triangles is

$$A(x) = \frac{x^2\sqrt{3}}{4} + \frac{(a-x)^2\sqrt{3}}{4},$$

$$A(x) = \frac{\sqrt{3}}{4}\left(2x^2 - 2ax + a^2\right).$$

$$A_{min} = \frac{a^2\sqrt{3}}{8}.$$

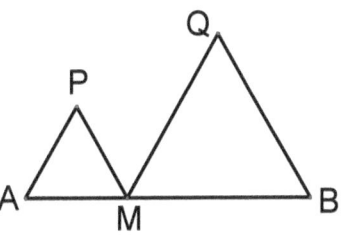

V. 67. Let a, b, and c be the side lengths of $\triangle ABC$. Let $MD = x$.
Applying Stewarts' theorem in $\triangle MBC$,

$$MB^2 \cdot CD + MC^2 \cdot BD - MD^2 \cdot BC = BD \cdot DC \cdot BC,$$

$$\frac{MB^2}{2} + \frac{MC^2}{2} - MD^2 = \frac{a^2}{4}.$$

$$S = MA^2 + 2MD^2 + \frac{a^2}{2} = (m_a - x)^2 + 2x^2 + \frac{a^2}{2}$$

$$S = 3x^2 - 2m_a x + m_a^2 + \frac{a^2}{2}. \text{ Minim of } S \text{ is for } x = \frac{a}{3}.$$

V. 68. Let $AM = 2x$. $A(x) = r^2\pi - x^2\pi - (r-x)^2\pi$,

$$A(x) = -2x^2\pi + 2rx\pi. \quad A_{max} = \frac{r^2\pi}{2}. \quad \textbf{V. 69.} \quad AM = 2x, \text{ the area of}$$

the circles is $A(x) = \pi x^2 + \pi(a-x)^2$, $A_{min} = \frac{\pi a^2}{2}$ and $AM = a$.

V. 70. Let y the width, x the length, and P the perimeter. We have
$xy = 100$ and $P = 2x + 2y$ or $2x^2 - xP + 200 = 0$. This equation
must have real roots, thus $\Delta = b^2 - 4ac \geq 0 \Rightarrow P \geq 40$, $P = 40$ is
minimum, then $x = 10$, and finally $y = 10$. **V. 71.** $A_{\triangle ABC} = \frac{10h}{2}$.

Let $AC = x$, $AD = y$, and $h = CD$, we have
$AC^2 - AD^2 = CB^2 - BD^2$ or $x^2 - y^2 = (10-x)^2 - (8-y)^2$,

$$y = \frac{5x-9}{4}, \quad A_{\triangle ABC} = \frac{5}{2}\sqrt{-9x^2 + 90x - 81} = \frac{5}{2}\sqrt{144 - 9(x-5)^2} \text{ thus}$$

the maximum areas is 30 for $x = 5$, therefore $h = 3$cm.

VI. 1. a) $z_1 = 5 + 2i$; **b)** $z_2 = 1 - 3i$; **c)** $z_3 = -5 + 14i$;

d) $z_4 = \dfrac{7}{5} - \dfrac{1}{5}i$;

e) $z_5 = \dfrac{1-i}{1+2i} = \dfrac{(1-i)(1+2i)}{1^2 - (2i)^2} = \dfrac{1 + 2i - i - 2i^2}{5} = \dfrac{3+i}{5}.$

VI. 2. a) $-32i$; **b)** $-1 - i$; **c)** 0; **d)** 0; **e)**

$\left(i^4\right)^{502} \cdot i^2 + \left(i^4\right)^{502} \cdot i^3 + \left(i^4\right)^{502} \cdot i^4 + \left(i^4\right)^{502} \cdot i^5 = i^2 + i^3 + i^4 + i^5 = 0.$

VI. 3. 0. **VI. 4. a)** $9 + i$; **b)** $-1 - 5i$; **c)** $26 + 2i$; **d)** $\dfrac{2+i}{10}$;

e) $28 + 8i$. **VI. 5.** yes.

VI. 6. a) $\mathrm{Re}(z_1) = -4$, $\mathrm{Im}(z_1) = 3$, $|z_1| = 5$;

b) $\mathrm{Re}(z_2) = 3$, $\mathrm{Im}(z_2) = -2$, $|z_2| = \sqrt{13}$; **c)** $\mathrm{Re}(z_3) = -2$,

$\mathrm{Im}(z_3) = 1$, $|z_3| = \sqrt{5}$; **d)** $\mathrm{Re}(z_4) = 1$, $\mathrm{Im}(z_4) = -2$, $|z_4| = \sqrt{5}$.

VI. 8. a) $2i$; **b)** $\dfrac{1+3i}{6}$; **c)** $\dfrac{3+3i}{2}$. **VI. 9. a)** $z = -2 + 4i$;

b) $|z| = 2\sqrt{5}$.

VI. 10. a) $-\dfrac{3}{29} + \dfrac{7}{29}i$; **b)** $-\dfrac{3}{13} + \dfrac{11}{13}i$; **c)** $-\dfrac{4}{5} + \dfrac{7}{5}i$, **d)** $\dfrac{11i}{2}$;

e) $\dfrac{\left(1 + 2i + i^2\right)^2 + \left(1 - 2i + i^2\right)^2}{\left(1 + 4i + 4i^2\right)^2 + \left(1 - 4i + 4i^2\right)^2} = \dfrac{4i^2 + 4i^2}{\left(-3 + 4i\right)^2 + \left(-3 - 4i\right)^2} = \dfrac{4}{7}.$

VI. 11. a) $\begin{cases} x + 2y = 3 \\ 3x - y = 2 \end{cases}$, $x = 1$, $y = 1$; **b)** $\begin{cases} x^2 + y = 2 \\ x + 3y^2 = 4 \end{cases}$, $x = 1$,

$y = 1$;

c) $\begin{cases} 3\sqrt{x^2 - 2y} + x^2 = 2y \\ -x^2 = 4y - 12 \end{cases}$, $x = \pm 2$, $y = 2$; **d)** $x = -6$, $y = 7$;

e) $x = 0, y = 3$; **f)** $x - 3 + (y - 3)i = (x + 2)i - y - 4$,

$$\begin{cases} x-3=-y-4 \\ y-3=x+2 \end{cases} \Rightarrow y=2,\ x=-3.$$

g) $x=1$, $y=4$. **VI. 12. a)** $a=0$, $b=1$; **b)** $a=0$, $b=1$.

VI. 13. a) $|z|=100$; **b)** $|z|=36$. **VI. 14.** $z=-8$.

VI. 15. 0. **VI. 16.** $z_1\bar{z}_2+z_2\bar{z}_1=4$.

VI. 17. $z=1+i$. **VI. 18.** $z=-2\pm i$.

VI. 19. $z_1=1-2i$, $z_2=-1+2i$,

$z_3=-1-2i$, $z_4=1+2i$, $E_1=E_2=0$. **VI. 20. a)** $x_{1,2}=\dfrac{-1\pm3i}{3}$;

b) $x_{1,2}=\dfrac{-5\pm i\sqrt{3}}{2}$; **c)** $x_{1,2}=\dfrac{3\pm i\sqrt{7}}{4}$; **d)** $x_{1,2}=\dfrac{3\pm i\sqrt{23}}{2}$.

VI. 21. The complex numbers $1+3i$ and $1-3i$ are the root of the equation $x^2-2x+10=0$, $x^3-4x^2+14x-20=\left(x^2-2x+10\right)(x-2)$
.

VI. 22. $\left(x+\dfrac{1}{2}-i\dfrac{\sqrt{3}}{2}\right)\left(x+\dfrac{1}{2}+i\dfrac{\sqrt{3}}{2}\right)=x^2+x+1$,

$x^4+x^2+1=\left(x^2+x+1\right)\left(x^2-x+1\right)$.

VI. 23. We have $x_1^2+x_1+1=0$ and $x_2^2+x_2+1=0$; $x_1+x_2=-1$ and $x_1x_2=1$. On the other hand

$x_1^3-1=\left(x_1-1\right)\left(x_1^2+x_1+1\right)=0 \Rightarrow x_1^3=1$, similarly $x_2^3=1$.

a) $\left(1+x_1^2-1\right)^{2002}+\left(1+x_2^2-1\right)^{2002}=\left(x_1^3\right)^{1334}x_1^2+\left(x_2^3\right)^{1334}x_2^2=$
$=x_1^2+x_2^2=\left(x_1+x_2\right)^2-2x_1x_2=1-2=-1$

b) -1; **c)** $\left[x_1^2\left(x_1^2+x_1+1\right)+1\right]^{2010}+\left[x_2^2\left(x_2^2+x_2+1\right)+1\right]^{2010}=2$.

VI. 24. $\varepsilon^3=1$, $\varepsilon^3-1=(\varepsilon-1)\left(\varepsilon^2+\varepsilon+1\right)=0 \Rightarrow \varepsilon^2+\varepsilon+1=0$

a) $z_1=\varepsilon^4+\varepsilon^2+1=\varepsilon\cdot\varepsilon^3+\varepsilon^2+1=\varepsilon\cdot1+\varepsilon^2+1=\varepsilon^2+\varepsilon+1=0$

b) $z_2=\dfrac{1+\varepsilon}{1-2\varepsilon+\varepsilon^2}+\dfrac{1-\varepsilon}{1+2\varepsilon+\varepsilon^2}=\dfrac{1+\varepsilon}{-3\varepsilon}+\dfrac{1-\varepsilon}{\varepsilon}=\dfrac{-2+4\varepsilon}{-3\varepsilon}$. If

$\varepsilon_1=\dfrac{-1+i\sqrt{3}}{2} \Rightarrow$

$$z_2 = \frac{-2+4\varepsilon}{-3\varepsilon} = \frac{2}{3\varepsilon} - \frac{4}{3} = \frac{2}{3} \cdot \frac{2}{-1+i\sqrt{3}} - \frac{4}{3} = \frac{-5-i\sqrt{3}}{3},$$

If $\varepsilon_2 = \dfrac{-1-i\sqrt{3}}{2} \Rightarrow$

$$z_2 = \frac{-2+4\varepsilon}{-3\varepsilon} = \frac{2}{3\varepsilon} - \frac{4}{3} = \frac{2}{3} \cdot \frac{2}{-1-i\sqrt{3}} - \frac{4}{3} = \frac{-5+i\sqrt{3}}{3}; \quad \textbf{c)} \ z_3 = 0;$$

d) $z_4 = a^3 + b^3 + c^3 - 3abc$. **VI. 26.** $z^2 - z + 1 = 0$,

$z^3 + 1 = (z+1)(z^2 - z + 1) = 0 \Rightarrow z^3 = -1$.

$$z^{13} + \frac{1}{z^{13}} = \left(z^3\right)^4 z + \frac{1}{\left(z^3\right)^4 z} = z + \frac{1}{z} = 1.$$

VI. 27. a) $z = \dfrac{\left(1-i\sqrt{3}\right)\left[a-(a+1)i\right]}{a^2 - \left[(a+1)i\right]^2} = \dfrac{a-(a+1)\sqrt{3} - \left(a+1+a\sqrt{3}\right)i}{a^2 + (a+1)^2}$,

$a + 1 + a\sqrt{3} = 0 \Rightarrow a = \dfrac{-1}{1+\sqrt{3}} = \dfrac{1-\sqrt{3}}{2}$; **b)** $b = x + yi$,

$$\left(x - \frac{1}{2}\right)^2 + \left(y - \frac{1}{2}\right)^2 = \frac{9}{2}.$$

VI. 28. $i^4 = 1$, $\dfrac{1+i}{1-i} = \dfrac{(1+i)(1+i)}{1^2 - i^2} = i$.

$z = \left(i^4\right)^{502} \cdot i^2 + \left(i^4\right)^{502} \cdot i^3 + \left(i^4\right)^{503} = -1 - i + 1 = -i$.

VI. 29. $x_1^2 + x_1 + 1 = 0$ and

$x_1^3 - 1 = (x_1 - 1)(x_1^2 + x_1 + 1) = 0 \Rightarrow x_1^3 = 1$, similarly $x_2^3 = 1$,

$$E = \frac{\left(-x_1^2\right)^{2011}}{-x_1} + \frac{\left(-x_2^2\right)^{2011}}{-x_2} =$$

$$= -\left(x_1^3\right)^{1340} x_1^2 - \left(x_2^3\right)^{1340} x_2^2 = -(x_1 + x_2)^2 + 2x_1 x_2 = 1.$$

VI. 30. Two solutions: **i)** A complex number is a real number if and only if $z = \bar{z}$. From $|z_1| = |z_2| = 1$ this means $z_1\bar{z}_1 = z_2\bar{z}_2 = 1$.

$$\overline{\left(\frac{z_1+z_2}{1+z_1z_2}\right)}=\frac{\overline{z_1}+\overline{z_2}}{1+\overline{z_1z_2}}=\frac{\overline{z_1}+\overline{z_2}}{1+\overline{z_1}\,\overline{z_2}}=\frac{\dfrac{1}{z_1}+\dfrac{1}{z_2}}{1+\dfrac{1}{z_1}\dfrac{1}{z_2}}=\frac{z_1+z_2}{1+z_1z_2}.$$

ii) Consider $z_1=\cos t_1+i\sin t_1$, $z_2=\cos t_2+i\sin t_2$, $|z_1|=|z_2|=1$.

$$\frac{z_1+z_2}{1+z_1z_2}=\frac{\cos t_1+\cos t_2+i(\sin t_1+\sin t_2)}{1+\cos(t_1+t_2)+i\sin(t_1+t_2)}=$$

$$=\frac{2\cos\dfrac{t_1+t_2}{2}\cos\dfrac{t_1-t_2}{2}+2i\sin\dfrac{t_1+t_2}{2}\cos\dfrac{t_1-t_2}{2}}{2\cos^2\dfrac{t_1+t_2}{2}+2i\sin\dfrac{t_1+t_2}{2}\cos\dfrac{t_1+t_2}{2}}=\frac{\cos\dfrac{t_1-t_2}{2}}{\cos\dfrac{t_1+t_2}{2}}\in R.$$

VI. 31. $\dfrac{1-z+z^2}{1+z+z^2}=\overline{\dfrac{1-z+z^2}{1+z+z^2}}\Rightarrow\dfrac{1-\overline{z}+\overline{z}^2}{1+\overline{z}+\overline{z}^2}=\dfrac{1-z+z^2}{1+z+z^2}\Rightarrow$

$(z-\overline{z})(1-z\cdot\overline{z})=0$, but $z\neq\overline{z}$, then $\overline{z}\cdot z=1$, which means $|z|=1$.

VI. 32. $z_2=\dfrac{1-\overline{z_1}}{1+\overline{z_1}}$, $x_2+iy_2=\dfrac{1-x_1^2-x_2^2}{(1+x_1)^2+x_2^2}+\dfrac{2y_1i}{(1+x_1)^2+x_2^2}$,

$y_2=\dfrac{2y_1}{(1+x_1)^2+x_2^2}$. Therefore $y_1y_2>0$. b) $z_1+z_2\in R\Rightarrow$

$y_1+y_2=0$, from $y_1y_2>0\Rightarrow y_1=0$ and $y_2=0$.

VI. 33. a) point $(-2,3)$; **b)** point $(3,-2)$; **c)** point $(-2,0)$;
d) point $(0,2)$; **e)** point $(4,-8)$; ; **f)** point $(0.5,2.5)$.
VI. 36. $|z_1-z_2|=3\sqrt{2}$, $|z_1-z_3|=2\sqrt{2}$, $|z_3-z_2|=\sqrt{2}$.

VI. 37. $x^2+y^2=5$. **VI. 38. a)** $z=-\dfrac{7}{3}-\dfrac{4}{3}i$;

b) $z=a+bi$, $|z|=\sqrt{a^2+b^2}$, $\left|\dfrac{z+1}{z}\right|=\dfrac{|z+1|}{|z|}=\dfrac{\sqrt{(a+1)^2+b^2}}{\sqrt{a^2+b^2}}$,

$\left|\dfrac{z-i}{1-i}\right|=\dfrac{|z-i|}{|1-i|}=\dfrac{|z-i|}{\sqrt{2}}=\dfrac{\sqrt{a^2+(b-1)^2}}{\sqrt{2}}$,

$$\begin{cases} \dfrac{\sqrt{(a+1)^2+b^2}}{\sqrt{a^2+b^2}}=2 \\[2mm] \dfrac{\sqrt{a^2+(b-1)^2}}{\sqrt{2}}=1 \end{cases} \Rightarrow \begin{cases} 3(a^2+b^2)=2a+1 \\ a^2+b^2=2b+1 \end{cases}, \quad z_1=1 \ \text{ and}$$

$z_2=\dfrac{-1-2i}{5}$, **c)** $z=2-i$.

VI. 39. a) $x^2+(y-1)^2=4$; $C(0,\ 1)$; $r=2$;

b) $\dfrac{x^2}{4}+\dfrac{y^2}{3}=1$ (ellipse); **c)** $x^2+(y+1)^2<4$;

d) $\left(\dfrac{x}{\sqrt{2}}\right)^2-\left(\dfrac{y}{\sqrt{2}}\right)^2-1<0$; **e)** $\dfrac{1}{4}<\left(x+\dfrac{3}{2}\right)^2+(y-1)^2\le 1$;

f) $x^2+(y-1)^2>16$. **VI. 40.** $z=x+iy$,

$z+i-2=(x-2)+i(y+1)\Rightarrow |z+i-2|=\sqrt{(x-2)^2+(y+1)^2}$

$\sqrt{(x-2)^2+(y+1)^2}\le 2\Rightarrow (x-2)^2+(y+1)^2\le 4$.

VI. 41. a) $x_1=1$, $x_{2,3}=\dfrac{-1\pm i\sqrt{3}}{2}$; **b)** $x_1=2$, $x_{2,3}=-1\pm i\sqrt{3}$;

c) $x_1=-\dfrac{1}{3}$, $x_{2,3}=\dfrac{1\pm i\sqrt{3}}{6}$; **d)** $x_1=\dfrac{3}{4}$, $x_{2,3}=\dfrac{-3\pm 3i\sqrt{3}}{8}$;

e) ± 2 and $\pm 2i$; **f)** ± 3, $\pm 3i$; **g)** $x_1=1$, $x_{2,3}=\dfrac{-1\pm i\sqrt{3}}{2}$; $x_4=\sqrt[3]{2}$,

$x_{5,6}=\dfrac{-\sqrt[3]{2}\pm i\cdot\sqrt[6]{108}}{2}$;

h) $x_1=-1$, $x_{2,3}=\dfrac{1\pm i\sqrt{3}}{2}$; $x_4=-\sqrt[3]{2}$, $x_{5,6}=\dfrac{\sqrt[3]{2}\pm i\cdot\sqrt[6]{108}}{2}$.

VI. 42. $|a+b|\le |a|+|b|$,

$|(1+2i)z+2i|z||\le |1+2i||z|+|2i||z|<\sqrt{5}\cdot\dfrac{1}{2}+2\cdot\dfrac{1}{2}<3$.

VI. 43. c) $\varepsilon^2+\varepsilon+1=0$, $\varepsilon^3=1$, $\overline{\varepsilon}=\varepsilon^2$, $\left(\overline{\varepsilon}\right)^2=\varepsilon$,

(1) $|z-1|^2=(z-1)(\overline{z}-1)=|z|^2-z-\overline{z}+1$,

(2) $|z-\varepsilon|^2 = |z|^2 - z\varepsilon^2 - \bar{z}\varepsilon + 1$,

(3) $|z-\varepsilon^2|^2 = |z|^2 - z\varepsilon - \bar{z}\varepsilon^2 + 1$. Add (1), (2), and (3).

VI. 44. $z = (a-b)^k$. **VI. 45. a)** $a = \pm 2$; **b)** $a = \pm\sqrt{6}$.

VI. 46. a) $f(-1) = -3$, $f(i) = -1 + 4i$, $f(2-i) = 7 - 6i$,

$f(\bar{z}) = \bar{z}^2 + 3\bar{z} - z$, $f(|z|) = |z|^2 + 2|z|$;

VI. 47. a) $\dfrac{z-2}{z-6} = \dfrac{x-2+iy}{x-6-iy} = \dfrac{(x-2+iy)(x-6+iy)}{(x-6)^2 - (iy)^2} =$

$= \dfrac{(x-2)(x-6)+y^2}{(x-6)^2+y^2} + i\dfrac{y(x-2)-y(x-6)}{(x-6)^2+y^2}$,

$\dfrac{(x-2)(x-6)+y^2}{(x-6)^2+y^2} = \dfrac{y(x-6)-y(x-2)}{(x-6)^2-(iy)^2} = 0 \Rightarrow (x-4)^2 + (y+2)^2 = 8$.

A circle with the center $C(4,-2)$ and the radios $r = 2\sqrt{2}$;

b) $\left(x-\dfrac{3}{2}\right)^2 + y^2 = \dfrac{1}{4}$ $C\left(\dfrac{3}{2},\ 0\right)$; $r = \dfrac{1}{2}$;

c) The points from the line $x = 0$.

VI. 48. $\dfrac{4}{9} < (x-3)^2 + y^2 < 100$.

VI. 49. $x \in \left(-\infty,\ \dfrac{-5-\sqrt{28}}{3}\right) \cup \left(\dfrac{-5+\sqrt{28}}{3},\ \infty\right)$.

VI. 50. a) $x(x-1)(x+1)\left(x+\dfrac{1}{2}-i\dfrac{\sqrt{3}}{2}\right)\left(x+\dfrac{1}{2}+i\dfrac{\sqrt{3}}{2}\right)$;

b) $(x+2)(x-2-i\sqrt{3})(x-2+i\sqrt{3})$; **c)** $x^2(x-i)(x+i)$;

d) $x^2(x+3)\left(x+\dfrac{3}{2}-i\dfrac{3\sqrt{3}}{2}\right)\left(x+\dfrac{3}{2}+i\dfrac{3\sqrt{3}}{2}\right)$.

VI. 51. a) $4 - 3i$; **b)** $5 - 2i$; **c)** $-3 - 4i$.

VI. 52. $0 + 0i$; $1 + 0i$; $-\dfrac{1}{2} \pm \dfrac{i\sqrt{3}}{2}$.

VI. 53. $A = \{(x;\ 0) \mid x \in \mathbf{R} \setminus \{1\}\}$.

VI. 54. a) $z_1 = 1 - 2i$, $z_2 = -1 + 2i$; **b)** $z_1 = 2 - \sqrt{3}i$, $z_2 = -2 + \sqrt{3}i$;

c) $z_1 = 2 + i$, $z_2 = -2 - i$; **d)** $z_1 = 4 + 3i$, $z_2 = -4 - 3i$.

VI. 55. a) $z_1 = i$, $z_2 = 3i$;

b) $z_{1,2} = \dfrac{1 - i \pm \sqrt{-8 - 6i}}{2} = \dfrac{1 - i \pm \sqrt{(1 - 3i)^2}}{2} = \dfrac{1 - i \pm (1 - 3i)}{2}$, $z_1 = i$,

$z_2 = 1 - 2i$; **c)** $z_{1,2} = \dfrac{2 - 6i \pm \sqrt{-48 + 14i}}{2 - i} = \dfrac{2 - 6i \pm (1 + 7i)}{2 - i}$,

$z_1 = 1 + i$, $z_2 = \dfrac{-11 + 27i}{5}$; **d)** $z_1 = -1 + 3i$, $z_2 = 1 + 2i$; **e)** $z_1 = -1 - i$

, $z_2 = i$;

f) $z_1 = 2$, $z_2 = 3 - 2i$; **g)** $z_1 = 1 - i$, $z_2 = 1 + 2i$;

h) $z_1 = 2 - i$, $z_2 = 1 + 2i$.

VI. 56. a) $(1 + i)(x + iy) + (5 - 3i)(x - iy) = 20 - 4i \Rightarrow$

$6x - 4iy - 2ix - 4y = 20 - 4i$. We get the system $\begin{cases} 6x - 4y = 20 \\ -4y - 2x = -4 \end{cases}$ whose

solution is $\begin{cases} x = 3 \\ y = -\dfrac{1}{2} \end{cases}$. Therefore $z = 3 - \dfrac{1}{2}i$; **b)** $z = 2 + 8i$. **VI. 57. a)**

$(1 + 3i)z - (3 - 2i)z = 11i + 13 \Rightarrow z(-2 + 5i) = 11i + 13 \Rightarrow z = 1 - 3i$.

b) $2(a + bi) + i(a - bi) = 9 + 3i$, $\begin{cases} 2a + b = 9 \\ 2b + a = 3 \end{cases}$, $z = 5 - i$.

VI. 58. a) From $(a + bi)^3 = -11 + 2i$ we get

$(a^3 - 3ab^2) + (3a^2b - b^3) = -11 + 2i$ or

$\begin{cases} a^3 - 3ab^2 = -11 \\ 3a^2b - b^3 = 2 \end{cases}$. In the equation $2(a^3 - 3ab^2) = -11(3a^2b - b^3)$ we

set $b = ta$ and it implies $2(1 - 3t^2) = -11(3t - t^3)$ or

$(t + 2)(11t^2 - 16t - 1) = 0$.

The only convenient solution of this equation is $t_1 = -2$; hence

$z = 1 - 2i$; **b)** $z = 3 - 2i$. **VI. 59. a)** $\dfrac{z - 5 + 12i}{z - 3 + 4i}$; **b)** $\dfrac{z + 1}{z - 1 + i}$;

c) $\dfrac{z - i}{z + i}$. **VI. 60. a)** $x = 1 - i$, $y = i$; **b)** $x = i$, $y = 2 + i$;

c) $x = 1 + i$, $y = 1 - i$; **d)** $z_1 = i$, $z_2 = 1 + 3i$.

VI. 61. Switching x and $\varepsilon^2 x$ we get $f(x) = \varepsilon^2 x$.

VI. 62. If $z = \cos t + i \sin t$, $z^n = \cos nt + i \sin nt$, $n \in N$ (De Moivre's theorem) also $z^{-n} = \cos nt - i \sin nt$. So

$$(\cos nt + i \sin nt) + (\cos nt - i \sin nt) = 2\cos nt = z^n + z^{-n} = \frac{z^{2n} + 1}{z^n}$$

and

$$(\cos nt + i \sin nt) - (\cos nt - i \sin nt) = 2i\sin nt = z^n - z^{-n} = \frac{z^{2n} - 1}{z^n}.$$

b) Let $z = \cos 20^\circ + i \sin 20^\circ$, we get
$z^9 = \cos 180^\circ + i \sin 180^\circ = -1$. Using a) we have

$$\sin 70^\circ = \cos 20^\circ = \frac{z^2 + 1}{2z}, \qquad \sin 50^\circ = \cos 40^\circ = \frac{z^4 + 1}{2z^2} \quad \text{and}$$

$$\sin 10^\circ = \cos 80^\circ = \frac{z^8 + 1}{2z^4}. \text{ So}$$

$$A = \frac{z^2 + 1}{2z} \frac{z^4 + 1}{2z^2} \frac{z^8 + 1}{2z^4} = \frac{1 + z^2 + z^4 + z^6 + z^8 + z^{10} + z^{12} + z^{14}}{8z^7} =$$

$$= \frac{\dfrac{1 - z^{16}}{1 - z^2}}{8z^7} = \frac{1 - z^7 z^9}{8(z^7 - z^9)} = \frac{1 + z^7}{8(z^7 + 1)} = \frac{1}{8}.$$

VI. 64. $z = \cos t + i \sin t$ then

$$Z_k = \sqrt[n]{z} = \sqrt[n]{\cos t + i \sin t} = \cos\frac{t + 2k\pi}{n} + i \sin\frac{t + 2k\pi}{n},$$

$k = 0, 1, \ldots, n - 1$.

We have $z^3 = \dfrac{3 + i}{2 - i} = 1 + i = \dfrac{\sqrt{2}}{2}\left(\cos\dfrac{\pi}{4} + i \sin\dfrac{\pi}{4}\right)$, then

$$Z_k = \sqrt[3]{\frac{\sqrt{2}}{2}\left(\cos\frac{\pi}{4} + iso\frac{\pi}{4}\right)} \Rightarrow$$

$$Z_k = \frac{1}{\sqrt[6]{2}}\left[\cos\frac{1}{3}\left(\frac{\pi}{4} + 2k\pi\right) + i\sin\frac{1}{3}\left(\frac{\pi}{4} + 2k\pi\right)\right], k = 1,2,3.$$

VI. 65. Let be $\dfrac{1 + ia}{1 - ia} = r(\cos t + i\sin t)$. But

$$\left|\frac{1 + ia}{1 - ia}\right| = \frac{|1 + ia|}{|1 - ia|} = \frac{\sqrt{1 + a^2}}{\sqrt{1 + a^2}} = 1 \Rightarrow r = 1. \text{ So}$$

$$\frac{1 + ia}{1 - ia} = \cos t + i\sin t \Rightarrow a = \tan\frac{t}{2}.$$

From $\left(\dfrac{1 + iz}{1 - iz}\right)^n = \cos t + i\sin t$ we get

$$\frac{1 + iz}{1 - iz} = \sqrt[n]{\cos t + i\sin t} = \cos\frac{t + 2k\pi}{n} + i\sin\frac{t + 2k\pi}{n}, k = 0,1,...,n-1$$

and

$$z = \tan\frac{t + 2k\pi}{2n}, k = 0,1,...,n-1 \text{ Since } t = 2\arctan a, \text{ therefore}$$

$z \in R$.

VI. 66. Let be (1) $\dfrac{x - i}{x + i} = z$, then the initial equation becomes

$z^{n-1} + z^{n-2} + ... + z + 1 = 0$.

$z^n - 1 = (z - 1)(z^{n-1} + z^{n-2} + ... + z + 1) = 0 \Rightarrow z^n = 1$, but $z \neq 1$. If

$z^n = 1$ we obtain, $z^n = \cos 0 + i\sin 0 \Rightarrow z = \sqrt{\cos 0 + i\sin 0} =$

$$= \cos\frac{0 + 2k\pi}{n} + i\sin\frac{0 + 2k\pi}{n},$$

From (1) we get $\dfrac{x - i}{x + i} = \cos\dfrac{2k\pi}{n} + i\sin\dfrac{2k\pi}{n} \Rightarrow x = \dfrac{i(1 + z)}{1 - z}$. So

$$x_k = i\left(\frac{1 + \cos\dfrac{2k\pi}{n} + i\sin\dfrac{2k\pi}{n}}{1 - \cos\dfrac{2k\pi}{n} - i\sin\dfrac{2k\pi}{n}}\right) =$$

$$= \frac{2i\cos^2\dfrac{2k\pi}{2n} - 2i\sin\dfrac{2k\pi}{2n}\cos\dfrac{2k\pi}{2n}}{2\sin^2\dfrac{2k\pi}{2n} - 2i\sin\dfrac{2k\pi}{2n}\cos\dfrac{2k\pi}{2n}} = -\cot\frac{k\pi}{n}, \quad k = 1,2,\ldots,n-1.$$

VI. 67. We have $\left(\dfrac{z + i\sqrt{1-z^2}}{z - i\sqrt{1-z^2}}\right)^n = -1 = \cos\pi + i\sin\pi \Rightarrow$

$$\frac{z + i\sqrt{1-z^2}}{z - i\sqrt{1-z^2}} = \cos\frac{\pi + 2k\pi}{n} + i\sin\frac{\pi + 2k\pi}{n} \Rightarrow$$

$$\frac{z}{\sqrt{1-z^2}} = \frac{i\left[\cos\dfrac{(2k+1)\pi}{2n} + i\sin\dfrac{(2k+1)\pi}{2n} + 1\right]}{\cos\dfrac{(2k+1)\pi}{2n} + i\sin\dfrac{(2k+1)\pi}{2n} - 1} =$$

$$= \frac{2i\cos^2\dfrac{(2k+1)\pi}{2n} - 2\sin\dfrac{(2k+1)\pi}{2n}\cos\dfrac{(2k+1)\pi}{2n}}{-2\sin^2\dfrac{(2k+1)\pi}{2n} + 2i\sin\dfrac{(2k+1)\pi}{2n}\cos\dfrac{(2k+1)\pi}{2n}} = \cot\frac{(2k+1)\pi}{2n}.$$

Finally $z = \cos\dfrac{(2k+1)\pi}{2n}$ $\quad k = 0,1,\ldots,n-1.$

VI. 68. It is known that a function is a *bijection* if it is a injection (one-to one) and a *surjection* (onto).

A function $f : A \to B$ is said to be *one-to-one* (*injective*), if and only if $f(z_1) = f(z_2)$ implies that $z_1 = z_2$ for all z_1 and z_2 in the domain of f.

A function $f : A \to B$ is called *onto* (*surjective*), if and only if for every element y in B there is an element x in A with $f(x) = y$ ($\forall y \in B \Rightarrow \exists x \in A$ such as $f(x)=y$).

For injection let be $z_1, z_2 \in C, z_1 = x_1 + iy_1$ and $z_2 = x_2 + iy_2$ if

$$f(z_1) = f(z_2) \Rightarrow 3z_1 + |z_1| = 3z_2 + |z_2| \Rightarrow$$

$$3(x_1 + iy_1) + \sqrt{x_2^2 + y_2^2} = 3(x_2 + iy_2) + \sqrt{x_2^2 + y_2^2} \Rightarrow$$

$$\begin{cases} 3x_1 + \sqrt{x_2^2 + y_2^2} = 3x_2 + \sqrt{x_2^2 + y_2^2} \\ y_1 = y_2 \end{cases} \Rightarrow \begin{cases} x_1 = x_2 \\ y_1 = y_2 \end{cases} \Rightarrow z_1 = z_2.$$

b) For surjection $\forall u \in C, \ u = a + ib \Rightarrow \exists z \in C$ such as $f(z) = u$,

$z = x + iy$. Since $f(z) = 3x + 3iy + \sqrt{x^2 + y^2}$. Therefore the function f is

a surjection (onto) if the system (1) $\begin{cases} 3x + \sqrt{x^2 + y^2} = a \\ 3y = b \end{cases}$ has a unique

solution. Substituting $y = \dfrac{b}{3}$ in the first equation we have

$\sqrt{9x^2 + b^2} = 3a - 9x$. This is an irrational equation, $3a - 9x \geq 0$

and we get $x \leq \dfrac{a}{3}$. Squaring the both sides of the irrational equation

we get $9x^2 + b^2 = (3a - 9x)^2 \Rightarrow 72x^2 - 54ax + 9a^2 - b^2 = 0$ whose

roots are $x_1 = \dfrac{9a + \sqrt{9a^2 + 8b^2}}{24}$ and $x_2 = \dfrac{9a - \sqrt{9a^2 + 8b^2}}{24}$. Notice

that $x_1 = \dfrac{9a + \sqrt{9a^2 + 8b^2}}{24} > \dfrac{9a + 3|a|}{24} > \dfrac{a}{3}$ and

$x_2 = \dfrac{9a - \sqrt{9a^2 + 8b^2}}{24} < x_1 = \dfrac{9a - 3|a|}{24} < \dfrac{a}{3}$. Finally

$x_2 = \dfrac{9a - \sqrt{9a^2 + 8b^2}}{24}$. So the solution of the system (1) is

$x = \dfrac{9a - \sqrt{9a^2 + 8b^2}}{24}$ and $y = \dfrac{b}{3}$. Since the function is injective

and surjective , it is a bijection. Therefore the function f is

inversible and its inverse is $f^{-1}(u) = z$. Finally

$f^{-1}(u) = \dfrac{9\,\mathrm{Re}\,u - \sqrt{9(\mathrm{Re}\,u)^2 + 8(\mathrm{Im}\,u)^2}}{24} + i\dfrac{\mathrm{Im}\,u}{3}.$

Chapter VII

EXPONENTIAL and LOGARITHMS

VII. 1. a) $x \in (-\infty, -4]$; **b)** $x \in (-\infty, -2]$; **c)** $x \in [-2, +\infty)$;
d) $x \in \mathbf{R}$; **e)** If $a > 1$, $x \in [0, +\infty)$, if $0 < a < 1$, $x \in (-\infty, 0]$,
if $a = 1$, $x \in \mathbf{R}$; **f)** $x \in \mathbf{R}$.
VII. 2. a) $f : [-1, \ 1] \rightarrow [1, 2]$;
b) $f_1 : (-\infty, \ -1] \rightarrow [1, \ \infty)$; **c)** $f_2 : [3, +\infty) \rightarrow [0, +\infty)$;
d) $f_3 : \mathbf{R} \rightarrow (0, 1]$; **e)** $f_4 : \mathbf{R} \rightarrow [1, +\infty)$.
VII. 3. $x \in [0, \infty) \setminus \{2\}$.
VII. 4. d) Vertical stretch by factor 2; Horizontal translation of 3 units to the right; Vertical translation by 1.5 units up.

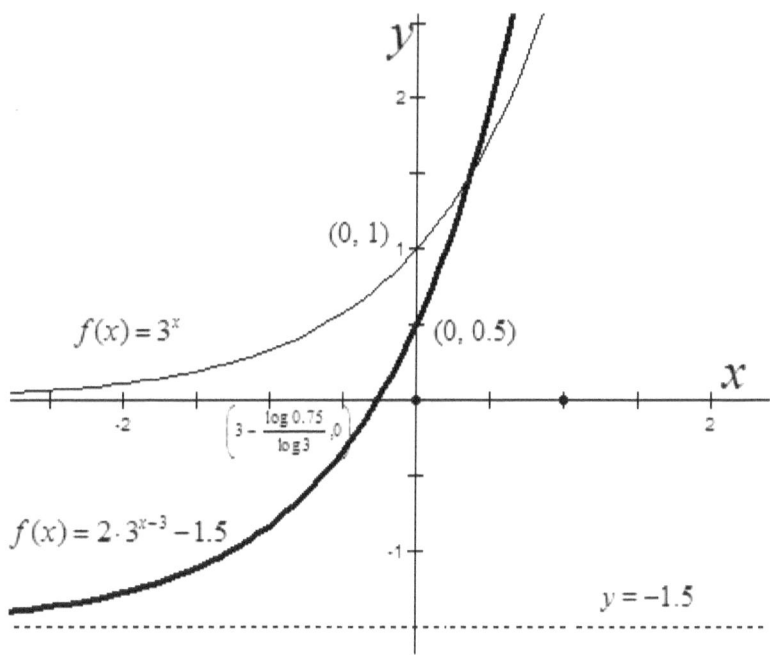

e) Reflexion in y-axis; Vertical stretch by factor 3; Vertical translation by 2 units up.

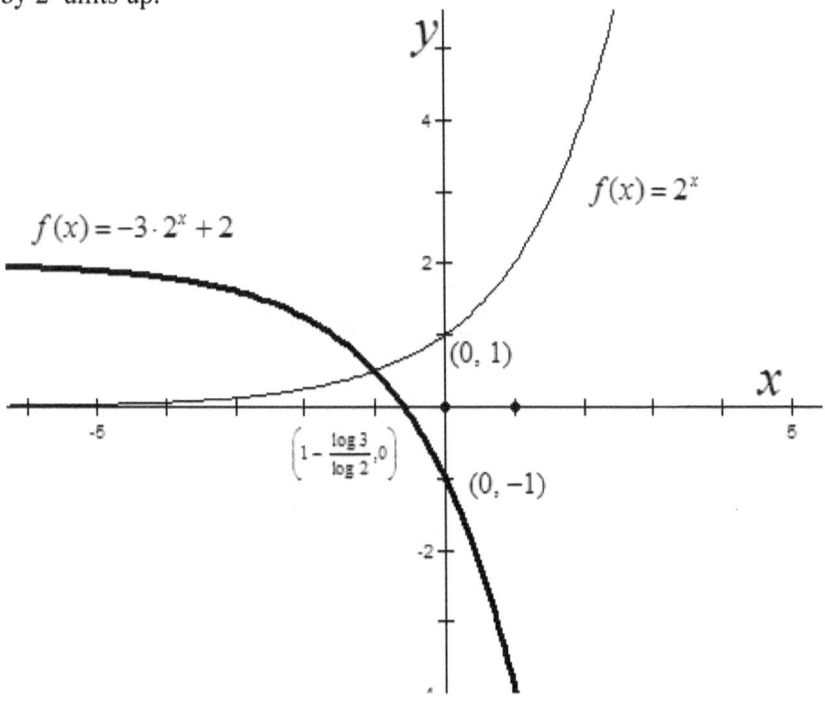

$f(x) = 2^x$

$f(x) = -3 \cdot 2^x + 2$

$(0, 1)$

$\left(1 - \dfrac{\log 3}{\log 2}, 0\right)$

$(0, -1)$

VII. 5. a) $x \neq -1$; **b)** $x \in (-1, 3)\backslash\{0\}$; **c)** $x \in (-\infty, -1) \cup (6, \infty)$;
d) $10 - x^2 > 0$, $\log_3(10 - x^2) > 0$, $x + 1 > 0$ and $x + 1 \neq 0$,
$x \in (-1, 3)\backslash\{0\}$; **e)** $x \in (1, 16)$; **f)** $x \in (1, \infty)$.

VII. 6.

	$y = 2^x$	$y = -2^{x-3} + 4$
Shape	increasing	decreasing
Shape factor	1	-1
point	$(0, 1)$	$(3, 3)$
domain	$x \in \mathbf{R}$	$x \in \mathbf{R}$
range	$y \in (0, \infty)$	$y \in (-\infty, 4)$
asymptote	$y = 0$	$y = 4$

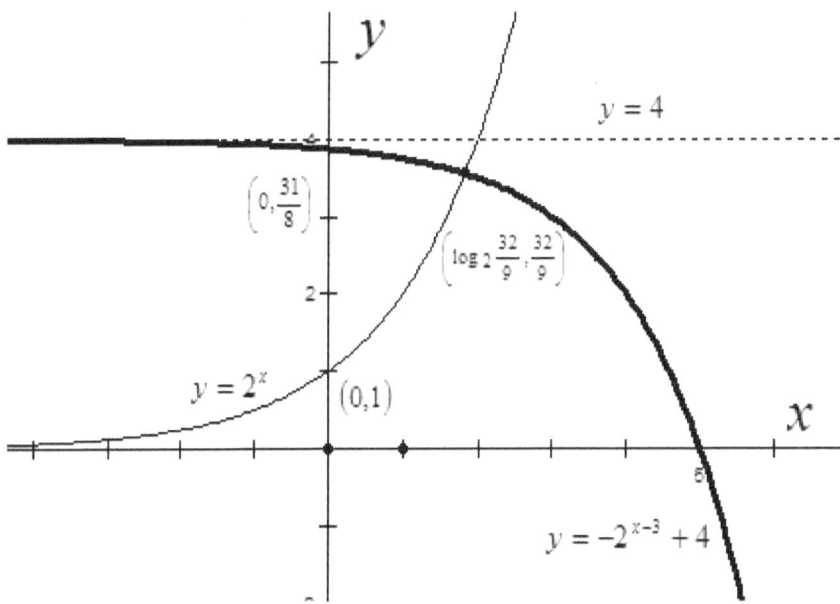

VII. 7. The basic function is $y = \left(\dfrac{1}{2}\right)^x$. It is translated 3 units to the right, a vertically streched by a factor 3, a reflection in the *x*-axis, and a translation up 2 units.

Function	$y = \left(\dfrac{1}{2}\right)^x$	$y = -3 \cdot \left(\dfrac{1}{2}\right)^{x-3} + 2$
shape	decreasing	increasing
domain	$x \in \mathbf{R}$	$x \in \mathbf{R}$
range	$y \in (0, \ \infty)$	$y \in (\infty, \ 2)$
asymptote	$y = 0$	$y = 2$
y-intercept	$(0, 1)$	$(0, -22)$

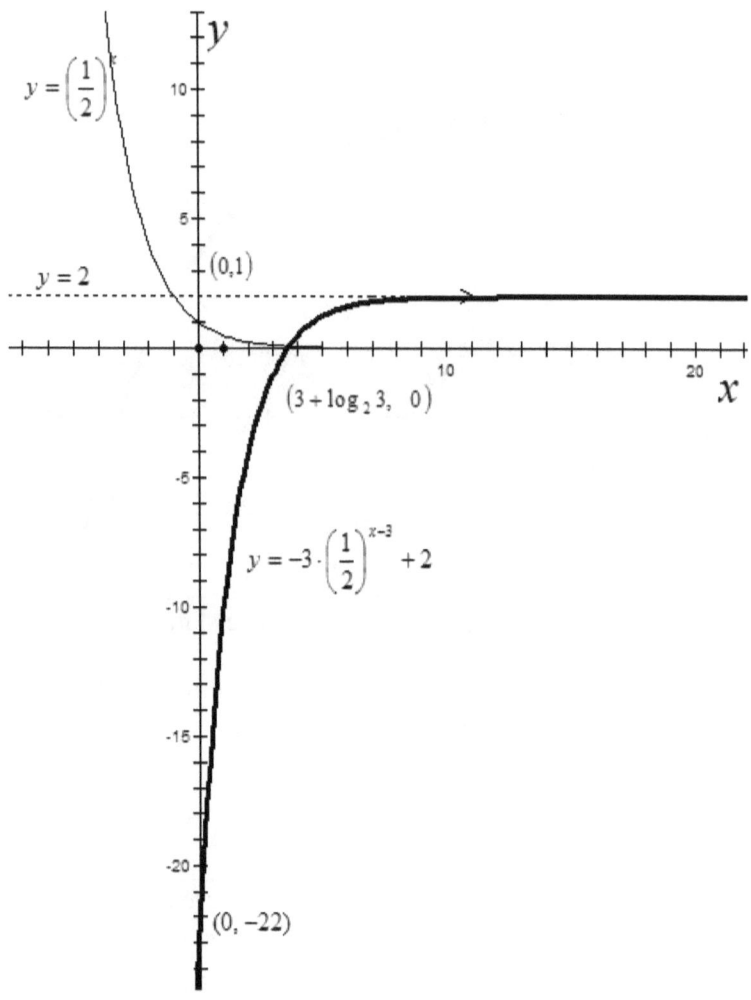

$y = \left(\dfrac{1}{2}\right)^x$

$y = 2$

$(0,1)$

$\left(3 + \log_2 3, \ 0\right)$

$y = -3 \cdot \left(\dfrac{1}{2}\right)^{x-3} + 2$

$(0, -22)$

VII. 8.

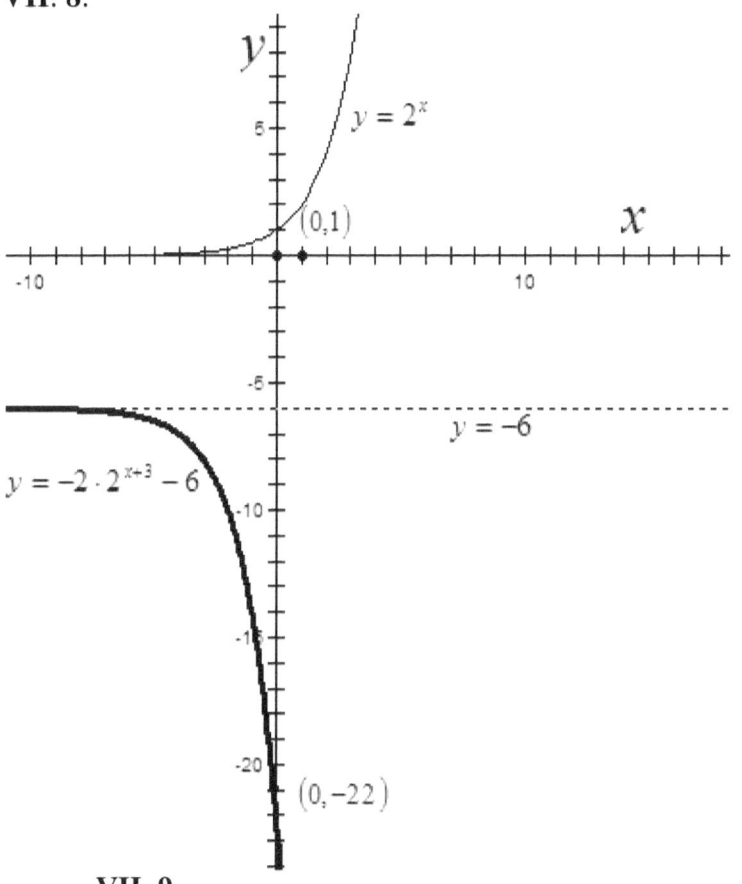

$y = 2^x$

$(0,1)$

$y = -6$

$y = -2 \cdot 2^{x+3} - 6$

$(0,-22)$

VII. 9.

Function	$y = \log_{\frac{1}{3}} x$	$y = 2\log_{\frac{1}{3}}(x-1) - 3$
Shape	decreasing	decreasing
Shape factor	1	2
point	$(1, 0)$	$(2, -3)$
domain	$x \in (0, \infty)$	$y \in (1, \infty)$
range	$y \in \mathbf{R}$	$y \in \mathbf{R}$
asymptote	$x = 0$	$x = 1$

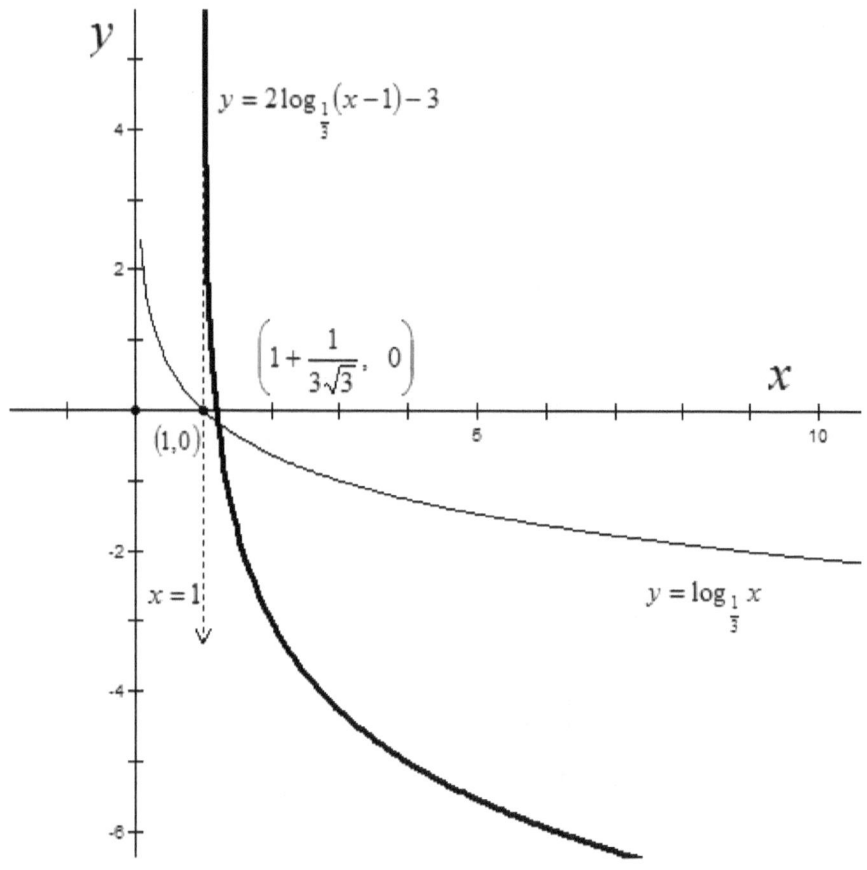

$$y = 2\log_{\frac{1}{3}}(x-1) - 3$$

$$\left(1 + \frac{1}{3\sqrt{3}},\ 0\right)$$

$(1, 0)$

$x = 1$

$$y = \log_{\frac{1}{3}} x$$

VII. 10. The basic function is $y = \log_3 x$. It is translated 2 unit to the left, a vertically streched by a factor 2, a reflection in the x-axis, and a translation down 1 unit.

Function	$y = \log_2 x$	$y = -2\log_2(x+2) - 1$
shape	increasing	decreasing
domain	$x \in (0,\ \infty)$	$x \in (-2,\ \infty)$
range	$y \in \mathbf{R}$	$y \in \mathbf{R}$
asymptote	$x = 0$	$x = -2$
x-intercept	$(1, 0)$	$(-1, -3)$

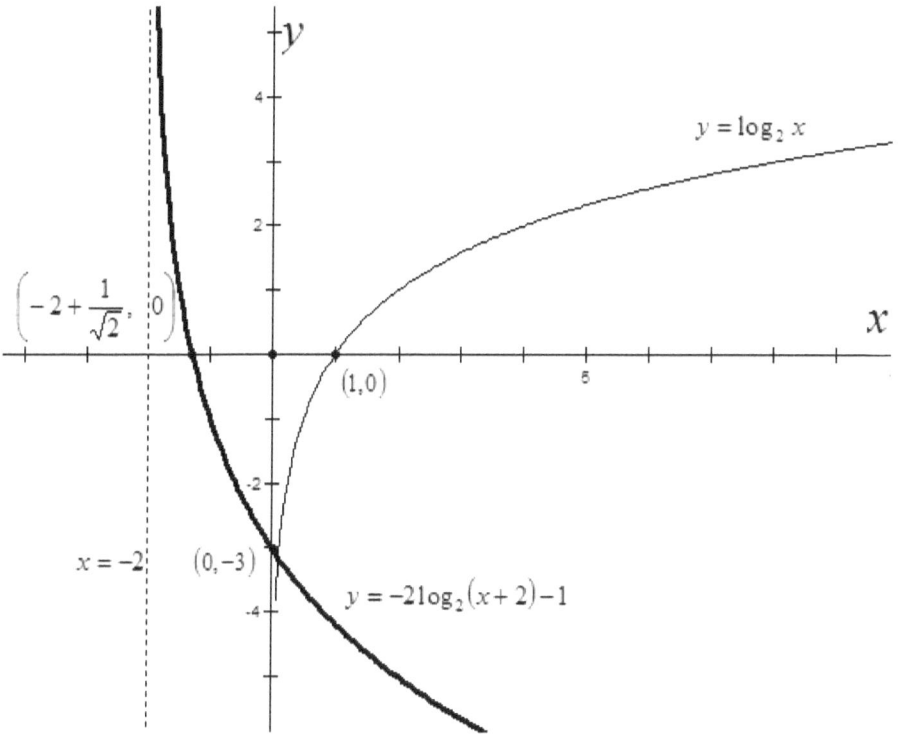

$y = \log_2 x$

$\left(-2+\dfrac{1}{\sqrt{2}},\ 0\right)$

$(1,0)$

$x = -2$ $(0,-3)$

$y = -2\log_2(x+2)-1$

VII. 11. $y = -2\log_2(x+3)+5$.

VII. 12. f) reflection in the x-axis, a vertical stretch by a factor of 5, translated 2 units right and 4 units down, and a dilatation (stretch) by a factor 3. **VII. 13. a)** 4; **b)** –7; **c)** –1; **d)** 1; **e)** $\dfrac{1}{2}$; **f)** 1; **g)** –3;

h) $-\dfrac{5}{2}$; **i)** $\log_2 5$; **j)** $-0.5\log_2 3$. **VII. 14. a)** $\log_6 5$; **b)**

$\log_a b = x$, $\left(x+\dfrac{1}{x}+2\right)\cdot\left(x-\dfrac{x}{x+1}\right)\dfrac{1}{x}-1 = x$, $\log_a b$; **c)** $\dfrac{9}{8}$;

d) $\log_a 2$; **e)** $\dfrac{\dfrac{1}{3}\log_a\left(a^2-1\right)\cdot\dfrac{1}{-4}\log_a\left(a^2-1\right)}{\dfrac{1}{4}\log_a\left(a^2-1\right)\cdot\dfrac{6}{4}\log_a\left(a^2-1\right)}=-\dfrac{2}{9}$; **f)**

$$\left(b^{\frac{\frac{1}{2}\log_{10}a}{\log_{10}a}}\times a^{\frac{\frac{1}{2}\log_{10}b}{\log_{10}b}}\right)^{2\log_{ab}(a+b)}=\left(\sqrt{b}\times\sqrt{a}\right)^{2\log_{ab}(a+b)}=(ab)^{\log_{ab}(a+b)}=a+b$$

g) $=\left[\left(9^{\log_9 5}\right)^2+\left(3^{\log_3 6}\right)^{\frac{3}{2}}\right]\left(25-\left(5^{\log_5 6}\right)^{\frac{3}{2}}\right)$

$=\left(25+6\sqrt{6}\right)\left(25-6\sqrt{6}\right)=409$; **h)**

$$\sqrt{x^{1+\frac{1}{\log_2 x}}+8^{\frac{1}{\frac{3}{2}\log_x 2}}+1}=\sqrt{x^{1+\log_x 2}+\left(2^{\log_2 x}\right)^2+1}=\sqrt{2x+x^2+1}=x+1$$

i) $\dfrac{\left[\left(3^{\log_3 2}\right)^3+5^{\log_5 7}\right]\left[2\cdot\left(3^{\log_3 2}\right)^4-\left(2^{\log_2 3}\right)^3\right]}{\left(3+5^{\log_5 4}\right)\cdot 5^{\log_5 3}}=\dfrac{\left(2^3+7\right)\left(2\cdot 2^4-3^3\right)}{(3+4)\cdot 3}=\dfrac{25}{7}$;

where $a^{\log_a x}=x$; **j)** $\dfrac{\log_a b}{\log_a b-1}$; **k)** $\left(\log_2 x+1\right)^3$;

l) $\begin{cases}1, & \text{if } 1<a\le b\\ \log_a b, & \text{if } 1<b<a\end{cases}$. **VII. 16.** $\dfrac{4(3-a)}{3+a}$. **VII. 17.** $\dfrac{a+3}{2}$;

VII. 18. $4(1-a-b)$. **VII. 19.** $abc+1$.

VII. 20. $\log_{30}8=\dfrac{\log_{10}8}{\log_{10}30}=\dfrac{3\log_{10}2}{\log_{10}2+\log_{10}3+\log_{10}5}$,

$\log_{10}2=\log_{10}\dfrac{10}{5}=\log_{10}10-\log_{10}5=1-\log_{10}5=1-a$,

$\log_{30}8=\dfrac{3(1-a)}{b+1}$. **VII. 21.** $\dfrac{(1-a-b)(2+a-2b)}{2a(1-b)}$.

VII. 22. $\dfrac{1}{1-\log_8 c}$. **VII. 23.** $\dfrac{6-6a+3b}{4a-2b}$.

VII. 25. a) $E=(2+4+...+512)^{\frac{1}{9}}=2$;

b) $F = 3^{\log_3 2} + 2\log_2 \dfrac{4}{\sqrt{3}+\sqrt{2}} - \log_2 \dfrac{1}{\left(\sqrt{3}+\sqrt{2}\right)^2} =$

$= 2 + \log_2 16 - \log_2 \left(\sqrt{3}+\sqrt{2}\right)^2 + \log_2 \left(\sqrt{3}+\sqrt{2}\right)^2 = 6$.

VII. 26. $\dfrac{1}{\log_{\sqrt{n}} 3} = \dfrac{1}{\frac{1}{2}\log_n 3} = 2\log_3 n$ and $\left(\sqrt{3}\right)^{\frac{1}{\log_{\sqrt{n}} 3}} = 3^{\log_3 n} = n$.

Therefore $E = 1 + 2 + 3 + \ldots + n = \dfrac{n(n+1)}{2} - 1$.

VII. 27. $\dfrac{3}{2}$. **VII. 31. a)** 1;

b) $x \in (0, \ \infty) \setminus \{1\}, \dfrac{1}{\sqrt{x}-1} - \dfrac{1}{\sqrt{x}+1} = \dfrac{2}{\sqrt{x}+1} \Leftrightarrow 4 = 2\sqrt{x} \Rightarrow x = 4$;

c) $x \in [5,\infty)$, $3^{\sqrt{x-5}} = a$, $a^2 - 6a - 27 = 0$, $a_1 = -3$ (impossible),

$a_2 = 9$, $x = 9$; **d)** $x < 9$, $\left(\dfrac{5}{2}\right)^{\frac{4+\sqrt{9-x}}{\sqrt{9-x}}} \times \left(\dfrac{5}{2}\right)^{-1+\sqrt{9-x}} = \left(\dfrac{5}{2}\right)^5$,

$\dfrac{4+\sqrt{9-x}}{\sqrt{9-x}} - 1 + \sqrt{9-x} = 5$, $13 - x = 5\sqrt{9-x}$, $x_1 = -7$, $x_2 = 8$;

e) $5^{\frac{1}{3}-\frac{1}{2x}} = 5^{2x-2}$, $x \neq 0$ $x_{1,2} = \dfrac{7 \pm \sqrt{13}}{12}$.

VII. 32. a) Since the LHS is a positive number, it follows that $x \in [0, \ \infty)$. From $x^2 + 3x - 4 = 0$, we get the solutions $x_1 = -4$ (impossible), $x_2 = 1$, also $x = 3$ is a solution;

b) $x \in \{-4, \ 1, \ 2, \ 4\}$;

c) If $x^2 - 5x + 6 = 1$ the solutions are $x_{1,2} = \dfrac{5 \pm \sqrt{5}}{2}$ and if

$0 < x^2 - 5x + 6 \neq 1$ the solutions are $x_3 = -3$, $x_4 = 0$.

VII. 33. a) $x = 2$; **b)** no solution; **c)** $x = 2$; **d)** $x \in \mathbf{R} \setminus \{2\}$;

e) $x = 3$; **f)** $x_1 = 1$, $x_2 = -\log_3 2$; **g)** $x = 2$; **h)** $x = 1$;

i) $x = 1$; **j)** $x = 1$;

VII. 34. a) $x_1 = -1$, $x_2 = 8$; **b)** $x_1 = 2$, $x_2 = -2^{-1}$; **c)** $x = 2$;

d) $5^{x^2} \cdot \dfrac{1}{5} \cdot 4^{x^2} = 4 \cdot 3^{x^2} \cdot \dfrac{1}{3}$, $\dfrac{5^{x^2} 4^{x^2}}{3^{x^2}} = \dfrac{5 \cdot 4}{3}$, $\left(\dfrac{20}{3}\right)^{x^2} = \dfrac{20}{3} \Rightarrow x^2 = 1$,

$x = \pm 1$; **e)** $x_1 = 1$, $x_2 = \dfrac{1}{4}$; **f)** $x_1 = 3$, $x_2 = -5^{-1}$; **g)** $x = \dfrac{-1}{2}$.

VII. 35. a) 2; **b)** 4; **c)** 4; **d)** 3; **e)** 1; **f)** 3; **g)** 4; **h)** 4; **i)** 1;
j) 3. **VII. 36. a)** $x = 0$; **b)** $x = -1$; **c)** $x = 2$; **d)** $x = 0$.

VII. 37. a) $5^{x-2012}\left(3 \cdot 5^4 + 5^3 + 2\right) = 2012$ but since 2012 does not
have 5 as a factor, it follows that $3 \cdot 5^4 + 5^3 + 2 = 2012$, and
$5^{x-2012} = 1$, $x = 2012$;

b) $6^{x-2003}\left(9 \cdot 6^3 + 6^2 + 3 \cdot 6 + 4\right) = 2003$. Through a similar reasoning
as in **a)**, it follows that $x = 2003$.

VII. 38. a) $x_1 = -1$, $x_2 = 0$; **b)** $2^x\left(2^x + 10\right) = 144$, $2^x = a > 0$,
$x = 3$;

c) $9^{x^2} - 12 \times 3^{x^2} + 27 = 0$, $3^{x^2} = a$; $a^2 - 12a + 27 = 0$,
$a_1 = 3 \Leftrightarrow x_{1,2} = \pm 1$ and $a_2 = 9 \Leftrightarrow x_{3,4} = \pm\sqrt{2}$; **d)** $x = 1$;

e) $x \in [-5, \infty)$, $3^{\sqrt{x+5}} = t$, $t^2 - 6t - 27 = 0$, $t_1 = -3$ (impossible) and
$t_2 = 9$, $\sqrt{x+5} = 2 \Rightarrow x + 5 = 4 \Rightarrow x = -1$; **f)** $x \in (-\infty, -2] \cup [2, \infty)$;
$2^{x+\sqrt{x^2-4}} = a$; $2a^2 - 5a - 12 = 0$; $x = 2.5$.

VII. 39. a) $6^{2x} \cdot 6 - 6^{2x} \div 6 = 210$ or $6^{2x}\left(6 - \dfrac{1}{6}\right) = 210$, $x = 1$;

b) $9 \cdot 9^{x^2} - \dfrac{9}{9^{x^2}} = 80$, $9^{x^2} = t > 0$, $9t^2 - 80t - 9 = 0$,

$t_1 = 9$, $t_2 = -1$, $x = \pm 1$; **c)** $5^{2x} \cdot 5^{-1} + 2 \cdot 5^x \cdot 5^{-1} = 5^x + 2$,
$5^x = t > 0$, $t^2 - 3t - 10 = 0 \Rightarrow t_1 = 5$ $t_2 = -2$, $x = 1$;

d) $3^{3x} - 13 \times 3^{2x} + 39 \times 3^x - 27 = 0$, $3^x = a$,
$(a-3)(a^2 - 10a + 9) = 0$ $a_1 = 3 \Rightarrow x = 1$, $a_2 = 9 \Rightarrow x = 2$ and

$a_3 = 1 \Rightarrow x = 0$; **e)** $x = 20$; **f)** $3^{3x} + \dfrac{1}{3^{3x}} + 3 \cdot 3^x + \dfrac{3}{3^x} = 8 \Rightarrow$

$\left(3^x + \dfrac{1}{3^x}\right)^3 = 2^3 \Rightarrow 3^{2x} - 2 \cdot 3^x + 1 = 0$, $\left(3^x - 1\right)^2 = 0$, $3^x = 1$ and

$x = 0$; **g)** $x = 0$.

VII. 40. a) $\left(2+\sqrt{3}\right)^x = t,\ \left(2-\sqrt{3}\right)^x = \dfrac{1}{t},\ t^2 - 14t + 1 = 0,$

$t_{1,2} = 7 \pm 4\sqrt{3} = \left(2 \pm \sqrt{3}\right)^2,\ x = \pm 2;$ **b)** $\pm 2;$ **c)** $\pm 2;$ **d)** $\pm 2;$

e) $x = 1;$ **f)** $x_{1,2} = 1 \pm \sqrt{2},\ x_3 = 1;$ **g)** $\pm 1;$ **h)** $\pm 1.$

VII. 41. a) $\pm 1;$ **b)** $-2;$ **c)** $-1, 1,$ and $0;$

d) $\dfrac{5^{2x}}{5} + 2^{2x} - 5^{2x} + 2^{2x} \times 2^2 = 0 \Leftrightarrow 5^{2x}\left(\dfrac{1}{5} - 1\right) + 2^{2x} \times 5 = 0,$

$\left(\dfrac{5}{2}\right)^{2x} = \left(\dfrac{5}{2}\right)^2 \Rightarrow x = 1;$ **e)** Dividing by $6^{\frac{1}{x+1}}$ we get

$6 \cdot \left(\dfrac{3}{2}\right)^{\frac{1}{x+1}} + 6 \cdot \left(\dfrac{2}{3}\right)^{\frac{1}{x+1}} - 13 = 0,\ \left(\dfrac{3}{2}\right)^{\frac{1}{x+1}} = t,\ 6t^2 - 13t + 6 = 0,$

$x_1 = 0$ and $x_2 = -2;$ **f)** $0;$ **g)** $-1;$ **h)** $0.5;$ **i)** 0 and $\log_{3/2} 2.$

VII. 42. Suppose that $\log_5 10$ is a rational number. $\log_5 10 = \dfrac{a}{b},$ where a, b are integer positive numbers and relative prime. The equality $5^a = 10^b$ is false, left side odd number, right side even number, contradiction.

VII. 43. a) $x \in (0,\ \infty) \setminus \{1\},\ x = 9;$ **b)** $x \in (-0.5,\ \infty),\ x = 3;$
c) $x \in (0,\ \infty),\ x = 1;$ **d)** $x \in (-\infty,\ 10),\ x = 1;$ **e)** $x \in (2.5,\ \infty),$
$x = 3;$ **f)** $x \in (2.\overline{3},\ \infty),\ x = 34;$ **g)** $x \in (-1.5,\ \infty),\ x_1 = -1,\ x_2 = 2;$
h) $x \in (-0.5,\ \infty) \setminus \{0\},\ x = 1;$ **i)** $x \in (-0.6,\ \infty) \setminus \{0.5\},\ x = 3.$

VII. 44. a) $x \in (3,\infty),\ x^2 - 3x = x,\ x_2 = 4;$ **b)** $x = 1;$ **c)** $x \in (2,\infty);$

$x = 5;$ **d)** $\begin{cases} x^3 + 1 > 0 \\ x^2 + 2x + 1 > 0 \end{cases},\ \begin{cases} x \in (-1,\infty) \\ x \neq -1 \end{cases},\ x \in (-1,\infty),$

$\log_3\left(x^2 + x + 1\right) = \log_3 7,\ x_1 = 2$ and $x_2 = -3;$ **e)** $x = \sqrt[10]{10};$

f) $x \in (0,\infty);\ \log_4 x + \dfrac{1}{2}\log_4 x = -\log_4 \sqrt{3} \Rightarrow x = \dfrac{1}{\sqrt[3]{x}};$

g) $x \in (-1,\ 0);\ x = -\dfrac{7}{8};$ **h)** $\log_2 \sqrt{\log_2 x} + \log_2 \dfrac{\log_2 x}{2} = 2 \Leftrightarrow$

$\log_2 x\sqrt{\log_2 x} = 8 \Leftrightarrow \log_2 x = 4 \Rightarrow x = 16.$

VII. 45. a) $x_1 = 2$; $x_2 = 2^{-2}$; **b)** $x_1 = 2\sqrt{2}$; $x_2 = -2\sqrt{2}$;
c) $x_1 = 2\sqrt{2}$; $x_2 = -2\sqrt{2}$; **d)** $x_1 = 25$; $x_2 = 5^{-1}$; **e)** $x_1 = 5$;
$x_2 = 10^{-2}$; **f)** $x = 0,99$; **g)** $x \in (0,\infty)$, $\log_a x = t$,
$(3t-2)t^2 = 2t - 3 \Rightarrow (t+1)(3t^2 - 5t + 3) = 0$ $t_1 = -1$, $t_{2,3} \notin \mathbf{R}$.
$x = \dfrac{1}{a}$; **h)** $x_1 = 7$, $x_2 = \dfrac{11}{7}$ **VII. 46. a)** $x \in (-2,\infty)$,
$\log_6 (2+x) + \log_6 (3+x) = 1 \Rightarrow \log_6 (2+x)(3+x) = 1$, $x_1 = 0$ and
$x_2 = -5$ (impossible); **b)** $x \in (4,\infty)$, $\log_2 (2+x)(x-4) = 0$,
$x_1 = 1 + \sqrt{10}$, $x_2 = 1 - \sqrt{10}$ (impossible); **c)** $x \in (0,\infty)$,
$\log_2 x + 2\log_2 x = 6$ $\log_2 x = 2$, $x = 4$; **d)** $x \in (2,\infty)$,
$\log_2 (x-2) = \log_2 \sqrt{2x-1} \Rightarrow x - 2 = \sqrt{2x-1}$, $x = 5$; **e)** $x = 2.5$;
f) $x = 3.5$.

VII. 47. a) $x \in (0,\infty) \setminus \{1\}$, $\dfrac{1}{2}\log_x 5 + \log_x 5 + 1 = \dfrac{1}{4}(\log_x 5)^2 + \dfrac{9}{4}$,

$\log_x 5 = t$, $t^2 - 6t + 5 = 0$, $x = \sqrt[5]{5}$, $x = 5$; **b)** $x \in (0,\infty)$,

$(-\log_3 9 - \log_3 x)^2 + 2\log_3 x - \log_3 27 = 8$, $\log_3 x = t$,

$t^2 + 6t - 7 = 0$, $x_1 = 3$, $x_2 = 3^{-7}$; **c)** $\sqrt{\log_a x} = y$, $y + \dfrac{1}{y} = \dfrac{10}{3}$,

$y_1 = 3 \Rightarrow x = a^9$ and $y_1 = \dfrac{1}{3} \Rightarrow x = a^{\frac{1}{9}}$; **d)** $x = 6$; **e)** $x = 7$;

f) $x = 4$; **g)** $x \in (-0.5 + 0.5\sqrt{33}, \infty)$, $\log_{x-1}(x^3 - 9x + 8) = 3$,
$3x^2 - 12x + 9 = 0$, $x = 6$.
VII. 48. a) $x = \pm\sqrt{\log_2 3}$; **b)** $x_1 = 2$, $x_2 = 3$;

c) $\log_8 (9^{x+1} - 1) - \log_8 (3^{x+1} + 1) = \dfrac{1}{3}$, $\log_8 \dfrac{(3^{x+1} + 1)(3^{x+1} - 1)}{3^{x+1} + 1} = \dfrac{1}{3}$,

$3^{x+1} - 1 = 8^{\frac{1}{3}}$, $x = 0$; **d)** $x = 5$; **e)** $x_1 = 2$, $x_2 = 4$; **f)** $x = 0$.

VII. 49. a) $\begin{cases} x \in (0,\infty) \setminus \{1\} \\ \log_x \sqrt{4x} \geq 0 \end{cases}$, $x \in \left(0, \dfrac{1}{4}\right] \cup (1,\infty)$. From

$\sqrt{\log_x \sqrt{4x}} = -\log_x 4$, $-\log_x 4 \geq 0$, $x \in \left(0, \dfrac{1}{4}\right]$,

$\sqrt{\dfrac{1}{2}\left(\log_x 4 + \log_x x\right)} = -\log_x 4$, $\log_x 4 = t < 0$, $\sqrt{\dfrac{1}{2}(t+1)} = -t$

$2t^2 - t - 1 = 0$, $x = \dfrac{1}{16}$; **b)** $x \in \left(0, \dfrac{1}{3}\right]$, $\sqrt{\dfrac{1}{2}\log_x 3 + \dfrac{1}{2}} \cdot \dfrac{1}{\log_x 3} = -1$

,

$\log_x 3 = -\dfrac{1}{2} \Rightarrow x = \dfrac{1}{9}$ and $\log_x 3 = 1$(impossible); **c)** $x = 16$;

d) $x_1 = 2^{-8}$, $x_2 = 2^{27}$; **e)** 50, $\dfrac{1}{200}$; **f)** $\log_5 x = a$, $x = 5$;

VII. 50. a) $x_1 = 2^{-\frac{4}{27}}$, $x_2 = 2^{\frac{1}{3}}$; **b)** $\log_6 x = t$,

$y^2 + \dfrac{1}{y^2} + 2 \cdot \left(y + \dfrac{1}{y}\right) + \dfrac{3}{4} = 0$, $y + \dfrac{1}{y} = z$, $4z^2 + 8z - 5 = 0$,

$x_1 = \dfrac{1}{36}$, $x_2 = \dfrac{1}{\sqrt{6}}$; **c)** $x_1 = \sqrt[5]{5}$; $x_2 = 5$; **d)** $x = \dfrac{a^4 + 1}{1 - a^4}$; **e)** $x_1 = \dfrac{1}{3}$,

$x_1 = 9$; **f)** $x = a^6$; **g)** $x \in (0,\infty) \setminus \{1\}$, $x_1 = \sqrt{2}$, $x_2 = \sqrt[4]{2}$.

VII. 51. a) $x \in (0,\infty) \setminus \{1\}$,

$\dfrac{1}{\log_3 3 + \log_3 x} - \dfrac{1}{\log_x 3 + \log_x x} + \log_3{}^2 x = 1$, $\log_3 x = t$,

$t^3 + t^2 - 2t = 0$, $x = 3^{-2}$, $x = 3$; **b)** $x_1 = \dfrac{1}{2}$, $x_2 = \dfrac{1}{8}$;

c) $x_1 = 3$, $x_2 = \sqrt{3}$; **d)** $x \in (0,\infty) \setminus \left\{\dfrac{1}{81}, \dfrac{1}{9}, 3\right\}$,

$\dfrac{10}{\log_x 9x} + \dfrac{21}{\log_x 81x} - \dfrac{6}{\log_x \dfrac{x}{3}} = 0$,

$$\frac{10}{2\log_x 3 + \log_x x} + \frac{21}{4\log_x 3 + \log_x x} - \frac{6}{\log_x x - \log_x 3} = 0,$$

$\log_x 3 = t$, $26t^2 - 3t - 5 = 0$, $x_1 = 3^3$, $x_2 = \dfrac{1}{\sqrt[5]{3^{26}}}$, also $x_3 = 1$;

e) $x_1 = a$, $x_2 = a^{\frac{9}{16}}$;

f) $x \in (0,\infty) \setminus \left\{ \dfrac{1}{2}, \sqrt{2}, 2 \right\}$, $\dfrac{1}{\log_x \dfrac{2}{x^2}} + \dfrac{2}{\log_x \dfrac{2}{x}} - \dfrac{8}{\log_x 2x} = 0,$

$\dfrac{1}{t-2} + \dfrac{2}{t-1} - \dfrac{8}{t+1} = 0$, $5t^2 - 22t + 21 = 0$, $\log_x 2 = t$, $x_1 = \sqrt[3]{2}$,

$x_2 = \sqrt[21]{32}$.

Notice: The initial equation has $x_3 = 1$ as a solution, too.

VII. 52. a) $\dfrac{1}{81}$;   **b)** $x^{\frac{\log_9 16}{\log_3 x}} = \left(x^{\log_x 3} \right)^{\log_3 4} = 4,$

$\log_2 3 + \log_2 x = 4$, $x = \dfrac{16}{3}$;

c) $x \in (0,\infty) \setminus \{1\}$, $x^{\log_{10} \sqrt{x}} = 100 \Leftrightarrow \log_{10} x^{\log_{10} \sqrt{x}} = \log_{10} 10^2 \Leftrightarrow$

$\log_{10} \sqrt{x} \log_{10} x = 2 \Leftrightarrow \dfrac{1}{2}(\log_{10} x)\log_{10} x = 2$, $x = 10^{\pm 2}$;

d) $x_1 = 10^{-4}$, $x_2 = 10$;   **e)** $\left(2\log^3 x - 1.5\log x \right)\log x = 0.5,$

$\log^2 x = t \geq 0$, $4t^2 - 3t - 1 = 0$; $x = 10^{\pm 1}$.

VII. 53. a) The equation becomes

$$(7^x)^3 + \left(8^x\right)^3 + \left(9^x\right)^3 = 3 \cdot (7 \cdot 8 \cdot 9)^x.$$

$\dfrac{a+b+c}{3} \geq \sqrt[3]{abc}$ is true for any $a,b,c \geq 0$, the equality holds for

$a = b = c$. Therefore $343^x = 512^x = 729^x$, thus $x = 0$;

b) Medium inequality gives

$9^x + 36^x + 32^x + 32^x \geq 4\sqrt[4]{9^x \cdot 36^x \cdot 32^x \cdot 32^x} =$

$= 4\sqrt[4]{3^{2x} \cdot 3^{2x} \cdot 2^{2x} \cdot 2^{5x} \cdot 2^{5x}} = 4\sqrt[4]{3^{4x} \cdot 2^{12x}} = 4\sqrt[4]{3^{4x} \cdot 8^{4x}} = 4 \cdot 24^x.$

Equality holds when the numbers are equal, then

$9^x = 36^x = 32^x \Rightarrow x = 0.$

VII. 54. a) $(1, 1)$, $(\dfrac{1}{\sqrt[3]{3}}$, $\sqrt[3]{9}$); **b)** $(2, 4)$; **c)** Multiplying both

equations $6^{x+y} = 6^3 \Rightarrow x + y = 3$. Dividing both equations

$\left(\dfrac{2}{3}\right)^{x-y} = \dfrac{3}{2} \Rightarrow x - y = -1$, $(1, 2)$; **d)** $(0, 0)$;

e) $\begin{cases} x^y \cdot x = 27 \\ \dfrac{\left(x^y\right)^2}{x^5} = \dfrac{1}{3} \end{cases} \Rightarrow \begin{cases} x^y = \dfrac{27}{x} \\ \left(x^y\right)^2 = \dfrac{x^5}{3} \end{cases}$, $(3, 2)$; **f)** $(1, 2)$, $(2, 1)$;

g) $2^{x+y} = t$, $3^{x-y} = z$, $\begin{cases} t^2 = 175 + z^2 \\ 5t + 6z = 134 \end{cases}$, $\begin{cases} x + y = 2 \\ x - y = 2 \end{cases}$, $(2, 0)$;

h) $x, y \in (0, \infty)$, $\begin{cases} \left(\sqrt[4]{x} + \sqrt{y}\right)\log x = \dfrac{8}{3}\log y \\ \left(\sqrt[4]{x} + \sqrt{y}\right)\log y = \dfrac{2}{3}\log x \end{cases}$,

$\left(\sqrt[4]{x} + \sqrt{y}\right)^2 = \dfrac{16}{9} \Rightarrow \sqrt[4]{x} + \sqrt{y} = \dfrac{4}{3}$, substituting the last relation in

the first equation $x^{\frac{4}{3}} = y^{\frac{8}{3}}$, $x = y^2$, $\begin{cases} x = y^2 \\ \sqrt[4]{x} + \sqrt{y} = \dfrac{4}{3} \end{cases}$,

$x = \dfrac{16}{81}$, $y = \dfrac{4}{9}$, $x = y = 1$ is a solution for the system, too.

i) $(-2, \infty)$ $x, y \in (0, \infty)$, $a = 4^{\sqrt[3]{x}}$, $b = 3^{\sqrt{y}}$, $\begin{cases} ab = 144 \\ a^2 + b^2 = 337 \end{cases} \Leftrightarrow$

$\begin{cases} ab = 144 \\ a + b = 25 \end{cases}$, $(8, 4)$, $\left(\left(\log_4 9\right)^3, \left(\log_3 16\right)^2 \right)$; **j)** $(8, 3)$, $\left(\left(\log_3 5\right)^3, \right.$

$2 + \left(\log_5 9\right)^4 \right)$; **k)** $(1, 1, 1)$, $(4, 2, \sqrt{2}$).

VII. 55. a) $x \in (0, \infty) \setminus \{1\}$, $\begin{cases} x + y = x^2 \\ 4x - 2y = x^2 \end{cases}$, $(2, 2)$;

b) $x, y \in (0, \infty) \setminus \{1\}$,

(4, 4); **c)** (3, 9); **d)** (2, 4); **e)** (3, 2); **f)** (3, 9), ($\frac{1}{9}$, $\frac{1}{3}$);

g) (3, 27), (27, 3); **h)** $\begin{cases} \sqrt{y}\left(2\sqrt[4]{x}-1\right)=3 \\ \sqrt[4]{x}\left(\sqrt{y}+1\right)=4 \end{cases}$, (1, 9), (6, 1);

i) (2, 3, 4); **j)** (3,4), (4,3); **k)** (3,9); **l)** (3,5), (5,3);

m) $\begin{cases} \log_3 x + \log_3 y = 3 \\ \log_3 y \log_3 x = 2 \end{cases}$, (3,9) or (9,3); **n)** $x, y > 0$,

$\begin{cases} x^2 + y^2 = 10^2 \\ x \cdot 2^2 = y \cdot 3 \end{cases}$, (6,8); **o)** (2, 6), (6, 2); **p)** $\log_a\left(x^{\log_a x} y^{\log_a y}\right) = 2m$,

$x^{\log_a x} y^{\log_a y} = a^{2m}$, $u = x^{\log_a x}$, $v = y^{\log_a y}$, $uv = a^{2m}$,

$u + v = 2\dfrac{p^2 + q^2}{p^2 - q^2} a^m$.

VII. 56. For $x > 1$, let $f(x) = 5^{\log_{10}\left(x^3 - x^2\right)}$. The three equations are
$f(x) = y^2$, $f(y) = z^2$ and $f(z) = x^2$. Since $x^3 - x^2 = x^2(x-1)$ is
increasing, f is an increasing function. If, say, $x < y$, then $y < z$ and $z < x$, yielding a contradiction. Thus, we can only have that $x = y = z$
and so $\log_{10}\left(x^3 - x^2\right) = \log_5 x^2$. Let $2t = \log_5 x^2$ so that $t > 0$,
$x^2 = 5^{2t}$ and so
$x = 5t$. Therefore $5^{3t} - 5^{2t} = 10^{2t} \Rightarrow 5^t - 1 = 4^t \Rightarrow 5^t - 4^t = 1$.

Since $5^t - 4^t = 4^t\left[\left(\dfrac{5}{4}\right)^t - 1\right]$ is an increasing function of t, we see

that the equation for t has a unique solution, namely $t = 1$.
Therefore $x = 5$.
VII. 57. a) $x \in [1, 2]$; **b)** $x \in (0, \infty)$; **c)** $x \in [0, 4]$;
d) $x \in \left(-0.\overline{3}, \infty\right)$; **e)** $3^{\sqrt[4]{x}-\sqrt{x}} = t$, $9t^2 + 8t - 1 \geq 0$, $x \in [0, 16]$;
f) $x \in (0, 1) \cup (1, 2)$; **g)** $x \in [-1,3)$; **h)** $x \in (-\infty,0) \cup (1,\infty)$;
i) $x \in (-\infty, 1]$; **j)** $x^{3x} < x^{2x^2 - 2x + 1}$ or $x^{(2x-1)(x-2)} > 1$; **I)** $x \in (0, 1)$,
$(2x-1)(x-2) < 0$, then $x \in (0.5, 1)$; **II)** $x \in (1, \infty)$,
$(2x-1)(x-2) > 0$, then $x \in (2, \infty)$. Finally, $x \in (0.5, 1) \cup (2, \infty)$.

VII. 58. a) $x \in (-\infty, -4) \cup (-1, \infty)$; **b)** $x \in (2^{-28}, 1)$;
c) j) $x \in (-1+\log_2 3, 2)$; **d)** $x \in (-1, 0) \cup [1, \infty)$;

e) $x \in (-\infty, 0) \cup \left(0, \dfrac{1}{2}\right) \cup \left(1, \dfrac{3}{2}\right) \cup (3, \infty)$;

f) $\dfrac{3}{\log_2 x - 1} - \dfrac{3}{\log_2 x - 2} < \dfrac{2\log_2 x}{\log_2 x - 2}$; $x \in (0, 2) \cup (4, \infty)$;

g) $2\log_3 \log_4 \dfrac{2x-1}{x-1} < 0$, $x \in (-\infty, 0) \cup \left(\dfrac{3}{2}, \infty\right)$;

h) $x \in (3, 4] \cup [6, \infty)$; **i)** $x = -1$; **j)** $x \in \left(\dfrac{1}{8}, \dfrac{1}{4}\right) \cup (4, 8)$.

VII. 59. For $x, y > 1$, then $\log_x y > 0$.

$\left(\log_c a + \log_c b\right) \geq 2\sqrt{\log_c a \log_c b} \Leftrightarrow$

$\sqrt{\log_a c \cdot \log_b c}\left(\log_a c + \log_b c\right) \geq 2 \Leftrightarrow (ab)^{\sqrt{\log_a c \cdot \log_b c}} \geq c^2$.

VII. 60. a) Notice that $x = 2$ is a solution of the equation.

$\left(\dfrac{3}{5}\right)^x + \left(\dfrac{4}{5}\right)^x = 1$. Assume that there is another solution $x_1 < 2$,

therefore $1 = \left(\dfrac{3}{5}\right)^{x_1} + \left(\dfrac{4}{5}\right)^{x_1} > \left(\dfrac{3}{5}\right)^2 + \left(\dfrac{4}{5}\right)^2 = 1$, contradiction.

Similarly for the interval $(2, \infty)$; **b)** $x = 4$;

c) $7 \cdot \left(\dfrac{4}{5}\right)^{2x-1} + 38 \cdot \left(\dfrac{1}{5}\right)^{2x-1} = 6$; $x = \dfrac{3}{2}$.

VII. 61. a) Consider the function $f : \mathbf{R} \to (0, \infty)$

$f(x) = \log_3(2^x + 1)$. The function is invertible (bijection),

$f^{-1} : (0, \infty) \to \mathbf{R}$ $f^{-1}(x) = \log_2(3^x - 1)$. The initial equation is
equivalent to the equation (1) $f(x) = f^{-1}(x)$. Since the two
graphs, are symmetric to the first bisector $y = x$, the equation (1) is
equivalent to one of the equations $f(x) = x$ or $f^{-1}(x) = x$,
$\log_3(2^x + 1) = x \Rightarrow 2^x + 1 = 3^x \Rightarrow$

(2) $\left(\dfrac{2}{3}\right)^x + \left(\dfrac{1}{3}\right)^x = 1$. The functions $\left(\dfrac{2}{3}\right)^x$ and $\left(\dfrac{1}{3}\right)^x$ are

monotonous decreasing also the sum $\left(\dfrac{2}{3}\right)^x + \left(\dfrac{1}{3}\right)^x$ is monotonous

decreasing. So the equation $\left(\dfrac{2}{3}\right)^x + \left(\dfrac{1}{3}\right)^x = 1$ has maximum one

solution , $x = 1$;

b) $x = 1$; **c)** $f : \mathbf{R} \to \mathbf{R},\ f(x) = \dfrac{e^x - e^{-x}}{2}$ is an odd function and

$f^{-1}(x) = \ln\left(x + \sqrt{x^2 + 1}\right)$; $x = 0$.

VII. 62. We have $2^{\log_5\left(2^{\log_5 x} + 3\right)} + 3 = x$. Consider the function
$f : (0, \infty) \to \mathbf{R}$ $f(x) = 2^{\log_5 x} + 3$ which, obviously is a strict increasing
function. Then the given equation becomes $f(f(x)) = x$. Denoting

$t = \log_5 x$ we get $2^t + 3 = 5^t \Rightarrow \left(\dfrac{2}{5}\right)^t + 3 \cdot \left(\dfrac{1}{5}\right)^t = 1$ with the solution

$t = 1$, because the left side function is decreasing, thus injective.
Finally $x = 5$ is the unique solution.

VII. 63. Denote $\alpha = \lg a, \beta = \lg b, \gamma = \lg c$ and we have

$\alpha \le \beta \le \gamma < 0,\ A = \dfrac{\alpha}{\beta} + \dfrac{\beta}{\gamma} + \dfrac{\gamma}{\alpha}$, and $B = \dfrac{\beta}{\alpha} + \dfrac{\gamma}{\beta} + \dfrac{\alpha}{\gamma}$

Therefore $A - B = \dfrac{(\alpha - \beta)(\beta - \gamma)(\gamma - \alpha)}{\alpha\beta\gamma} \le 0 \Rightarrow A \le B$.

VII. 64. The equation becomes $5^{[x]} + 5^{\{x\}} = 6$. For $x < -1 \Rightarrow$
$[x] < -1 \Rightarrow 5^{[x]} + 5^{\{x\}} < 5^{-1} + 5 < 6$; For $x \ge 2 \Rightarrow [x] \ge 2 \Rightarrow$
$5^{[x]} + 5^{\{x\}} \ge 5^2 + 1 > 6$; For $x \in [1;2) \Rightarrow [x] = 1 \Rightarrow 5^{\{x\}} = 1 \Rightarrow \{x\} = 0$
$\Rightarrow x = 1$; For $x \in [0;1) \Rightarrow [x] = 0 \Rightarrow 5^{\{x\}} = 5 \Rightarrow \{x\} = 1 \Rightarrow x \in \varnothing$;
For $x \in [-1;0) \Rightarrow [x] = -1 \Rightarrow 5^{\{x\}} = \dfrac{29}{5} > 5 \Rightarrow \{x\} > 1 \Rightarrow x \in \varnothing$;

Therefore the solution is $x = 1$.

Chapter VIII

ARITHMETIC AND GEOMETRIC PROGRESSIONS

VIII. 1. a) 2, 4, 6, 8, 10; **b)** 9, 10, 11, 12, 13; **c)** 3, 6, 9, 12, 15;
d) 2, 4, 8, 16, 32; **e)** 3, 5, 7, 9, 5; **f)** 11, 22, 33, 44, 55;

g) 4, 7, 10, 13, 16; **h)** -1, 2, 7, 14, 23; **i)** $-1, \dfrac{1}{2}, -\dfrac{1}{6}, \dfrac{1}{24}, -\dfrac{1}{120}$;

j) $0, \dfrac{3}{2}, -\dfrac{8}{3}, \dfrac{15}{4}, -\dfrac{24}{5}$; **k)** $\dfrac{1}{2}, \dfrac{2}{5}, \dfrac{3}{10}, \dfrac{4}{17}, \dfrac{5}{26}$; **l)** $\dfrac{3}{2}, \dfrac{9}{4}, \dfrac{16}{9}, \dfrac{25}{16}, \dfrac{36}{25}$;

m) $1, \dfrac{1}{2}, \dfrac{1}{3}, \dfrac{1}{4}, \dfrac{1}{5}$. **VIII. 2. a)** $(-1)^{n+1}\dfrac{1}{n}$; **b)** $a_n = \dfrac{1}{n^2+1}$;

c) $a_n = \dfrac{2n \cdot 2(n+1)}{(2n-1)(2n+1)}$; **d)** $a_n = \dfrac{2 \cdot 3^{n-1}}{5^n}$;

e) $b_1 = 1,\ b_2 = \dfrac{1}{2}, b_3 = \dfrac{1}{3}, b_4 = \dfrac{1}{4},\ldots,\ b_n = \dfrac{1}{n}$; **f)** $x_n = n!$;

g) $x_n = \dfrac{n+1}{n+2}$. **VIII. 3. a)** $t_n = 2n-1$; **b)** $t_n = 6n$; **c)** $t_n = 5n-1$

d) $t_n = 10^n$; **e)** $t_n = \dfrac{1}{2} \cdot 10^n$. **VIII. 4. a)** 32, 36; **b)** 19, 37; **c)** 12,

77; **d)** 4, -32; **e)** $\dfrac{4}{7}, \dfrac{7}{13}$; **f)** $\dfrac{28}{27}, \dfrac{82}{81}$. **VIII. 5.** $a_n = n(2n-1)$.

VIII. 6. $S_n = n(2^{n-1} - n + 1)$. **VIII. 7. a)** 3, 3, 3; **b)** 2, 8, 14;
c) -1, 3, 7; **d)** 2, 5, 5; **e)** 3, 5, 7. **VIII. 8. a)** -4; **b)** 16; **c)** 32; **d)** 40.
VIII. 9. a) i) $n^2 - 3n + 2$; **ii)** $2n - 2$; **b) i)** $3n^2 - 11n + 8$;
ii) $6n - 8$; **c) i)** $2^{n-1} - 1$; **ii)** 2^{n-1}; **d) i)** $2n^2 - 7n + 5$; **ii)** $4n - 5$;
e) i) $2(3^{n-1} - 1)$; **ii)** $4(3^{n-1})$; **f) i)** $n^2 + 1$; **ii)** $2n - 1$.
VIII. 10. a) i) $n^2 - 3n + 2$; **ii)** $2n - 2$; **b) i)** $3n^2 - 11n + 8$; **ii)** $6n - 8$;
c) i) $2^{n-1} - 1$; **ii)** 2^{n-1}; **d) i)** $2n^2 - 7n + 5$; **ii)** $4n - 5$;
e) i) $2(3^{n-1} - 1)$; **ii)** $4(3^{n-1})$; **f) i)** $n^2 - 4n + 3$; **ii)** $6n - 3$.
VIII. 11. a) i) 79, $4n - 1$; **ii)** 820, $2n^2 + n$; **b)** 125.
VIII. 12. a) 72, 152; **b)** t_{51}.
VIII. 13. a) 20, 14, -4; **b)** 2, 23, 30; **c)** 17, 27, 32;
d) 10, 17, 31; **e)** 1, -9, -14; **f)** 6, -3, -12.

VIII. 14. $-10, -6, -2$. **VIII. 15. a)** no; **b)** no; **c)** a_{12}; **d)** a_{37}.

VIII. 16. a) ii; **b)** vi; **c)** v; **d)** iii.

VIII. 17. $a_n = 2^n + 1$, for any $n > 1$. **VIII. 18.** $a_n = n^2$.

VIII. 19. $-3, 4, 11, 18, \ldots$ **VIII. 20.** $-\dfrac{57}{2}$. **VIII. 21.**

$n = 1: \qquad a_2 - a_1 = 1^2 - 1 + 1$

$n = 2: \qquad a_3 - a_2 = 2^2 - 2 + 1$

$\ldots\ldots\ldots\ldots\ldots\ldots\ldots\ldots\ldots\ldots$

$n = 99: \qquad a_{100} - a_{99} = 99^2 - 99 + 1$. Adding all these 99 equalities

$a_{100} - a_1 = 1^2 + 2^2 + \ldots + 99^2 - (1 + 2 + \ldots + 99) + 99 = 323499$;

$1 + 2 + \ldots + n = \dfrac{n(n+1)}{2}$; $1^2 + 2^2 + \ldots + n^2 = \dfrac{2(n+1)(2n+1)}{2}$.

$a_{100} = 323{,}500$. **VIII. 22. a)** $2, -1, -4, -7, -10$, $a_n = -3n + 5$;

b) $a_1 = 2$, $a_2 = 4$, $a_3 = 6$, $a_4 = 8$, $a_5 = 10$, $a_n = 2n$;

c) $a_1 = 21$, $a_2 = 22\dfrac{1}{2}$, $a_3 = 24$, $a_4 = 25\dfrac{1}{2}$, $a_5 = 27$,

$a_n = \dfrac{3}{2}(n+13)$; **d)** $a_1 = 4$, $a_2 = 9$, $a_3 = 14$, $a_4 = 19$, $a_5 = 24$,

$a_n = 5n - 1$. **VIII. 23.** 1380. **VIII. 24. a)** 31; **b)** 37; **c)** 41; **d)** 21;

e) 32; **f)** 100. **VIII. 25. a)** 179; **b)** 2760. **VIII. 26.** $a_{10} = 30$,

$S_{12} = 570$. **VIII. 27.** 85, 80, 75. **VIII. 28.** $a_1 = 1$, $d = 4$, $n = 26$,

$S_{26} = 1326$. **VIII. 29.** 735. **VIII. 30.** 2. **VIII. 31.**

$2x^2 = (x - 2) + (x + 6)$, $x_1 = -1$, $x_2 = 2$. **VIII. 32.** $x - 5y + 2 = 0$.

VIII. 33. $a_{50} = 2a_{49} - a_{48} = 2\sqrt{27 - 10\sqrt{2}} - \sqrt{18 - 8\sqrt{2}} =$

$= 2(5 - \sqrt{2}) - (4 - \sqrt{2}) = 6 - \sqrt{2}$.

VIII. 34. a) 345; **b)** 14.5; **c)** -2625.50.

VIII. 35. $\begin{cases} a_1 + a_5 = 8 \\ a_3 \cdot a_5 = 80 \end{cases} \Rightarrow \begin{cases} 2a_1 + 4d = 8 \\ (a_1 + 2d)(a_1 + 4d) = 80 \end{cases}$, $d = 8$,

$a_1 = -12$, $a_{20} = 140$, $S_{20} = 1280$. **VIII. 36.** $18 + 22 + 26 + \ldots$

VIII. 37. $a_1 = 0$, $d = 14/3$, and $S_{10} = 210$. **VIII. 38.** $a_1 = 4$, $d = 3$.

VIII. 39. a) $a_1 = 3$, $d = 5$; **b)** $a_1 = 3$, $d = 5$ or $a_1 = 13$, $d = -5$.

VIII. 40. 5, 7, 9, …, 23 and reverse. **VIII. 41.** $a_1 = a - 3x$,

$a_2 = a - x$, $a_3 = a + x$, $a_4 = a + 3x$, $a = 9$ and $x = \pm 3$.

VIII. 42. 5. **VIII. 43.** 1.5, 405. **VIII. 44.** $a_n = S_n - S_{n-1} = 6n - 8$.

VIII. 45. a) $a_1 = 11$ and $d = 10$; **b)** $a_1 = 7$ and $d = 6$.

VIII. 46. Let $a - d$, a, and $a + d$ be the numbers, where d is the ratio of the arithmetic progression. Then we have

$$\begin{cases} (a-d) + a + (a+d) = 6 \\ (a-d)^2 + a^2 + (a+d)^2 = 62 \end{cases}.$$ From the first equation we obtain

$a = 2$ and the second equation becomes $d^2 = 25$. Therefore the numbers are $-3, 2, 7$ or $7, 2, -3$.

VIII. 47. $x_1 = x_2 = 1$. **VIII. 50.** $r = \dfrac{\alpha}{2}$, $a_n = \alpha(2n - 1)$.

VIII. 51. $2(a^2 + b^2) = (a + b)^2 + (a - b)^2$.

VIII. 52. Prove that if the numbers

$\dfrac{1}{a + 2b}, \dfrac{1}{3c + a}, \dfrac{1}{3c + 2b} (a \neq -2b \neq -3c)$ are in arithmetic

progression then the numbers $a^2, 4b^2, 9c^2$ are also in arithmetic progression.

VIII. 53. $m_a^2 = \dfrac{2(b^2 + c^2) - a^2}{4}$. **VIII. 55.** $m = 3$.

VIII. 57. $\dfrac{1 + 2 + \ldots + (x - 1)}{x} = 10$. $1 + 2 + \ldots + (x - 1)$ is an arithmetic

progression with $x - 1$ terms, where $a_1 = 1$, $d = 1$ and $a_n = x - 1$

then $S_n = \dfrac{(a_1 + a_n)n}{2} = \dfrac{(1 + x - 1)(x - 1)}{2}$; $\dfrac{x(x - 1)}{2} = 10x$, $x_2 = 21$.

VIII. 58. $a_1 = 3$; $d = 4$; $x = a_n$, $a_n = a_1 + (n - 1)d = 3 + (n - 1)4$.

$S_n = \dfrac{(a_1 + a_n)n}{2}$, then $210 = \dfrac{(3 + x)n}{2}$; $x = 39$.

VIII. 59. $a_{2k} = \dfrac{a_{2k-1} + a_{2k+1}}{2} \geq \sqrt{a_{2k-1}a_{2k+1}} \Rightarrow \dfrac{a_{2k-1}}{a_{2k}} \leq \dfrac{\sqrt{a_{2k-1}}}{\sqrt{a_{2k+1}}}$,

$k \geq 1$ and multiplying these inequalities side by side we obtain the result.

VIII. 60. a) Use the identity $\dfrac{1}{\sqrt{a_k}+\sqrt{a_{k+1}}}=\dfrac{\sqrt{a_{k+1}}-\sqrt{a_k}}{d}$, or by

induction. **VIII. 61.** $S_1=\dfrac{1}{d}\left(\dfrac{1}{a_1}-\dfrac{1}{a_n}\right),S_2=\dfrac{1}{d}\left(\dfrac{1}{a_1^2}-\dfrac{1}{a_n^2}\right),$

$S_3=\dfrac{a_1a_2}{a_{n-1}a_n}\cdot\dfrac{a_{n-1}a_n+d^2}{a_1a_2+d^2};\ S_4=\dfrac{1+(n-1)\sqrt{q}}{\sqrt{q}-1}.$

VIII. 62. a) Setting

$x_n=a_na_{n+1}...a_{n+k}\Rightarrow x_{n+1}-x_n=(k+1)da_{n+1}a_{n+2}...a_{n+k}$, d is the ratio.

$A_n=a_1a_2...a_k+\left[(x_2-x_1)+(x_3-x_2)+...+(x_{n+1}-x_n)\right]\dfrac{1}{(k+1)r}$

therefore $A_n=a_1a_2...a_k+\dfrac{a_{n+1}a_{n+2}...a_{n+k-2}-a_1a_2...a_{k+1}}{(k+1)r};$

b) $y_n=\dfrac{1}{a_na_{n+1}...a_{n+k-2}}$ using y_n-y_{n+1} we obtain

$B_n=1\dfrac{1}{(k-1)d}\left(\dfrac{1}{a_1a_2...a_{k-1}}-\dfrac{1}{a_{n+1}a_{n+2}...a_{n+k-1}}\right)$, d is the ratio of the

progression.

VIII. 64. a) $b_1=2,\ b_2=6,\ b_3=18,\ b_4=54,\ b_5=162;$

b) $b_1=3,\ b_2=-6,\ b_3=12,\ b_4=-24,\ b_5=48;$ **c)** $b_1=\sqrt{2},\ b_2=\sqrt{6}$

, $b_3=3\sqrt{2}$, $b_4=3\sqrt{6}$, $b_5=9\sqrt{2};$ **d)** $b_1=5,\ b_2=10,\ b_3=20,\ b_4=$

$40,\ b_5=80.$ **VIII. 65.** $r_{a_n}=\dfrac{1}{2}$ and $r_{b_n}=3.$

VIII. 66. $A_n(-n,\ n),\ A_1A_2:\ y=-x.$ **VIII. 67.** $y_n=\dfrac{10}{3}2^{n-1}$ and

$x_n=\dfrac{5\cdot2^n+3}{5\cdot2^n-3}.$ **VIII. 68.** $b_4=b_1\cdot r^3\Rightarrow r=\sqrt{3}.$ Therefore

$b_2=\sqrt{6},b_3=3\sqrt{2},b_5=9\sqrt{2},b_6=9\sqrt{6}.$

VIII. 69. $\begin{cases} x^2=y\sqrt{3} \\ y^2=x\cdot18\sqrt{2} \end{cases}$, $x=3\sqrt{2}$ and $y=6\sqrt{3}.$

VIII. 70. $b_5 = -81$, $S_{10} = \dfrac{1-3^{10}}{2}$. **VIII. 71.** $\dfrac{b_1 r + b_1 r^2}{b_1 r^2 + b_1 r^3} = \dfrac{14}{28}$,

$r = 2$, $b_1 = \dfrac{7}{3}$, $S_{10} = \dfrac{7}{3}\left(2^{10} - 1\right)$. **VIII. 72.** 15, 45. **VIII. 73.** 3.

VIII. 74. a) $b_1 = 1$, $r = 3$; $b_1 = -81$, $r = \dfrac{1}{3}$; **b)** $b_1 = 2$, $r = 3$.

VIII. 75. $\dfrac{1+\sqrt{5}}{2}$. **VIII. 76.** $\dfrac{-1+\sqrt{5}}{2}$.

VIII. 79. $\begin{cases} b_1 + b_2 + b_3 = 21 \\ b_1 b_2 b_3 = 64 \end{cases}$; $\begin{cases} b_1\left(1+r+r^2\right) = 21 \\ b_1^3 r^3 = 64 \end{cases}$; $1, \dfrac{1}{4}, \dfrac{1}{16}$ or

16,4,1. **VIII. 80.** $\dfrac{1}{2}$. **VIII. 81.** $b_1, b_2, b_3, \ldots, b_n$

$\begin{cases} b_1 b_2 b_3 = 3\sqrt{3} \\ b_1^3 + b_2^3 + b_3^3 = 28 + 3\sqrt{3} \end{cases} \Rightarrow \begin{cases} b_1^3 r^3 = 3\sqrt{3} \\ b_1^3\left(1+r^3+r^6\right) = 28 + 3\sqrt{3} \end{cases}$;

$3\sqrt{3} r^6 - 28 r^3 + 3\sqrt{3} = 0$,

I) $r = \sqrt{3}$ and $b_1 = 1$, $b_2 = \sqrt{3}$, $b_3 = 3$;

II) $r = \dfrac{1}{\sqrt{3}}$ and $b_1 = 27$, $b_2 = 9\sqrt{3}$, $b_3 = 9$.

VIII. 82. $a = 5$, $b = 10$, $c = 20$, $d = 40$.

VIII. 83. $b^2 = ac$, $a+b+c = 13$, $(b+1)^2 = (a+1)(c-1)$; 1, 3, 9;

$\dfrac{-55}{7}$, $\dfrac{121}{7}$, $\dfrac{25}{7}$. **VIII. 84.** 4 or $-\dfrac{85}{26}$. **VIII. 85.** $\dfrac{2}{3}$.

VIII. 86. a) 4; **b)** -3; **c)** 5; **d)** $-\dfrac{8}{3}$; **e)** 119; **f)** $\dfrac{500}{67}$.

VIII. 87. \$165,984. **VIII. 88.** $b^2 = ac$, $b + 8 = \dfrac{a+c}{2}$,

$(b+8)^2 = a(c+64)$; $(4, 12, 64)$ or $\dfrac{4}{9}, -\dfrac{20}{9}, \dfrac{100}{9}$. **VIII. 89.**

$\begin{cases} b_1 - b_4 = b_2 \\ b_3 - 6 = b_4 \end{cases} \Rightarrow \begin{cases} b_1(1-r) = 54 \\ b_1 q^2(1-r) = 6 \end{cases} \Rightarrow \dfrac{b_1(1-r)}{b_1 q^2(1-r)} = \dfrac{54}{6} \Rightarrow r = \pm\dfrac{1}{3}$.

Therefore $q = \dfrac{1}{3}$, $b_1 = 81$, or $q = -\dfrac{1}{3}$, $b_1 = \dfrac{81}{2}$.

VIII. 90. The sum of the first nine terms of an arithmetic progression is equal to 9/2 the sum of the first and ninth terms, from which it is seen that the ninth term is 81. Let r be the common ratio of the geometric progression whose first term is 1 and whose ninth term is 81. Then $r^8 = 81$, whence $r = \pm\sqrt{3}$. The sum of the first twenty terms of the geometric progression is $\dfrac{1}{2}\left(3^{10}-1\right)\left(\pm\sqrt{3}+1\right)$.

VIII. 91. 32. **VIII. 92.** $a_1, a_2, a_3, \ldots, a_n; d, \ b_1, b_2, b_3, \ldots, b_n; r.$

$a_1 + a_2 + a_3 + a_4 + a_5 = 62$ and $a_5 = b_1$, $a_8 = b_2$, and $a_{11} = b_{10}$.

$a_1 + (a_1 + d) + (a_1 + 2d) + (a_1 + 3d) + (a_1 + 4d) = 62$ and

$a_1 + 4d = b_1$, $a_1 + 7d = b_1 r$, $\ a_1 + 10d = b_1 r^9$. $b_1 = 2$ or $b_1 = \dfrac{62}{5}$.

VIII. 93. $b_1 + b_2 + b_3 = 65$ and $\log_{15} b_1 + \log_{15} b_2 + \log_{15} b_3 = 3$,

(5, 3) and (45, 1/3). **VIII. 94.** $b_3 = 2$, $b_5 = 6$, $S_9 = \dfrac{2\left(1 - 81\sqrt{3}\right)}{3\left(1 - \sqrt{3}\right)}$.

VIII. 95. $2a_1 + 9b_1 = 31$, $b_1(1 + a_1) = 9$, $a_1 = \dfrac{25}{2}, b_1 = \dfrac{2}{3}$ or

$a_1 = 2$, $b_1 = 3$.

VIII. 96. $\begin{cases} b_1 + b_5 = 164 \\ b_3 \cdot b_5 = 2916 \end{cases}$; $r = 3$; $b_1 = 2$, $b_{10} = 2 \cdot 3^9$.

VIII. 98. We have $\quad S = 1 + 3 + 3^2 + 3^3 + \ldots + 3^{98} + 3^{99} +$
$$+ 3 + 3^2 + 3^3 + \ldots + 3^{98} + 3^{99} +$$
$$+ 3^2 + 3^3 + \ldots + 3^{98} + 3^{99} +$$
$$\ldots\ldots\ldots\ldots\ldots\ldots\ldots\ldots\ldots\ldots\ldots$$
$$+ 3^{98} + 3^{99} +$$
$$+ 3^{99} \quad .$$

Therefore

$$S = \frac{1\left(3^{100}-1\right)}{3-1} + \frac{3\left(3^{99}-1\right)}{3-1} + \frac{3^2\left(3^{98}-1\right)}{3-1} + \ldots + \frac{3^{98}\left(3^2-1\right)}{3-1} + \frac{3^{99}\left(3^1-1\right)}{3-1} =$$

$$= \frac{1}{2}\left[\left(3^{100}-1\right) + \left(3^{100}-3\right) + \left(3^{100}-3^2\right) + \ldots + \left(3^{100}-3^{98}\right) + \left(3^{100}-3^{99}\right)\right] =$$

$$= \frac{1}{2}\left[100 \cdot 3^{100} - \left(1 + 3 + 3^2 + \ldots + 3^{99}\right)\right] = \frac{1}{2}\left(199 \cdot 3^{100} + 1\right).$$

VIII. 99. $3 = \dfrac{10-1}{3}$, $\quad 33 = \dfrac{10^2 - 1}{3}$, ..., $\underbrace{33...3}_{n \text{ times}} = \dfrac{10^n - 1}{3}$;

$S_n = \dfrac{10\left(10^n - 1\right)}{27} - \dfrac{n}{3}$. **VIII. 100.** $S_1 = \dfrac{n q^{n+1}}{q-1} - \dfrac{q^{n+1} q}{(q-1)^2}$;

$S_2 = 2\dfrac{q^{n+1} - q}{(q-1)^3} - \dfrac{(2n-1)q^{n+1} + q}{(q-1)^2} - \dfrac{n^2 q^{n+1}}{q-1}$.

VIII. 101. $S_1 = \dfrac{\left(a^{2n} - 1\right)\left(a^{2n+2} - 1\right)}{a^{2n}\left(a^2 - 1\right)} + 2n$;

$S_2 = \dfrac{\left(a^{2n} - 1\right)\left(a^{2n+2} - 1\right)}{a^{2n}\left(a^2 - 1\right)} - 2n$.

VIII. 102. $A_n = \dfrac{n-1}{\sqrt{r} - 1}$, b) $B_n = b_1 \dfrac{r^{n+1} - r}{r^2 - 1}$.

VIII. 103. $A = b_1 \dfrac{r^n - 1}{r - 1}$, $\quad B = \dfrac{1}{b_1} \dfrac{r^n - 1}{r^{n-1}(r-1)}$, $\quad \dfrac{A}{B} = b_1^2 r^{n-1}$,

$C = \left(b_1 r^{\frac{n-1}{2}}\right)^n = \left(b_1^2 r^{n-1}\right)^{\frac{n}{2}} = \left(\dfrac{A}{B}\right)^{\frac{n}{2}}$. **VIII. 104.** $a_1 = x^4 + 1$,

$r = \sqrt{x^4 + 1}$, $S = \dfrac{n^2 + 3n + 1}{4}$.

VIII. 105. $b_1^2 \cdot (1 - r^2 + r^4 - r^6 + ... + r^{4n-4} - r^{4n-2})(1 - r^{4n}) =$

$b_1^2 \cdot \dfrac{r^{4n} - 1}{-r^2 - 1} \cdot (1 - r^{4n}) = b_1^2 \cdot \dfrac{(1 - r^{4n})^2}{1 + r^2} \geq 0$.

VIII. 107. $S_1 = \dfrac{1}{b_1^p \left(q^p + 1\right)}\left(1 + \dfrac{1}{q^p} + ... + \dfrac{1}{q^{2p}}\right)$,

$S_2 = \dfrac{1}{b_1^p \left(q^p - 1\right)}\left(1 + \dfrac{1}{q^p} + ... + \dfrac{1}{q^{2p}}\right)$, therefore $\dfrac{S_1}{S_2} = \dfrac{q^p - 1}{q^p + 1}$,

whre q is the ratio lf the geometric progression.

VIII. 108. Let $b_1, b_2, ..., b_n$ be a geometric progression

and $r \in \mathbf{R} \setminus \{0,1\}$ the ratio. Use $S_n = b_1 \dfrac{r^n - 1}{r - 1}$.

VIII. 110. For $n = 2$ we have $2\left(\sqrt{x_1} + \sqrt{x_2}\right) = 3\sqrt{x_2} \Rightarrow x_2 = 4a$,

for $n = 3$ we have $2\left(\sqrt{x_1} + \sqrt{x_2} + \sqrt{x_3}\right) = 4\sqrt{x_3} \Rightarrow x_3 = 9a$.

Next, using mathematical induction $x_n = n^2 a$. Therefore

$$\sum_{k=1}^{n} x_k = a\left(1^2 + 2^2 + \ldots + n^2\right) = \frac{n(n+1)(2n+1)a}{2}.$$

VIII. 111. $|x_0| \le 2p \Rightarrow \left|\dfrac{x_0}{2p}\right| \le 1$. If we denote $y_0 = \dfrac{x_0}{2p}$ in this case

there exists a real number $\alpha \in \left[-\dfrac{\pi}{2}, \dfrac{\pi}{2}\right]$ such that $y_0 = \sin \alpha$.

Denoting $y_1 = \dfrac{x_1}{2p}$ we get $y_1 = \sin 3\alpha$ and using mathematical

induction, we get $y_n = \sin 3^n \alpha$. Finally $x_n = 2p \sin 3^n \alpha$ where

$\alpha = \arcsin \dfrac{x_0}{2p}$. **VIII. 112.** The equation $r^2 - \dfrac{5}{7}r - \dfrac{2}{7} = 0$

has the roots $r_1 = 1$ and $r_2 = -\dfrac{2}{7}$. Therefore $x_n = A \cdot 1^n + B \cdot \left(-\dfrac{2}{7}\right)$,

$\forall n \ge 2$.

For $n = 0$ and 1 $\quad \begin{cases} A - \dfrac{2}{7}B = b \\ A + B = a \end{cases}$. Thus $A = \dfrac{2a + 7b}{9}$ and

$B = \dfrac{7(a-b)}{9}$. Consequently $x_n = \dfrac{2a + 7b}{9} + \dfrac{7(a-b)}{9} \cdot \left(-\dfrac{2}{7}\right)^n$.

Chapter IX

POLYNOMIALS AND ALGEBRAIC EQUATIONS

IX. 1.

	x^4	x^3	x^2	x^1	x^0
-1	1	1	2	0	1
	1	0	2	-2	3

$P(x) = x^4 + x^3 + 2x^2 + 1 = (x^3 + 2x - 2)(x + 1) + 3$.

IX. 2. a) $q(x) = x^2 + 5x + 3$, $r = 0$;

b) $q(x) = 2x^3 + 9x^2 + 22x + 63$, $r = 190$;

c) $q(x) = 3x^3 + 9x^2 + x + 4$, $r = 1$;

d) $q(x) = x^3 - 2x^2 + 7x - 13$, $r = 16$;

e) $q(x) = 2x^2 + 2x - 4.5$, $r = 13.5$;

f) $q(x) = x^4 - 3x^3 + 6x^2 - 9x + 21$, $r = -44$;

g) $q(x) = x^4 - x^3 - x^2 + 2x + 2$, $r = -3$;

h) $q(x) = x^7 - x^6 - 3x^5 + 3x^4 + 3x^3 - 3x^2 + x - 1$, $r = 3$.

IX. 3. $f(1) = 0$. **IX. 4.** 13. **IX. 5. a)** $P(1) = 0$; **b)** The quotient is $x^2 - x - 6$ and the reminder is zero; **c)** $x_1 = -2$, $x_2 = 1$, $x_3 = 3$.

IX. 6. $f(x) = 2(x^2 - 1)(x + 3)(x - 2)$.

IX. 7. $f(x) = 3(x^2 - 1)(x + 2)$.

IX. 8. $a(a + 1)(a + 2) = -1320$, -12, -11, -10.

IX. 9. $(x - 1)(x - 2)(x + 3) = 42$, 3, 2 and 7.

IX. 10. $V(r) = \frac{4}{3}\pi(r^3 + 12r^2 + 48r + 64) = \frac{4}{3}\pi(r + 4)^3$. Therefore the radius is $(r + 4)$. **IX. 11.** $(x - 5)(x^2 - 1)$.

IX. 12. $P(x) = x(x-2)(x-3)$, $x_1 = 0$, $x_2 = 2$, $x_3 = 3$.

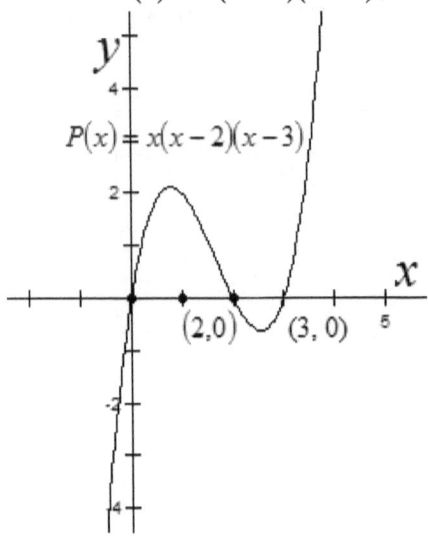

IX. 13. a)

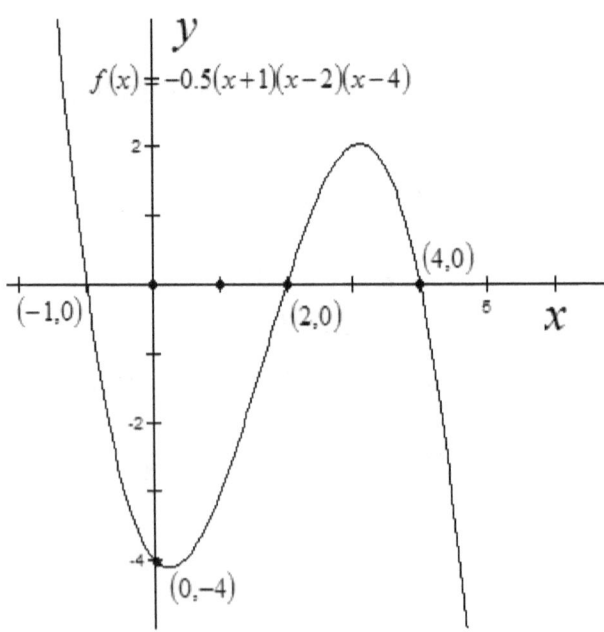

b)

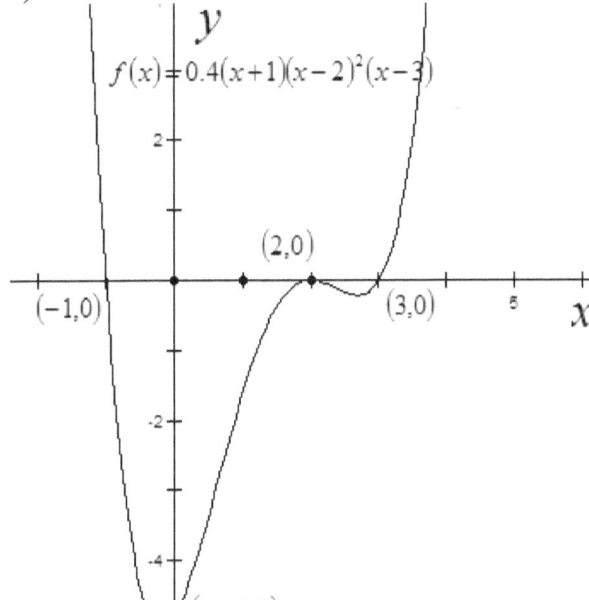

$f(x) = 0.4(x+1)(x-2)^2(x-3)$

$(-1,0)$ $(2,0)$ $(3,0)$

$(0,-4.8)$

c)

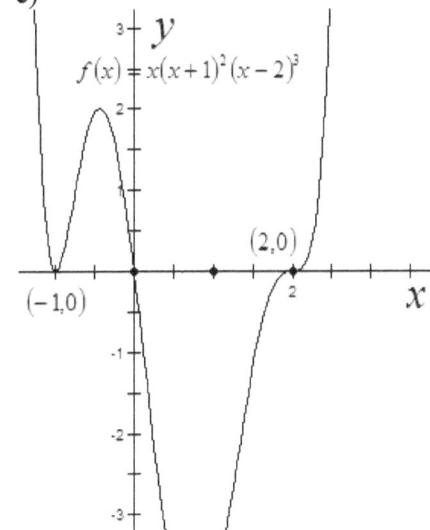

$f(x) = x(x+1)^2(x-2)^3$

$(-1,0)$ $(2,0)$

IX. 14. $f(x) = (x-2)(x-1)(x+2)$.

IX. 15. a) Vertical stretch by factor 2; Horizontal translation of 2 units to the right.

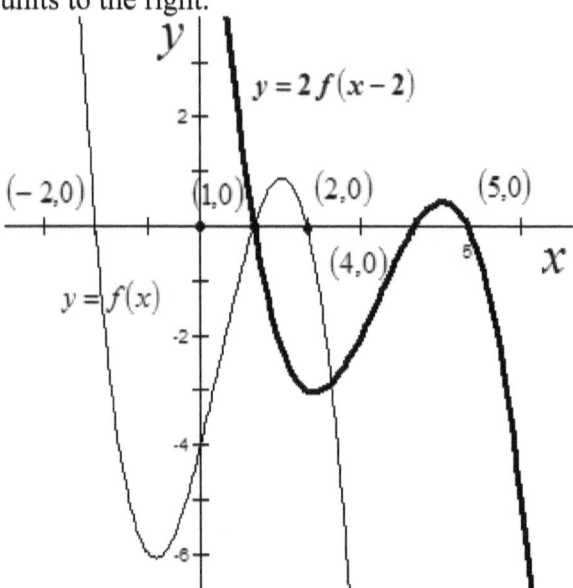

b) Reflexion in the x-axis; Vertical translation up 3 units.

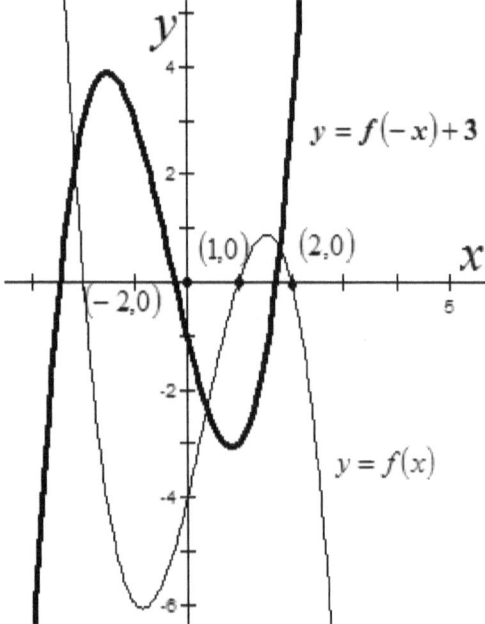

c) Reflection in the y-axis; Horizontal compression by factor 3.

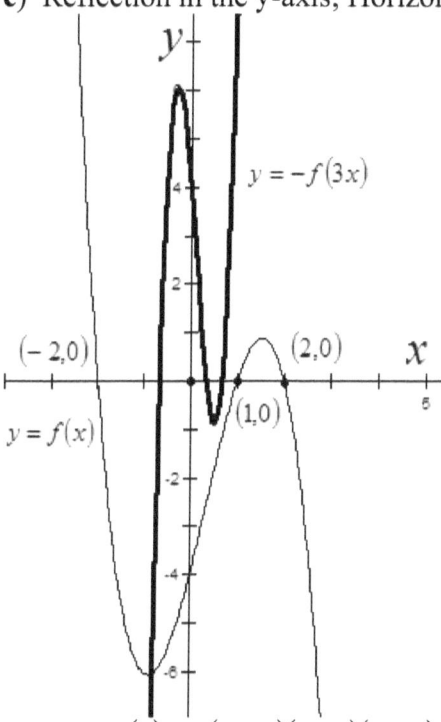

IX. 16. $f(x)=2(x-4)(x-1)(x+3)$.

IX. 17. a) $f(x)=ax(x-4)(x-2)(x+3)$;

b) $f(x)=-2x(x-4)(x-2)(x+3)$.

IX. 18. $2x-1$. **IX. 19.** $2x+8$.

IX. 20. a) $f(1)=0, f(-1)=0$; **b)** The quotient is x^2-5x+6 and the reminder is 0; **c)** $x_1=-1$, $x_2=1$, $x_3=2$, $x_4=3$.

IX. 21. $P(x)=ax+b$, $P(-1)=-a+b=2$, $P(1)=a+b=-3$,

$P(x)=-\dfrac{5}{2}x-\dfrac{1}{2}$. **IX. 22.** $f(x)=(x+1)(x-1)(x+5)$,

$f(x)=(x^2-1)(x+5)$. **IX. 23.** $(x^2-x-2)(x-5)$. **IX. 24.** $f(2)=0$

,

$m=5$. **IX. 25.** $p=3$. **IX. 26.** $f(m)=0$ or $(m^2-m+6)(m-1)$,

$m=1$. **IX. 27.** $f(1)=1-m+m^2+4-2=5\Rightarrow m^2-m-2=0$,

$m=-1$ or 2. **IX. 28.** $P(-1)=(-1)^3+k(-1)^2+(-1)+2=3$, $k=3$.

IX. 29. a) -1; **b)** 1; **c)** -60. **IX. 30.** $f(x) = 2x^2 + x - 1$.

IX. 31. $f(x) = 3x^2 - 9x + 6$.

IX. 32. a) $f(-1) = 0$, $f(1) = 4$, $a = 1$ and $b = 2$;

b) $x_1 = 1$, $x_2 = \dfrac{1 + i\sqrt{7}}{2}$, $x_3 = \dfrac{1 - i\sqrt{7}}{2}$; **c)** $x \in (-\infty,\ -1)$.

IX. 33. a) $q(x) = 2x$, $r = -10x + 1$; **b)** $q(x) = x^3 - x^2 + 6x - 11$,

$r = 34$; **c)** $q(x) = x + 6$, $r(x) = 9x^2 + 5x - 5$;

d) $q(x) = x^2 + 2x + 8$, $r = 0$;

e) $q(x) = x^3 - x - 4$, $r = 8$; **f)** $q(x) = x^3 + 3$, $r(x) = 3x - 4$;

g) $q(x) = x^5 + x^3 - x^2 + 9x - 5$, $r(x) = 33x^2 - 29x + 6$.

IX. 34. $k = 7$. **IX. 35.** We have

$$\begin{cases} P(-2) = 4(-2)^3 + a(-2)^2 - 2b + 11 = -7 \\ P(1) = 4(1)^3 + a(1)^2 + b + 11 = 14 \end{cases} . \begin{cases} 4a - 2b = 14 \\ a + b = -1 \end{cases} . \text{ From the}$$

second equation $a = -1 - b$. Substituting this in the first equation we

get $4(-1 - b) - 2b = 14 \Rightarrow b = -3$ and $a = 2$. **IX. 36.** $a = -\dfrac{7}{3}$,

$b = \dfrac{1}{3}$. **IX. 37.** $\begin{cases} f(1) = 1 + 6 + k - 4 = 3 + k \\ f(-2) = -8 + 24 - 2k - 4 = 12 - 2k \end{cases}$. They have the

same remainder, so they are equal to each other.

$3 + k = 12 - 2k \Rightarrow k = 3$. **IX. 38.** $m = 6$, $n = -5$.

IX. 39. $P(x) = -\dfrac{1}{3}x^3 + \dfrac{4}{3}x$.

IX. 40. The sum of the coefficients is $f(1)$,

$$f(1) = 1 + (1 + a) + (1 + a)^2 + \ldots + (1 + a)^n == \dfrac{(1 + a)^{n+1} - 1}{a}.$$

IX. 41. For $m \neq -1$ and $m \neq 2$ the degree of f is four; for $m = -1$ the degree of f is zero; for $m = -2$ the degree of f is two.

IX. 42. a) $\alpha = \sqrt[3]{7 + \sqrt{41}} + \sqrt[3]{7 - \sqrt{41}}$,

$\alpha^3 = \left(\sqrt[3]{7 + \sqrt{41}}\right)^3 + \left(\sqrt[3]{7 - \sqrt{41}}\right)^3 + 3\left(\sqrt[3]{7 + \sqrt{41}} \cdot \sqrt[3]{7 - \sqrt{41}}\right)\alpha =$

$= 14 + 6\alpha$, $P(\alpha) = \alpha^3 - 6\alpha + 5 = 14 + 5 = 19$; **b)** 13; **c)** 17.

IX. 43. $m = -1$. **IX. 44.** $m = 1$. **IX. 45.** $a = 1$; $a = -\dfrac{1}{3}$.

IX. 46. $P(x) = (x - a)(x - b)q(x) + mx + n$.

IX. 47. $P(x) = (x - 1)(x + 1)(x - 2)Q(x) + ax^2 + bx + c$, $P(1) = 0$,

$P(-1) = 0$, $P(2) = 0$, $\dfrac{7}{6}x^2 - \dfrac{1}{2}x - \dfrac{2}{3}$. **IX. 48.** $m = -5$.

IX. 49. $a = -3$. **IX. 50.** $a = -40$, $b = -288$, $c = 544$.

IX. 51. $P(-1) = 2$, $P(2) = -1$, $P(x) = (x^2 - x - 2)q(x) + ax + b$,

$-a + b = 2$ and $2a + b = -1$. The remainder is $-x + 1$.

IX. 52. a) $f(1) = 0$ and $f'(1) = 0$. *Another solution* using synthetic division (Horner's Algorithm)

	x^{n+1}	x^{n-1}	x^{n-1}	. . . . .	x^2	x^1	x^0
1	1	0	0		0	$-n-1$	n
1	1	1	1		1	$-n$	0
	1	2	3		n	0	

In this case the quotient is

$q(x) = x^{n-1} + 2x^{n-2} + 3x^{n-3} + \ldots + (n-1)x + n$.

IX. 53. Let ε be a root of the equation $g(x) = 0 \Rightarrow \varepsilon^2 + \varepsilon + 1 = 0$,

$\varepsilon^3 - 1 = (\varepsilon - 1)(\varepsilon^2 + \varepsilon + 1) = 0 \Rightarrow \varepsilon^3 = 1$ and $\varepsilon \neq 1$. $f(\varepsilon) = 0$,

$f(\varepsilon) = \varepsilon(\varepsilon^2 + 2\varepsilon + 1)^n (\varepsilon + 1) + (m - 1)\varepsilon^n =$

$= (\varepsilon^2 + \varepsilon)\varepsilon^n + (m - 1)\varepsilon^n = (m - 2)\varepsilon^n = 0 \Rightarrow m = 2$.

IX. 55. Let $x_1 = \alpha$ be a real root, then

$\alpha^3 - 4\alpha^2 + (6 - i)\alpha - 3 + i = 0$ or $\alpha^3 - 4\alpha^2 + 6\alpha - 3 + i(-\alpha + 1) = 0$

$\Rightarrow -\alpha + 1 = 0$ and $\alpha^3 - 4\alpha^2 + 6\alpha - 3 = 0 \Rightarrow \alpha = 1$. Therefore

$f(x) = (x - 1)(x^2 - 3x + 3 - i)$. x_2 and x_3 is found from

$x^2 - 3x + 3 - i = 0$ such that $x_2 = 2 + i$ and $x_3 = 1 - i$.

IX. 56. The roots of the equation $3x^2 + 3x + 1 = 0$ are x_1 and x_2,

therefore $3x_1^2 + 3x_1 + 1 = 0$, and $3x_2^2 + 3x_2 + 1 = 0$.

$f(x_1) = (3x_1 + 1)^{6n+1} + 3x_1^2 = [9(3x_1^2 + 3x_1 + 1) + 1]^{2n}(3x_1 + 1) + 3x_1^2 =$

$= 3x_1 + 1 + 3x_1^2 = 0$, thus f is divisible by $x - x_1$. Similarly f is divisible by $x - x_2$. Finally f is divisible by
$(x - x_1)(x - x_2) = 3x^2 + 3x + 1$.

IX. 57. $f(x) = (x-1)^3 \left(-\dfrac{5}{16} x^2 - \dfrac{13}{16} x - \dfrac{3}{8} \right) - 1$.

IX. 58. $f(x) = ax^5 + bx^4 + cx^3 = (x+1)^3 (ax^2 + px + q) + 2$,
$f(x) = -12x^5 - 30x^4 - 20x^3$.

IX. 59. $b + d - c = a \Rightarrow -\dfrac{b}{a} + \dfrac{c}{a} - \dfrac{d}{a} = -1 \Rightarrow$

$S_1 + S_2 + S_3 + 1 = 0 \Rightarrow x_1 + x_2 + x_3 + x_1 x_2 + x_2 x_3 + x_3 x_1 + x_1 x_2 x_3 + 1 = 0 \Rightarrow$
$x_1 (1 + x_2 + x_3 + x_2 x_3) + (1 + x_2 + x_3 + x_2 x_3) = 0 \Rightarrow$
$(1 + x_2 + x_3 + x_2 x_3)(x_1 + 1) = 0$
$\Rightarrow (x_1 + 1)(x_2 + 1)(x_3 + 1) = 0 \Rightarrow x_1 = x_2 = x_3 = -1$.

IX. 60. $x_1 + x_2 = 0 \Leftrightarrow (x_1 + x_2)(x_2 + x_3)(x_3 + x_1) = 0 \Leftrightarrow$

$(x_1 + x_2 + x_3)(x_1 x_2 + x_2 x_3 + x_3 x_1) - x_1 x_2 x_3 = 0 \Leftrightarrow$

$S_1 S_2 = S_3 \Leftrightarrow -\dfrac{b}{a} \cdot \dfrac{c}{a} = -\dfrac{d}{a} \Leftrightarrow bc = ad$.

IX. 61. **a)** $x_1 = -4$, $x_2 = -2$, $x_3 = 1$; **b)** $x_1 = 3$, $x_2 = 1$, $x_3 = 2$, $x_4 = 4$; **c)** $m = 5$, $x_1 = 2$, $x_2 = 1$, $x_3 = 5$; **d)** $m = 7$ $x_1 = 1$, $x_2 = 4$, $x_3 = 2$, $x_4 = 3$; **e)** $m = -8$, $x_1 = 1 + \sqrt{3}$, $x_2 = 1 - \sqrt{3}$, $x_3 = 1 + 2\sqrt{2}$, $x_4 = 1 - 2\sqrt{2}$; **f)** $x_1 = 7$, $x_2 = 4$, $x_3 = -2$; **g)** $m = 11$, $x_1 = 2$, $x_2 = -5$, $x_3 = 1$, $x_4 = -1$; **h)** $x_1 = -2$, $x_2 = 1$, $x_3 = \sqrt{3}$; **i)** $m = 0$, $x_1 = 1 + 2i$, $x_2 = 1 - 2i$, $x_3 = 1 + \sqrt{2}$, $x_4 = 1 - \sqrt{2}$. **IX. 62.** $2 - \sqrt{2}$. **IX. 63.** $(x+1)^2 = \left(\sqrt{2} + \sqrt{3} \right)^2$, then $\left(x^2 + 2x - 4 \right)^2 = 24$ or $x^4 + 4x^3 - 4x^2 - 16x - 8 = 0$. **IX. 64.** $m = 1$.

IX. 66. $x_1 = -\dfrac{2}{3}$, $x_{2,3} = -1 \pm i\sqrt{2}$. **IX. 67. a)**

$f(-1) \cdot f(1) = (m+1)^2 < 0$; **b)** $m = -1$, $x_1 = x_2 = -1$, $x_3 = 1$; **c)**

$$\left(x_1x_2 + x_3x_1 + x_2x_3\right)^2 - 2x_1x_2x_3\left(x_1 + x_2 + x_3\right) = 4x_1x_2x_3, \ m = \pm\sqrt{2}.$$

IX. 68. $0 < \sum_{i=1}^{4} x_i^2 = \left(\sum_{i=1}^{4} x_i\right)^2 - 2\sum_{1\le i \le j \le 4} x_i x_j =$

$= \left[-2(a+1)\right]^2 - 2\left(2a^2 + 4a + 3\right) = -2 < 0$

IX. 69. $x_1 + x_2 + x_3 = \dfrac{1}{2}$, $x_1 + x_2 = 1 \Rightarrow x_3 = -\dfrac{1}{2}$; $m = -3$.

IX. 70. $a = -3$, $b = -6$. **IX. 71.** $a = 11$, $b = -9$.

IX. 72. $x_2^2 = x_1x_3$ and Vieta's Formulas $x_1 + x_2 + x_3 = -a$,

$x_1x_2 + x_2x_3 + x_3x_1 = b$, $x_1x_2x_3 = -c$. Solving this system

$b^3 - a^3c = 0$.

IX. 73. Setting $x_1 = a$, $x_2 = qa$, $x_3 = q^2a$, and $x_4 = q^3a$.

Using Vieta's formulae,

$q = 2$, $x_1 = 1$, $x_2 = 2$, $x_3 = 4$, $x_4 = 8$, $m = 64$.

$$\begin{cases} x_1 + x_2 + x_3 + x_4 = 15 \\ x_1x_2 + x_1 x_3 + x_1x_4 + x_2x_3 + x_2x_4 + x_3x_4 = 70 \\ x_1x_2 x_3 + x_1x_2x_4 + x_1x_3x_4 + x_2 x_3x_4 = 121 \\ x_1x_2 x_3x_4 = m \end{cases}$$

IX. 74. x_1, x_2, x_3, and x_4 are in geometric progression $\Rightarrow$

$x_2^2 = x_1x_3$ and $x_3^2 = x_2x_4 \Rightarrow x_2x_3 = x_1x_4$. Denote $x_2 + x_3 = s_1$,

$x_2x_3 = p_1$ and $x_1 + x_4 = s_2$, $x_1x_4 = p_2$. Then $s_1 + s_2 = \dfrac{15}{2}$,

$s_1 + s_2 + p_1 + p_2 = \dfrac{35}{2}$, $s_1p_2 + s_2p_1 = 15$, $p_1p_2 = \dfrac{m}{2}$, and $p_1 = p_2$.

Therefore m = 8 and $\left(\dfrac{1}{2},\ 1,\ 2,\ 4\right)$, $\left(4,\ 2,\ 1,\ \dfrac{1}{2}\right)$.

IX. 75. $x_1 = -1$, $x_2 = 2$, $E = -\dfrac{7}{2}$.

IX. 76. **a)** $S(m) = m^3 + 6m - 6$; **b)** $m \in [0,\ 1]$; **c)** $m = 0$, $x_1 = -1$, $x_2 = -1$, $x_3 = 2$. **IX. 77.** $a = 4$, $b = -6$, $c = 4$, $d = -1$.

IX. 78. $\alpha^3 = \left(\sqrt[3]{\sqrt{5}+2}\right)^3 + \left(\sqrt[3]{\sqrt{5}-2}\right)^3 + 3\left(\sqrt[3]{\sqrt{5}+2}\cdot\sqrt[3]{\sqrt{5}-2}\right)\alpha =$

$= 2\sqrt{5}+3\alpha$. **IX. 79.** Denote $a = \sqrt[5]{27}$ and $b = \sqrt[5]{9}$.

$a^3 - b^3 = (a-b)^3 - 3ab(a-b) = x^3 - 9x$,

$(a-b)^5 = a^5 - b^5 - 5ab(a^3 - b^3) + 10a^2b^2(a-b)$ or

$x^5 = 18 - 15(x^3 - 9x) + 90x$, thus $x^5 + 15x^3 - 225x - 18 = 0$.

IX. 80. Let α be the real root, then

$\alpha^3 - (4+i)\alpha^2 + (7+3i)\alpha - 2i - 6 = 0$ or

$\alpha^3 - 4\alpha^2 + 7\alpha - 6 - i(\alpha^2 - 3\alpha + 2) = 0$. Therefore $\alpha = 2$ and

$z_2 = 1-i$, $z_3 = 1+2i$.

IX. 81. (*Three solutions*) **i)** $x_2 = \dfrac{1+i\sqrt{3}}{2}$ is also a root, therefore there is

$p \in \mathbf{R}$ such that

$f(x) = (x^2 - x + 1)(x + p) = x^3 + (p-1)x^2 + (1-p)x + p$, then
identifying the coefficients $p - 1 = a$, $1 - p = 4$ and $p = b$.
Finally $p = -3$, $a = -4$ and $b = -3$.

ii) Using long division

$x^3 + ax^2 + 4x + b = (x^2 - x + 1)(x + a + 1) + x(a + 4) + a + b + 1$. Since
the reminder is zero, $a = -4$ and $b = -3$.

iii) x_1 is a root, then $x_1^2 - x_1 + 1 = 0 \Rightarrow$

$x_1^3 + 1 = (x_1 + 1)(x_1^2 - x_1 + 1) = 0 \Rightarrow x_1^3 = -1$. Therefore

$f(x_1) = x_1^3 + ax_1^2 + 4x_1 + b = -1 + a(x_1 - 1) + 4x_1 + b = 0 \Rightarrow$

$-1 + a\cdot\left(\dfrac{1-i\sqrt{3}}{2} - 1\right) + 4\cdot\dfrac{1-i\sqrt{3}}{2} + b = 0$ or

$b + 1 - \dfrac{a}{2} + i\cdot\left(\dfrac{-\sqrt{3}}{2}a - 2\sqrt{3}\right) = 0 \Rightarrow a = -4$ and $b = -3$.

IX. 82. $a = -6$ and $b = -20$.

IX. 83. (*Four solutions*) **i)** Solving the system $x_1 = x_2$,

$x_1 + x_2 + x_3 = -\dfrac{m}{2}$, $x_1x_2 + x_1x_3 + x_2x_3 = 2$, and $x_1x_2x_3 = -2$, we get

$m = -7$ and $x_1 = x_2 = 2$.

ii) Obviously $x = 0$ is not a solution of

$f(x) = 0 \Rightarrow 2x + m + \dfrac{4}{x} + \dfrac{4}{x^2} = g(x)$ has the same roots as f. Therefore g

has a double root and $g'(x) = 0$, $g(x) = 0$. $g'(x) = 2 - \dfrac{4}{x^2} - \dfrac{8}{x^3}$,

$g'(x) = 0 \Rightarrow x^3 - 2x - 4 = 0 \Rightarrow x = 2$. From $g(2) = 0 \Rightarrow m = -7$.

iii) Let $\alpha \in \mathbb{R}$ be a double root. Using synthetic division (Horner's algorithm) two times

	x^3	x^2	x^1	x^0
α	2	m	4	4
α	2	$2\alpha + m$	$2\alpha^2 + m\alpha + 4$	$2\alpha^3 + m\alpha^2 + 4\alpha + 4$
	2	$4\alpha + m$	$6\alpha^2 + 2m\alpha + 4$	

we get the system $2\alpha^3 + m\alpha^2 + 4\alpha + 4 = 0$, $6\alpha^2 + 2m\alpha + 4 = 0$.
Therefore $\alpha = 2$ and $m = -7$. iv) Let $\alpha \in \mathbb{R}$ be a double root.
$f(x) = (x - \alpha)^2 (2x + p) = 2x^3 + (p - 4\alpha)x^2 + (2\alpha^2 - 2p\alpha)x + \alpha^2 p$.
Identifying the coefficients we get the system $p - 4\alpha = m$,
$2\alpha^2 - 2p\alpha = 4$ and $\alpha^2 p = 4$. Finally $\alpha = 2$ and $m = -7$.

IX. 84. $a = -2$, $b = 1$, $c = -2$. The roots are $x_2 = \dfrac{1 - i\sqrt{3}}{2}$, $x_3 = 3$,

$x_4 = -1$. **IX. 85.** $x_2 = 1 + i$ is also a root, therefore
$f(x) = (x^2 - 2x + 2)(x^2 + ax + b)$. Equating the coefficients
$a = -4$, $b = 3$, $m = -6$, $n = 6$, $x_3 = 1$, $x_4 = 3$.

IX. 86. $x_2 = 2 + i$, $x_3 = 1$, $x_4 = 0$, $a = 9$, $b = -5$, $c = 0$.

IX. 87. a) $x_2 = 2 + \sqrt{3}$, $x_3 = -1$; **b)** $x_2 = 1 + 2\sqrt{2}$, $x_3 = 1$;

c) $x_2 = 1 + \sqrt{2}$, $x_3 = i$, $x_4 = -i$; **d)** $x_2 = 3 - i$, $x_3 = 1$;

e) $x_2 = 2 - 3i$, $x_3 = \sqrt{2}$, $x_4 = -\sqrt{2}$;

f) $x_2 = 2 - i$, $x_3 = 1$, $x_4 = 2$, $m = -23$, $n = 10$; **g)** $x_2 = -1 - i$,

$x_3 = 1$, $x_4 = -1$, $m = 1$, $n = -2$; **h)** $x_2 = -\sqrt{3} + \sqrt{5}$, $x_3 = \sqrt{3} - \sqrt{5}$, $x_4 = \sqrt{3} + \sqrt{5}$, $x_5 = -1$; **i)** $x_2 = -\sqrt{3} + \sqrt{7}$, $x_3 = \sqrt{3} - \sqrt{7}$, $x_4 = \sqrt{3} + \sqrt{7}$, $x_5 = -\sqrt{3}$, $x_6 = \sqrt{3}$.

IX. 88. a) $x_1 = -1$, $x_2 = \dfrac{-1 + i\sqrt{15}}{4}$, $x_3 = \dfrac{-1 - i\sqrt{15}}{4}$; **b)** $x_1 = -1$,

$x_2 = -2$, $x_3 = -\dfrac{1}{2}$; **c)** $x_1 = \dfrac{1 + i\sqrt{3}}{2}$, $x_2 = \dfrac{1 - i\sqrt{3}}{2}$, $x_3 = \dfrac{-3 + i\sqrt{7}}{4}$,

$x_4 = \dfrac{-3 - i\sqrt{7}}{4}$; **d)** $x_1 = 2$, $x_2 = \dfrac{1}{2}$, $x_3 = i$, $x_4 = -i$; **e)** $x_1 = -1$,

$x_2 = i$, $x_3 = -i$, $x_4 = \dfrac{-1 + i\sqrt{3}}{2}$, $x_5 = \dfrac{-1 - i\sqrt{3}}{2}$; **d)** $x_1 = -1$,

$x_2 = -1$, $x_3 = -1$, $x_4 = \dfrac{1 + i\sqrt{3}}{2}$, $x_5 = \dfrac{1 - i\sqrt{3}}{2}$.

IX. 89. $P(x) = (x - 1)^2 q(x) + ax + b$. Obviously $P(1) = a + b$ and $P'(1) = a$. Therefore $r(x) = (n + 3)x - n - 1$.

IX. 90. $f(x) = g(x)x^2(x - 1) + r(x)$ where $r(x) \le 2$. Let $r(x) = ax^2 + bx + c$ be the reminder with a, b, $c \in \mathbf{R}$ and $a \neq 0$. For $x = 0 \Rightarrow c = 1$; for $x = 1 \Rightarrow$ $a + b + c = n + 1$. Differentiate in both members of the equation from the division theorem, we get $nx^{n-1} + (n - 1)x^{n-1} + \ldots + 2x + 1 = = g'(x)x^2(x - 1) + g(x)(3x^2 - 2x) + 2ax + b$. For $x = 0 \Rightarrow b = 1$. Finally $r(x) = (n - 1)x^2 + x + 1$.

IX. 91. a) $f(x) = x^{n-1} + x^{n-2} + \ldots + 1 = (x - x_1)(x - x_2)\ldots(x - x_{n-1})$,

we have $\displaystyle\sum_{i=1}^{n} \dfrac{1}{1 - x_i} = \dfrac{f'(1)}{f(1)}$. **IX. 92.** $f(x) = g(x)x^2(x - 1) + r(x)$

where $r(x) \le 2$. Let $r(x) = ax^2 + bx + c$ with a, b, $c \in \mathbf{R}$ and $a \neq 0$. For $x = 0 \Rightarrow c = 1$; for $x = 1 \Rightarrow a + b + c = n + 1$. Differentiating in both sides of the equation from the division theorem, we obtain $nx^{n-1} + (n - 1)x^{n-1} + \ldots + 2x + 1 = = g'(x)x^2(x - 1) + g(x)(3x^2 - 2x) + 2ax + b$. For $x = 0 \Rightarrow b = 1$.

Finally $r(x) = (n-1)x^2 + x + 1$.

IX. 93. $P(x) = (x-1)^2 q(x) + ax + b$. Obviously $P(1) = a + b$ and $P'(1) = a$. Therefore $r(x) = (m+3)x - n - 1$.

IX. 94. a) $f(x) = (x^2 + x + 1)\left[(m-1)x^2 - 2(m+1)x + m - 4\right]$.

b) $(m-1)x^2 - 2(m+1)x + m - 4 = 0$ has a double root

$\Delta = 4(m+1)^2 - 4(m-1)(m-4) = 0$, thus $m = \dfrac{3}{7}$.

c) Using Viete formulae, we get $\dfrac{m+6}{m-4} \geq \dfrac{5}{4}$, $4 < m \leq 44$.

d) $x_0 = -1 + i\tan\beta$ cannot be a root for $x^2 + x + 1$.

$\overline{x_0} = -1 - i\tan\beta$ is also a root for $(m-1)x^2 - 2(m+1)x + m - 4 = 0$.

Therefore $-2 = \dfrac{2(m+1)}{m-1}$ and $1 + \tan^2\beta = \dfrac{m-4}{m-1}$, $m \neq 1$.

Finally $m = 0$ and $\beta = \pm\dfrac{\pi}{3} + 2k\pi$, $k \in \mathbf{Z}$ or $\beta = \pm\dfrac{2\pi}{3} + 2k'\pi$,

$k' \in \mathbf{Z}$. The roots are $x_0 = -1 - i\sqrt{3}$ and $x_0' = -1 + i\sqrt{3}$.

IX. 95. a) From the stated equation, we have $P_n(i) - P_n(i-1) = (i-1)^n$, $(\forall) i = 1, 2\ldots k$; **b).** From $P_n(1) - P_n(0) = 0 \Rightarrow P_n(1) = 0$.

IX. 96. Let α be a real root of the given equation. We have

$\alpha^{2n+1} - \alpha^{2n} + \alpha^{2n-1} + 2n\alpha^n - n^2 = 0 \Rightarrow \alpha^{2n-1}(\alpha^2 + 1) = (\alpha^n - n)^2$ or

$\alpha^{2n-1} = \dfrac{(\alpha^2 - n)^2}{\alpha^2 + 1} \geq 0$. Then $\alpha \geq 0$. Since 0 is not a solution for the given equation, it follows that $\alpha > 0$.

IX. 96. We assume that there is $\lambda \in \mathbf{Z}$ such that $f(\lambda) = 0$. We have the cases $\lambda = 2k$ and $\lambda = 2k + 1$ and we get $f(2k)$ and $f(2k+1)$ are odd numbers. Therefore f does not have any integer root.

IX. 97. The degree of the polynomial in the left side is n^2 but the degree of the polynomial in the right side is $2n$. Then $n^2 = 2n \Rightarrow n = 2$. Finally $f(x) = a(x^2 - x)$ where $a \in \mathbf{R}^*$.

IX. 98. Let $x_1, x_2 \ldots x_n$ be the roots of the polynomial P. Therefore $x_1 x_2 \ldots x_n = 1$ and $|x_1| \cdot |x_2| \cdot \ldots \cdot |x_n| = 1 \Rightarrow |x_1| = |x_2| = \ldots = |x_n| = 1$, since

$$P(x) = (x - x_1)(x - x_2)...(x - x_n) \Rightarrow$$

$$P(-1) = (-1)^n (1 + x_1)(1 + x_2)...(1 + x_n) \text{ as } |x_i| = 1, \ (\forall) i = 1, 2...k \Rightarrow$$

$$x_i \cdot \overline{x_i} = 1 \Rightarrow \overline{xi} = \frac{1}{x_i}. \text{ Then } \overline{P(-1)} = (-1)^n (1 + \overline{x_1})(1 + \overline{x_2})...(1 + \overline{x_n}) =$$

$$= (-1)^n \frac{(1 + x_1)(1 + x_2)...(1 + x_n)}{x_1 \cdot x_2 \cdot ... \cdot x_n} = P(-1), \text{ therefore } P(-1) \text{ is real.}$$

IX. 99. It is obviously that $\varepsilon^n = 1$.

Therefore $\displaystyle\sum_{k=1}^{n} f\left(\frac{1}{\varepsilon^k}\right) = \sum_{k=1}^{n}\sum_{j=1}^{n} a_j \left(\frac{1}{\varepsilon^k}\right)^j = \sum_{k=1}^{n} a_j \sum_{j=1}^{n} \frac{1}{\varepsilon^{kj}} =$

$$= na_0 + a_1 \frac{1}{\varepsilon} \frac{1 - \left(\dfrac{1}{\varepsilon}\right)^n}{1 - \dfrac{1}{\varepsilon}} + a_2 \frac{1}{\varepsilon^2} \frac{1 - \left(\dfrac{1}{\varepsilon}\right)^{2n}}{1 - \dfrac{1}{\varepsilon^2}} + ... + na_n = n(a_0 + a_n).$$

Chapter X

PERMUTATIONS AND COMBINATIONS.
NEWTON'S BINOMIAL.

X. 1. $5! = 120$. **X. 2.** $6! = 720$. **X. 3. a)** $6! - 5! = 600$;
b) $5! = 120$; **c)** $5! - 4! = 96$; **d)** $4! = 24$; **e)** $600 - 24 = 576$.

X. 4. $\dfrac{7!}{3!\cdot 2!\cdot 1!\cdot 1!}$, $\dfrac{13!}{3!(2!)^2}$, $\dfrac{11!}{2!(4!)^2}$. **X. 5.** $\dfrac{15!}{6!\cdot 5!\cdot 4!} = 630,630$.

X. 6. $P(14,3) = 2184$.

X. 7. $P(7,5) = 2520$, $7! = 5040$. **X. 8.** Five digit numbers are
$P(5,5) - P(4,4) = 96$, four digit numbers are $P(5,4) - P(4,3) = 96$,
three digit numbers are $P(5,3) - P(4,2) = 48$, two digit numbers are
$P(5,2) - P(4,1) = 16$, and one digit number are five. In total 261
numbers.

X. 9. $\binom{7}{4} + \binom{7}{5} + \binom{7}{6} + \binom{7}{7} = 35 + 21 + 7 + 1 = 64$. **X. 10.** $\binom{20}{4} = 4845$.

X. 11. a) $\binom{4}{1} \cdot \binom{8}{5}$; **b)** $\binom{12}{6} - \binom{8}{6}$; **c)** $\binom{11}{5}$. **X. 12.** $\binom{4}{2} \cdot \binom{10}{5} = 1512$.

X. 13. a) $\binom{25}{5} = 53,130$, **b)** $\binom{24}{4} = 10,626$. **X. 14.** $\binom{6}{2} = 15$.

X. 15. $\binom{5}{2} = 10$. **X. 16.** $\binom{9}{5} = 126$; $\binom{10}{5} = 252$.

X. 17. $\binom{6}{3} \cdot \binom{9}{4} + \binom{6}{4} \cdot \binom{9}{3} + \binom{6}{5} \cdot \binom{9}{2} + \binom{6}{6} \cdot \binom{9}{1} C_9^1 = 4,005$.

X. 18. $\binom{7}{4} \cdot P(9,4)$.

X. 19. a) $\binom{15}{4} \cdot \binom{10}{6}$; **b)** 785,565. **X. 20.** See the previous problem.

X. 21. $\binom{9}{7} = 36$. **X. 22. a)** $\dfrac{(x+9)!}{(x+4)!\cdot 5!} = 5\dfrac{(x+7)!}{(x+4)!} \Rightarrow \dfrac{(x+9)(x+8)}{5!} = 5$

$\Rightarrow x^2 - 17x - 528 = 0$, $x = 33$; **b)** $x = 7$;
c) $x = 10$; **d)** $x = 5$; **e)** $x = 19$; **f)** $x = 5$; **g)** $x = 2$; **h)** $x = 10$;
i) $x = 4$; **j)** $x = 3$; **k)** $x = 2$; **l)** $n = 10$; **m)** $n = 12$.

X. 24. a) 15; **b)** 84; **c)** $2\dfrac{6}{7}$; **d)** $8\dfrac{13}{14}$; **e)** $4k^2$; **f)** $\dfrac{(n-1)!}{(k-1)!}$;

g) $\dfrac{k+1}{n+1}$; **h)** $\dfrac{2(n-k)}{n-1}$; **i)** $\dfrac{k}{(n+1)(n+2)}$.

X. 25. Use $\binom{n}{k} = \binom{n}{n-k}$. **X. 26. a)** $n > 6$ $(n \in \mathbf{N})$;

b) $n \in [0, 5]$, $n \in \mathbf{N}$; **c)** $n \in [4, 10]$, $n \in \mathbf{N}$; **d)** $1 \le n \le 9$ $(n \in \mathbf{N})$; **e)** $4 \le n \le 13$ $(n \in \mathbf{N})$; **f)** $9 \le n \le 15$ $(n \in \mathbf{N})$;
g) $5 \le n \le 20$ $(n \in \mathbf{N})$; **h)** $0 \le n \le 5$ $(n \in \mathbf{N})$;
i) $1 \le n \le 4$ $(n \in \mathbf{N})$. **X. 27. a)** $x = 7$, $y = 6$; **b)** $x = 3$, $y = 8$;
c) $x = 5$, $y = 5$; **d)** $n = 10$, $k = 4$; **e)** $x = 3$, $y = 5$ or $x = 5$, $y = 3$; **f)** $x = 6$, $y = 3$; **g)** $x = 5k - 1$, $y = 3k$. **X. 29. a)**

$$S_1 = \sum_{k=1}^{n}\left[(k+1)!-k!\right] = (2!-1!) + (3!-2!) + ... + (n+1)!-n! = (n+1)!-1;$$

b) $S_2 = \displaystyle\sum_{k=2}^{n}\left(\dfrac{1}{k!} - \dfrac{1}{(k+1)!}\right) = 1 - \dfrac{1}{(k+1)!}$;

c) $S_3 = \displaystyle\sum_{k=2}^{n}\dfrac{k(k+1)}{k!} - \sum_{k=2}^{n}\dfrac{k+2}{k!} = \sum_{k=2}^{n}\dfrac{k+1}{(k-1)!} - \sum_{k=2}^{n}\dfrac{k+2}{k!} = 3 - \dfrac{n+2}{n!}$;

d) $S_4 = \displaystyle\sum_{k=1}^{n}\dfrac{1}{k(k+1)} = 1 - \dfrac{1}{n+1}$;

e) $S_5 = \displaystyle\sum_{k=2}^{n}\left(\dfrac{1}{k!\,k} - \dfrac{1}{(k+1)!(k+1)}\right) = \dfrac{1}{4} - \dfrac{1}{(n+1)!(n+1)}$;

f) $S_6 = \displaystyle\sum_{k=1}^{n}\dfrac{k+1}{(k-1)!+k!+(k+1)!} = \sum_{k=1}^{n}\dfrac{k+1}{(k-1)!\left[1 + k + k(k+1)\right]} =$

$= \displaystyle\sum_{k=1}^{n}\dfrac{k+1}{(k-1)!(k+1)^2} = \sum_{k=1}^{n}\dfrac{k}{(k+1)!} = \sum_{k=1}^{n}\left(\dfrac{1}{k!} - \dfrac{1}{(k+1)!}\right) = \dfrac{n}{n+1}$.

X. 30. Use the formula $\binom{n}{k} = \binom{n-1}{k} + \binom{n-1}{k-1}$.

X. 31. a) $\dfrac{S_1}{3!} = \dfrac{1\cdot2\cdot3}{3!} + \dfrac{2\cdot3\cdot4}{3!} + ... + \dfrac{n\cdot(n+1)\cdot(n+2)}{3!} =$

$= \binom{3}{3} + \binom{4}{3} + ... + \binom{n+2}{3} = \binom{n+3}{4}$. $S_1 = \dfrac{n\cdot(n+1)\cdot(n+2)\cdot(n+3)}{4}$;

b) Similarly $\dfrac{S_2}{k!} = \binom{k}{k} + \binom{k+1}{k} + ... + \binom{n+k-1}{k} = \binom{n+k}{k+1}$. Therefore

$S_2 = \dfrac{(n+k)\cdot(n+k-1)...(n+1)n}{k+1}$. **X. 32. a)** $\dfrac{n(n+1)(n-1)}{6}$;

~ 330 ~

b) $\sum_{k=2}^{n} \dfrac{n^2(n-1)^2}{4} = \dfrac{1}{4}\sum_{k=2}^{n} k^4 - \dfrac{1}{2}\sum_{k=2}^{n} k^3 + \dfrac{1}{4}\sum_{k=2}^{n} k^2 = \dfrac{n\cdot(n^2-1)(3n^2-2)}{60}.$

We used the equalities $\sum_{k=1}^{n} k = \dfrac{n\cdot(n+1)}{2}$; $\sum_{k=1}^{n} k^2 = \dfrac{n\cdot(n+1)(2n+1)}{6}$;

$\sum_{k=1}^{n} k^3 = \dfrac{n^2\cdot(n+1)^2}{4}$; $\sum_{k=1}^{n} k^4 = \dfrac{n\cdot(n+1)(2n+1)(3n^2+3n-1)}{30}.$

X. 33. $a = 2$, $b = 2$, and $c = 4$. **X. 34.** Use $\binom{n-1}{k} = \dfrac{n-k}{k}\binom{k-1}{n-1}\binom{n-1}{k-1}$

and $\binom{n}{k} = \dfrac{n}{k}\binom{n-1}{k-1}$. **X. 36.** $\dfrac{\binom{2n}{n}}{n+1} = \dfrac{(2n)!}{n!\cdot(n+1)!} = \dfrac{(2n)!}{n!\cdot(n-1)!}\cdot\dfrac{1}{n(n+1)} =$

$= \dfrac{(2n)!}{n!\cdot(n-1)!}\left(\dfrac{1}{n} - \dfrac{1}{n+1}\right) = \binom{2n}{n} - \binom{2n}{n-1}$. **X. 37.** Use the identity

$\dfrac{k\binom{n}{k}}{\binom{n}{k-1}} = n-k$. **X. 38.** $\binom{n-1}{n+1} = \dfrac{n(n+1)}{2} = 1+2+3+...+n$. Therefore

$\dfrac{\binom{n-1}{n+1}}{\prod_{k=1}^{n} k!} = \dfrac{(1+2+3+...+n)!}{1!\cdot 2!\cdot 3!\cdot...\cdot n!} =$

$= \binom{n}{1+2+3+...+n}\cdot\binom{n-1}{1+2+3+...+(n-1)}\cdot\binom{n-2}{1+2+3+...+(n-2)}\cdot...\cdot\binom{2}{1+2}.$

X. 39.

$\sum_{k=0}^{2n}(-1)^k\cdot k!(2n-k)! = (2n)!\left(\sum_{k=0}^{2n}\dfrac{(-1)^k\cdot k!(2n-k)!}{(2n)!}\right) = (2n)!\left(\sum_{k=0}^{2n}\dfrac{(-1)^k}{\binom{2n}{k}}\right)\cdot$

$\cdot\left[\sum_{k=0}^{2n}(-1)^k\left(\dfrac{1}{\binom{2n+1}{k}} + \dfrac{1}{\binom{2n+1}{k+1}}\right)\right] = (2n)!\cdot\dfrac{2n+1}{2(n+1)} = $ (we used **a**))

$= \left[\left(\dfrac{1}{\binom{2n+1}{0}} + \dfrac{1}{\binom{2n+1}{1}}\right) - \left(\dfrac{1}{\binom{2n+1}{1}} + \dfrac{1}{\binom{2n+1}{2}}\right) + ... + \left(\dfrac{1}{\binom{2n+1}{2n}} + \dfrac{1}{\binom{2n+1}{2n+1}}\right)\right]$

where the last square bracket is 2.

X. 40. a) $\binom{4}{0} + \binom{4}{1}\cdot(2x) + \binom{4}{2}\cdot(2x)^2 + \binom{4}{3}\cdot(2x)^3 + \binom{4}{4}\cdot(2x)^4$;

b) $2\binom{6}{0}x^6 + 2\binom{6}{2}\cdot x^4(x^2-1) + 2\binom{6}{4}\cdot x^2(x^2-1)^2 + 2\binom{6}{6}\cdot(x^2-1)^3.$

X. 41. $t_5 = \binom{9}{4}\left(\sqrt[3]{x}\right)^5 \cdot \left(\dfrac{1}{\sqrt[3]{x}}\right)^4 = 126\sqrt[3]{x}$.

X. 42. $\dfrac{x^5}{32} + \dfrac{5x^4 y}{8} + 5x^3 y^2 + 20x^2 y^3 + 40xy^4 + 32y^5$

X. 43. a) $256 - 3072x + 16128x^2$; **b)** 24.

X. 44. $t_{k+1} = \binom{2010}{k}x^k \left(\dfrac{2}{\sqrt[5]{x}}\right)^{2010-k} = \binom{2010}{k} \cdot 2^{2010-k} \cdot x^{k + \frac{k-2010}{5}}$,

$k + \dfrac{k-2010}{5} = 0 \Rightarrow k = 370$. Therefore $t_{371} = \binom{2010}{370} \cdot 2^{1640}$.

X. 45. a) 198; **b)** $1\dfrac{2}{7}$. **X. 46.** 152.

X. 47. $\dfrac{1}{x}\left[(1+x)^{21} - 1\right]$, 210 each.

X. 48. $x = \dfrac{1}{3}$, $n = 9$, $\dfrac{28}{9}$. **X. 49.** 2300.

X. 50. $n = 10$, $a = 2$, $b = 960$.

X. 51. $-\dfrac{448}{3\sqrt{3}}x^3$. **X. 52.** 11. **X. 53.** $n = 15$. **X. 54.** $n = 20$.

X. 55. 2. **X. 56.** $a = \dfrac{1}{4}$.

X. 57. $t_4 = \binom{8}{3} \cdot \left(x^{\frac{4}{5} - \frac{n-1}{n}}\right)^5 \left(x^{1 + \frac{n-1}{n+1}}\right)^3 = 56x^{\frac{11}{5}}$, $x = 2$.

X. 58. $t_5 = 792$. **X. 59.** $(-1)^n \dfrac{(3n)!}{n! \cdot (2n)!}$. **X. 60.** $t_6 = 210x^2 y^3$.

X. 61. $t_2 = 153x^{\frac{13}{2}}$. **X. 62.** $2^{n-1} = 2048$, $t_7 = -264x^3 y^7$.

X. 64. a) $n = 6$; **b)** $n = 10$.

X. 65. $t_{10} = \binom{21}{5} \cdot \left(\sqrt[3]{\dfrac{x}{\sqrt{y}}}\right)^{21-k} \left(\sqrt{\dfrac{y}{\sqrt[3]{x}}}\right)^k = \binom{21}{5} \cdot x^{\frac{21-k}{3} - \frac{k}{6}} \cdot y^{\frac{k-21}{6} + \frac{k}{2}}$.

Since x and y have equal powers, we obtain $k = 9$, therefore
$t_{10} = \binom{21}{5}\sqrt{x^5 y^5}$.

X. 66. $x_1 = \dfrac{1}{100\sqrt{10}}$, $x_2 = 10$. **X. 67.** $m = 12$.

X. 68. $\binom{n}{0} + \binom{n}{1} + \binom{n}{2} = 22$, $n = 6$.

$t_3 + t_5 = \binom{6}{2}\left(\sqrt{2^x}\right)^4 \left(\sqrt{2^{1-x}}\right)^2 + \binom{6}{4}\left(\sqrt{2^x}\right)^2 \left(\sqrt{2^{1-x}}\right)^4$,

$15 \cdot 2^{x+1} + 15 \cdot 2^{2-x} = 135$, $x_1 = -1$, $x_2 = 2$.

X. 69. $t_3 = \binom{9}{2}\left(x^{\log\sqrt{x}}\right)^2 \cdot \left(\dfrac{1}{\sqrt[7]{x^2}}\right)^7 = 36x^{\log x - 2}$. Therefore

$36x^{\log x - 2} = 36000 \Rightarrow x^{\log x - 2} = 1000 \Rightarrow (\log x - 2)\log x = 3$, $x_1 = 10^{-1}$

and $x_2 = 10^3$. **X. 70.** $n = 6$, $x_1 = -2$, $x_2 = 1$.

X. 71. $x = 2$. **X. 72.** $x \in (0, \infty) \setminus \{10^{-1}\}$, $x_1 = 10$, $x_2 = 10^{-4}$.

X. 73. $n = 8$, t_0, t_4, t_8.

X. 74. $10 - 3^x > 0$, $\binom{n}{1} + \binom{n}{3} = 2\binom{n}{2} \Rightarrow n = 7$. From $t_6 = 21 \Rightarrow x_1 = 0$

and $x_2 = 2$. **X. 75.** $n = 8$; t_6.

X. 76. $\binom{n}{1} + \binom{n}{3} + \binom{n}{5} + \ldots = 2^{n-1}$, $n = 9$, $t_7 = \binom{9}{3}x^{-3}$. **X. 77. a)**

$t_{k+1} = \binom{5}{k}\left(\sqrt[3]{3}\right)^{5-k} \cdot \left(\sqrt{2}\right)^k = \binom{5}{k} \cdot 3^{\frac{5-k}{3}} \cdot 2^{\frac{k}{2}}$ where $\dfrac{5-k}{3} \in Z$, $\dfrac{k}{2} \in Z$ and

$0 \le k \le 5$ we get $k = 2$ and $t_3 = \binom{2}{5} \cdot 3 \cdot 2 = 60$; **b)** t_6 and t_{23};

c) t_{4k+1}, $0 \le k \le 50$, $k \in N$; **d)** $t_{15} = -\binom{24}{10} \cdot 36$.

X. 78. t_1 and t_{15} are the rational terms.

X. 79. a) 161,700; **b)** t_{91}; **c)** $(2.5)^{100}$; **d)** t_{50}; **e)** 7.

X. 80. a) 1; **b)** 2^{22}.

X. 81. a) $t_{k-1} < t_k > t_{k+1}$, $t_{50} = \binom{100}{50} \cdot \left(\dfrac{1}{2}\right)^{100}$, **b)** t_{10}; **c)** $314925 \cdot 10^5$.

X. 82. $a = \dfrac{1}{4}$, **X.83. a)** $(x-1)^9 = 1$, $x = 2$; **b)** $x = 4$;

c) $\sin x = -\dfrac{1}{2}$, thus $x_1 = -\dfrac{\pi}{6} + 2k\pi$ $k \in Z$ and

$x_2 = \dfrac{\pi}{6} + (2k'+1)\pi$ $k' \in Z$.

X. 83. a) 1; **b)** $(-1)^n$. **X. 85. a)** $z = 1+i$,

$$z^n = 2^{\frac{n}{2}} \cdot \left(\cos\frac{n\pi}{4} + i\sin\frac{n\pi}{4} \right),$$

$$(1+i)^n = 1 - \binom{n}{2} + \binom{n}{4} - \binom{n}{6} + \ldots + i\left(\binom{n}{1} - \binom{n}{3} + \binom{n}{5} - \binom{n}{7} + \ldots \right);$$

d) $z = 1 + i\sqrt{3}$, $z^n = 2^n \cdot \left(\cos\frac{n\pi}{3} + i\sin\frac{n\pi}{3} \right).$

X. 86. a) $\varepsilon = \dfrac{-1+i\sqrt{3}}{2}$ is the root of the equation $\varepsilon^2 + \varepsilon + 1 = 0$,

also $\varepsilon^3 = 1$. $\varepsilon = \cos\dfrac{2\pi}{3} + i\sin\dfrac{2\pi}{3}$, $1 + \varepsilon = \cos\dfrac{\pi}{3} + i\sin\dfrac{\pi}{3}$,

$1 + \varepsilon^2 = \cos\dfrac{\pi}{3} - i\sin\dfrac{\pi}{3}$. Finally add the equalities

$$2^n = \binom{n}{0} + \binom{n}{1} + \binom{n}{2} + \binom{n}{3} + \binom{n}{4} + \ldots,$$

$$(1+\varepsilon)^n = \binom{n}{0} + \binom{n}{1}\varepsilon + \binom{n}{2}\varepsilon^2 + \binom{n}{3} + \binom{n}{4}\varepsilon + \ldots,$$

$$\left(1+\varepsilon^2\right)^n = \binom{n}{0} + \binom{n}{1}\varepsilon^2 + \binom{n}{2}\varepsilon + \binom{n}{3} + \binom{n}{4}\varepsilon^2 + \ldots.$$

X. 87. $z^n = \left(1 + i\dfrac{1}{\sqrt{3}}\right)^n = \left(\dfrac{2}{\sqrt{3}}\right)^n \left(\cos\dfrac{n\pi}{6} + i\sin\dfrac{n\pi}{6} \right).$

X. 88. $z^{6n} = \left(1 - i\sqrt{3}\right)^{6n} = \left[2\left(\cos\dfrac{5\pi}{3} + i\sin\dfrac{5\pi}{3} \right) \right]^{6n} = 2^{6n}.$

X. 89. Use $z = (1+i)^n$. **X. 90. a)** $S_n = 2^n$; **b)** Denote

(1) $S_n = \binom{n}{1} + 2\binom{n}{2} + 3\binom{n}{3} + \ldots + n\binom{n}{n}$. From $\binom{n}{k} = \binom{n}{n-k}$, we get

(2) $S_n = n\binom{n}{0} + (n-1)\binom{n}{1} + (n-2)\binom{n}{2} + \ldots + \binom{n}{n-1}$. Adding (1) and (2) we

get $2S_n = n\left[\binom{n}{0} + \binom{n}{1} + \binom{n}{2} + \ldots + \binom{n}{n} \right]$. Finally $S_n = 2^{n-1}n$.

Another way: Using $\binom{n}{k} = \dfrac{n}{k}\binom{n-1}{k-1}$,

$S_n = n\binom{n-1}{0} + n\binom{n-1}{1} + n\binom{n-1}{2} + \ldots + n\binom{n-1}{n-1} = 2^{n-1}n$;

c) $S_n = 2^{n-1}(n+2)$; **d)** $S_n = (n-2)2^{n-1} + 1$; **e)** $S_n = (n+1)2^n$;

f) $S_n = 2^{n-1}(2k+n)$;   **g)** $S_n = 0$;   **h)** $S_n = 2^{n-1}n - 2^n + 1$;   **i)** Use

$\binom{n}{k} = \binom{n-1}{k-1} + \binom{n-1}{k}$, 0;   **j)** Use $\binom{n}{k} = \dfrac{k+1}{n+1}\binom{n+1}{k+1}$, $S_n = \dfrac{2^{n+1}-1}{n+1}$;

k) Use $\binom{n}{k} = \dfrac{(k+2)(k+1)}{(n+2)(n+1)}\binom{n+2}{k+2}$, $S_n = \dfrac{2^{n+1}n+1}{(n+1)(n+2)}$;

l) $S_n = \dfrac{2^{n+2}-n-3}{n+2}$;   **m)** $S_n = \dfrac{1}{n+1}$.

X. 91. a) The coefficient of x^n in the identity

$x^n(1-x)^n + x^{n-1}(1-x)^n + ... + x^{n-m}(1-x)^n = x^{n-m}(1-x)^{n-1} - x^{n+1}(1-x)^{n-1}$

is $(-1)^m \binom{n-1}{m}$ when $m \le n-1$ and it is zero if $m = n$;   **b)** Similarly,

the coefficient of x^k from the expression

$(1+x)^n + (1+x)^{n+1} + ... + (1+x)^{n+m} = \dfrac{1}{x}\left[(1+x)^{n+m+1} - (1+x)^n\right]$ is

$\binom{n+m+1}{k+1} - \binom{n}{k+1}$ when $k \le n-1$ and it is $\binom{n+m+1}{n+1}$ when $k = n$.

X. 92. a) The coefficient of x^n in the identity

$(1+x)^n(1+x)^n = (1+x)^{2n}$ is $\binom{2n}{n}$;   **b)** The coefficient of x^n in the

identity $(1+x)^n(1+x)^n = (1-x^2)^n$ is zero if $n = 2k+1$ and

$(-1)^k\binom{2k}{k}$ if $n = 2k$;   **c)** Use $\binom{n}{k} = \binom{n-1}{k} + \binom{n-1}{k-1}$.

The sum is 1 for $n = 3k$,

0 for $n = 3k+1$, and -1 for $n = 3k-1$, $k \in N$;

d) 2^{2n};   **e)** $\dfrac{(2n-1)!}{[(n-1)!]^2}$.   **X. 93.** Consider the equality

$(1+x)^p(x+1)^{n-p} = (1+x)^n$. $\displaystyle\sum_{k=0}^{p}\binom{p}{k}P(m,k)P(n-p,m-k) = \binom{n}{m}$ or

$m!\displaystyle\sum_{k=0}^{p}\binom{p}{k}P(m,k)P(n-p,m-k) = P(n,m)$.

X. 94. The number $\binom{3n}{n}$ is the coefficien of x^n from the expansion

$(1+x)^{3n}$. In the equality $(1+x)^{3n} = (1+x)^n(1+x)^{2n}$ we have

$(1+x)^{2n} = \displaystyle\sum_{k=0}^{2n}\binom{2n}{k}x^k$ and $(1+x)^n = \displaystyle\sum_{l=0}^{n}\binom{n}{l}x^l$. Therefore

$$(1+x)^{3n} = \sum_{k=0}^{2n}\sum_{l=0}^{n}\binom{2n}{k}\binom{n}{l}x^{k+l} \text{ and the coefficient of } x^n \text{ in } (1+x)^{3n} \text{ is}$$

$$\sum_{k+l=n}\binom{2n}{k}\binom{n}{l} = \sum_{k=0}^{n}\binom{2n}{k}\binom{n}{k}, \text{ thus } \binom{3n}{n} = \sum_{k=0}^{n}\binom{2n}{k}\binom{n}{k}.$$

X. 95. $\left(1+\sqrt{2}\right)^{2n} = \sum_{k=0}^{n} 2^k \binom{2n}{2k} + \sqrt{2}\cdot\sum_{k=0}^{n} 2^{k+1}\binom{2n}{2k+1}$ (1),

$$\left(1-\sqrt{2}\right)^{2n} = \sum_{k=0}^{n} 2^k \binom{2n}{2k} - \sqrt{2}\cdot\sum_{k=0}^{n} 2^{k+1}\binom{2n}{2k+1} \text{ (2).}$$

Multiplying (1) and (2) side by side we get the initial identity.

X. 96. $S_1 + iS_2 = 1 + \binom{n}{1}(\cos x + i\sin x)^1 + \binom{n}{2}(\cos x + i\sin x)^2 + \ldots +$
$+ \binom{n}{n}(\cos x + i\sin x)^n =$

$$= (1 + \cos x + i\sin x)^n == \left(2\cos^2\frac{x}{2} + 2i\sin\frac{x}{2}\cos\frac{x}{2}\right)^n =$$

$$= 2^n \cos^n\frac{x}{2}\left(\cos\frac{nx}{2} + i\sin\frac{nx}{2}\right).$$

Bibliography

*Canadian Mathematics Competition, Problems Problems Problems, Waterloo, On, Canada, Volume 1, 2, 3, 4, 5.

*James Stewart, Lothar Redlin, Saleem Watson, *Precalculus Mathematics for calculus 4th*, Brooks/Cole.

*Wesner/Nustad, *Elementary Algebra with applications*, Iowa1983.

*Elliot Mendelson, *Theory and problems of beginning calculus,* 2nd edition, McGra-Hill.

*Titu Andreescu, Dorin Andrica, *Complex Numbers from A to...Z*, Birkhäuser, 2006.

*A. Tsypkin, A. Pinsky, *Methods of solving problems in high school math*, Mir Publishers Moscow, 1983.

*Călugăriţa Gh., Mangu V., *Probleme de matematică pentru treapta I şi a II-a de liceu*, Ed. Albatros, Bucureşti, 1977.

*Comissaire H., Anzemberger E., *Exercises d'Algebre et de Trigonometrie*, Paris, 1923.

* Iaglom I. M., Iaglom A. M., *Challenging Mathematical Problems with elementary Solutions*, Dover Publications, 1964.

*Ioachimescu A. G., *Culegere de probleme de algebră*, ediţia a V-a, Ed. Didactică şi Pedagogică, Bucureşti, 1968.

*Ikramov H. D., *Zadacinik po lineinoi algebre*, Moskva, 1975.

*Kuterov A., Rubanov A., *Zadacinik po algebre i elementarnîm funcţiiam*, Moskva, 1974.

*M. I Skanavi, Sbornik Zadach po matematike, Moskva, 1996.

*Vîşenski V. A., Kartaşov N. V., Mihailovski B. I., Iardenko M. I., *Sbornik zadaci kievskih matematiceskih olimpiad*, Kiev, 1984.

*Nesterenko I. V., Olenik S. N., Potapov M. K., *Zadaci vstupitelnîh eczamenov po matematike*, Moskva, 1986.

*Dorofeev G. V., Potapov M. K., Rozov N. H., *Posobie pe matematike dlia postupaiuşcih v vuzî*, Moskva, 1976.

*Sklearski D. O., Cenţov N. N., Iaglom I. M., *Izbranîe zadacii teoremî elementarnoi matematiki (Arifmetica i algebra)*, Moskva, 1965.
*Stamate I., Stoian I., *Culegere de exerciţii şi probleme de algebră*, Ed. Didactică şi Pedagogică, Bucureşti, 1979.

*C. Nastasescu, M. Brandiburu, C. Nita, D. Joita, *Exercitii si probleme de algebra*, Editura Didactica si pedagogica, Bucuresti.

Matematică - Algebră (high school textbooks, grades IX - X, 1980 edition), Ed. Didactică şi Pedagogică, Bucureşti, 1980.

*Collection of "Gazeta Matematică - seria B", Bucharest.

www.ingramcontent.com/pod-product-compliance
Lightning Source LLC
Chambersburg PA
CBHW031820170526
45157CB00001B/127